Lecture Notes in Computer Science 4848

Commenced Publication in 1973
Founding and Former Series Editors:
Gerhard Goos, Juris Hartmanis, and Jan van Leeuwen

T0223148

Max H. Garzon   Hao Yan (Eds.)

DNA Computing

13th International Meeting on DNA Computing, DNA13
Memphis, TN, USA, June 4-8, 2007
Revised Selected Papers

 Springer

Volume Editors

Max H. Garzon
The University of Memphis
Computer Science
209 Dunn Hall, TN 38152-3240, U.S.A.
E-mail: mgarzon@memphis.edu

Hao Yan
Arizona State University
The Biodesign Institute
1001 S. McAllister Ave, Tempe, AZ 85287-5601, USA
E-mail: hao.yan@asu.edu

Library of Congress Control Number: 2008920717

CR Subject Classification (1998): F.1, F.2.2, I.2.9, J.3

LNCS Sublibrary: SL 1 – Theoretical Computer Science and General Issues

ISSN   0302-9743
ISBN   3-540-77961-2 Springer Berlin Heidelberg New York
ISBN   978-3-540-77961-2 Springer Berlin Heidelberg New York

Springer is a part of Springer Science+Business Media

springer.com

© Springer-Verlag Berlin Heidelberg 2008

Typesetting: Camera-ready by author, data conversion by Scientific Publishing Services, Chennai, India
Printed on acid-free paper   SPIN: 12225927   06/3180   5 4 3 2 1 0

Preface

Biomolecular/DNA computing is now well established as an interdisciplinary field where chemistry, computer science, molecular biology, physics, and mathematics come together with the common purpose of fundamental scientific understanding of biology and chemistry and its applications. This international meeting has been the premier forum where scientists with different backgrounds and a common focus meet to present their latest results and entertain visions of the future. In this tradition, about 100 participants converged in Memphis, Tennessee to hold the 13th International Meeting on DNA Computing during June 4–8, 2007, under the auspices of the International Society for Nanoscale Science, Computation and Engineering (ISNSCE) and The University of Memphis.

The call for papers encouraged submissions of original, recent, and promising experimental and theoretical results in the field. The Call for Papers elicited some 62 submissions, almost perfectly balanced among the major theoretical and experimental categories. It is evidence of how well the interdisciplinary nature of the conference has truly matured that the major criterion of quality, agreed upon in advance by the Program Committee (PC), produced a nearly balanced program as well across the two major categories, full papers and talks with an abstract only. The program with the greatest perceived impact consisted of 24 papers for plenary oral talks; in addition, 15 full-paper posters and 10 poster abstracts were accepted, of which 5 authors were invited to give five short demos in a new submission category this year.

The conference program retained the structure now customary for this meeting. It began with four *tutorials* on Monday June 4. The customary introductory tutorials to biochemistry and computation were delivered by Thom LaBean (Duke University, "Basic Bioscience for Computer Scientists") and Vinhthuy Phan (University of Memphis, "Basic Computer Science for Bioscientists") in the morning, while Ned Seeman (New York University, "Structural DNA Nanotechnology"), and Byoung-Tak Zhang (Seoul National University, "Molecular Evolutionary Computation *In Vitro* and *In Silico*") delivered the advanced tutorials in the afternoon. An exciting set of four invited talks by Charles Bennett (IBM Research, "Brownian Molecular Computers and the Thermodynamics of Computation"), David Harel (Weizmann Institute, "*In Silico* Biology"), Paul Rothemund (Caltech, "DNA Origami") and Steve Skienna (SUNY- Stony Brook, "Designing Useful Viruses") for the *main conference* provided a most appropriate setting for the recurring themes of self-assembly, encodings, as well as intriguing new trends in what may be termed the design and bio-engineering of robust biocomputers. Perhaps one day we will regard these budding new ideas as the onset of true interdisciplinary outcomes of the DNA conference. On Friday, we held the *Third Symposium on Nanotechechnology*, with four pointed lectures on cutting-edge developments in the field by Bob Austin (Princeton University, "The City of Cells: Adaptation and Evolution on a Chip"), William Shih (Harvard Medical School, "From Structural DNA Nanotechnology to Membrane-Protein NMR Structure"), Bernie Yurke (Lucent Technologies, "The Dynamic DNA Nanoworld") and Todd Yeates (UCLA, "Progress and Challenges in Designing

Proteins for Self-Assembly"). Complementary poster reception, tours of Memphis's major attractions, and a banquet at the Botanic Gardens facilitated the discussion of ideas arising from this program.

This volume consists of a selection of invited papers accepted for presentation at the conference and revised after feedback at the event. We can identify roughly six major themes in the contributions, as described in the table of contents. They appear to convey a much clearer focus and purpose in the realization of the potential in the field, despite their apparent similarity to topics addressed early in the history of the conference over a decade ago. Robustness and scalability in analyses, tools, and applications seem to characterize well the overall nature of these contributions, both experimental and theoretical, across all the major themes.

Many people generously provided much of their time and energy in organizing this meeting. We thank all 12 invited speakers for offering thought-provoking talks in the main conference, the symposium, and the tutorials, as well as the authors for their quality submissions. Authors and reviewers also deserve kudos for submissions and their positive attitude and patience in proving the new manuscript submission system at http://dna13.memphis.edu/subs effective in alleviating the burden on the PC in the various submission, review and feedback rounds upto camera-ready copy. The invaluable help from the Organizing Committee, assistants Makram Raboudi, Sujoy Roy, Jennifer Grazier, as well as Cheryl Hayes and Yolanda Feifer in the computer science department, made possible a very smooth meeting that everyone seemed to enjoy. We are also indebted to the Steering Committee, led by Lila Kari, for the terse and timely reminders that made possible the content on the following pages. Not least, DNA13 would simply not have been possible without the support of our sponsors. We are grateful to all of them for their contributions.

October 2007   Max H. Garzon
                                                                            Hao Yan

Organization

Program Committee

Alessandra Carbone	Université Pierre et Marie Curie, France
Mark Daley	University of Western Ontario, Canada
Max Garzon (*Co-chair*)	University of Memphis, USA
Ashish Goel	Stanford University, USA
Hendrik Jan Hoogeboom	Leiden University, The Netherlands
Thomas LaBean	Duke University, USA
Giancarlo Mauri	University of Milan-Bicocca, Italy
Satoshi Murata	Tokyo Instute of Technology, Japan
Andrei Paun	Louisiana Tech University, USA
Ion Petre	University of Turku, Finland
John A. Rose	Ritsumeikan APU, Japan
Yasubumi Sakakibara	Keio University, Japan
Dipankar Sen	Simon Fraser University, Canada
Ehud Shapiro	Weizmann Institute, Israel
William Shih	Harvard Medical School, USA
Friedrich C. Simmel	University of Munich, Germany
Lloyd Smith U	Wisconsin-Madison, USA
David Soloveichik	CalTech, USA
Petr Sosik	Universidad Politechnica de Madrid, Spain
Milan Stojanovic	Columbia University, USA
Ron Weiss	Princeton University, USA
Masahito Yamamoto	Hokkaido University, Japan
Masayuki Yamamura	University of Tokyo, Japan
Hao Yan (*Co-chair*)	Arizona State University, USA

Steering Committee

Lila Kari, Computer Science (*Chair*)	University of Western Ontario, Canada
Leonard Adleman (Honorary member)	Computer Science, University of Southern California, USA
Anne Condon	Computer Science University of British Columbia, Canada
Masami Hagiya	Computer Science, University of Tokyo, Japan
Natasha Jonoska	Mathematics, University of South Florida, USA
Chengde Mao	Chemistry, Purdue University, USA
Giancarlo Mauri	Computer Science, University of Milan-Biccoca Italy
Satoshi Murata	Computer Science, Tokyo Institute of Technology, Japan

Gheorge Paun	Computer Science, RomanianAcademy/ University of Seville, Romania/Spain
John Reif	Computer Science, Duke University, USA
Grzegorz Rozenberg	Computer Science, University of Leiden, The Netherlands
Nadrian Seeman	Chemistry, New York University, USA
Andrew Turberfield	Physics, University of Oxford, UK
Erik Winfree	Computer Science/Computation and Neural Systems, Caltech, USA

Organizing Committee

Derrel Blain	Computer Science, The University of Memphis, USA
Russell Deaton	Computer Science and Engineering, The University of Arkansas, USA
Susannah Gal	Biological Sciences, Binghamton University, USA
Max Garzon (*Chair*)	Computer Science, The University of Memphis, USA
Andrew Neel	Computer Science, The University of Memphis, USA
Vinhthuy Phan	Computer Science, The University of Memphis, USA

Sponsors

The University of Memphis
    (The Provost Office, Bioinformatics Program, and Computer Science Department)
Air Force Office of Scientific Research (AFOSR)
The Army research Lab (ARO/ARL)
The University of Arkansas, College of Engineering (Dean Saxena's Offfice)
An anonymous sponsor (by choice).

Table of Contents

Self-assembly

Biomolecular Machines and Automata

Codes for DNA Memories and Computing

Novel Techniques for DNA Computing *in vitro*

Novel Techniques for DNA Computing *in silico*

Models and Languages

Staged Self-assembly:
Nanomanufacture of Arbitrary Shapes with $O(1)$ Glues

Erik D. Demaine[1], Martin L. Demaine[1], Sándor P. Fekete[2], Mashhood Ishaque[3],
Eynat Rafalin[4], Robert T. Schweller[5], and Diane L. Souvaine[3]

[1] MIT Computer Science and Artificial Intelligence Laboratory, 32 Vassar St., Cambridge,
MA 02139, USA*
{edemaine,mdemaine}@mit.edu
[2] Institut für Mathematische Optimierung, Technische Universität Braunschweig, Pockelsstr. 14,
38106 Braunschweig, Germany
s.fekete@tu-bs.de
[3] Department of Computer Science, Tufts University, Medford, MA 02155, USA**
{mishaq01,dls}@cs.tufts.edu
[4] Google Inc.,***
erafalin@cs.tufts.edu
[5] Department of Computer Science, University of Texas Pan American, 1201 W. University
Drive, Edinburg, Texas 78539, USA
schwellerr@cs.panam.edu

Abstract. We introduce *staged self-assembly* of Wang tiles, where tiles can be added dynamically in sequence and where intermediate constructions can be stored for later mixing. This model and its various constraints and performance measures are motivated by a practical nanofabrication scenario through protein-based bioengineering. Staging allows us to break through the traditional lower bounds in tile self-assembly by encoding the shape in the staging algorithm instead of the tiles. All of our results are based on the practical assumption that only a constant number of glues, and thus only a constant number of tiles, can be engineered, as each new glue type requires significant biochemical research and experiments. Under this assumption, traditional tile self-assembly cannot even manufacture an $n \times n$ square; in contrast, we show how staged assembly enables manufacture of arbitrary orthogonal shapes in a variety of precise formulations of the model.

1 Introduction

Self-assembly is the process by which an organized structure can form spontaneously from simple parts. It describes the assembly of diverse natural structures such as crystals, DNA helices, and microtubules. In nanofabrication, the idea is to co-opt natural self-assembly processes to build desired structures, such as a sieve for removing viruses from serum, a drug-delivery device for targeted chemotherapy or brachytherapy, a magnetic device for medical imaging, a catalyst for enzymatic reactions, or a

---

* Partially supported by NSF CAREER award CCF-0347776 and DOE grant DE-FG02-04ER25647.
** Partially supported by NSF grant CCF-0431027.
*** Work performed while at Tufts University. Partially supported by NSF grant CCF-0431027.

M.H. Garzon and H. Yan (Eds.): DNA 13, LNCS 4848, pp. 1–14, 2008.
© Springer-Verlag Berlin Heidelberg 2008

biological computer. Self-assembly of artificial structures has promising applications to nanofabrication and biological computing. The general goal is to design and manufacture nanoscale pieces (e.g., strands of DNA) that self-assemble uniquely into a desired macroscale object (e.g., a computer).

Our work is motivated and guided by an ongoing collaboration with the Sackler School of Graduate Biomedical Sciences that aims to nanomanufacture sieves, catalysts, and drug-delivery and medical-imaging devices, using protein self-assembly. Specifically, the Goldberg Laboratory is currently developing technology to bioengineer (many copies of) rigid struts of varying lengths, made of several proteins, which can join collinearly to each other at compatible ends. These struts occur naturally as the "legs" of the *T4 bacteriophage*, a virus that infects bacteria by injecting DNA. In contrast to nanoscale self-assembly based on DNA [WLWS98, MLRS00, RPW04, BRW05, See98, SQJ04, Rot06], which is inherently floppy, these nanorod structures are extremely rigid and should therefore scale up to the manufacture of macroscale objects.

The traditional, leading theoretical model for self-assembly is the two-dimensional *tile assembly model* introduced by Winfree in his Ph.d. thesis [Win98] and first appearing at STOC 2000 [RW00]. The basic building blocks in this model are *Wang tiles* [Wan61], unrotatable square tiles with a specified glue on each side, where equal glues have affinity and may stick. Tiles then self-assemble into supertiles: two (super)tiles nondeterministically join if the sum of the glue affinities along the attachment is at least some threshold τ, called *temperature*. This basic model has been generalized and extended in many ways [Adl00, ACGH01, ACG$^+$02, SW04, ACG$^+$05, RW00, KS06]. The model should be practical because Wang tiles can easily simulate the practical scenario in which tiles are allowed to rotate, glues come in pairs, and glues have affinity only for their unique mates. In particular, we can implement such tiles using two unit-length nanorods joined at right angles at their midpoints to form a plus sign.

Most theoretical research in self-assembly considers the minimum number of distinct tiles—the *tile complexity t*—required to assemble a shape uniquely. In particular, if we allow the desired shape to be scaled by a possibly very large factor, then in most models the minimum possible tile complexity (the smallest "tile program") is $\Theta(K/\lg K)$ where K is the Kolmogorov complexity of the shape [SW04]. In practice, the limiting factor is the number of distinct glues—the *glue complexity g*—as each new glue type requires significant biochemical research and experiments. For example, a set of DNA-based glues requires experiments to test whether a collection of codewords have a "conflict" (a pair of noncomplementary base sequences that attach to each other), while a set of protein-based glues requires finding pairs of proteins with compatible geometries and amino-acid placements that bind (and no other pairs of which accidentally bind). Of course, tile and glue complexities are related: $g \leq t \leq g^4$.

We present the *staged tile assembly model*, a generalization of the tile assembly model that captures the temporal aspect of the laboratory experiment, and enables substantially more flexibility in the design and fabrication of complex shapes using a small tile and glue complexity. In its simplest form, staged assembly enables the gradual addition of specific tiles in a sequence of stages. In addition, any tiles that have not yet attached as part of a supertile can be washed away and removed (in practice, using a weight-based filter, for example). More generally, we can have any number of *bins* (in reality, batches of liquid solution stored in separate containers), each containing tiles and/or supertiles that self-assemble as in the standard tile assembly model. During a stage, we can perform any collection of operations of two types: (1) add (arbitrarily many copies of) a new tile to an existing bin; and (2) pour one bin into another bin, mixing the contents of the former bin into the latter bin, and keeping the former bin intact. In both cases, any pieces that do not assemble into larger structures are washed away and removed. These operations let us build intermediate supertiles in isolation and then combine different supertiles as whole structures. Now we have two new complexity measures in addition to tile and glue complexity: the number of stages—or *stage complexity s*—measures the time required by the operator of the experiment, while the number of bins—or *bin complexity b*—measures the space required for the experiment.[1] (When both of these complexities are 1, we obtain the regular tile assembly model.)

Our results. We show that staged assembly enables substantially more efficient manufacture in terms of tile and glue complexity, without sacrificing much in stage and bin complexity. All of our results assume the practical constraint of having only a small constant number of glues and hence a constant number of tiles. In contrast, an information-theoretic argument shows that this assumption would limit the traditional tile assembly model to constructing shapes of constant Kolmogorov complexity.

For example, we develop a method for self-assembling an $n \times n$ square for arbitrary $n > 0$, using 16 glues and thus $O(1)$ tiles (independent of n), and using only $O(\log \log n)$ stages, $O(\sqrt{\log n})$ bins, and temperature $\tau = 2$ (Section 4.2). Alternatively, with the minimum possible temperature $\tau = 1$, we can self-assemble an $n \times n$ square using 9 glues, $O(1)$ tiles and bins, and $O(\log n)$ stages (Section 4.1). In contrast, the best possible self-assembly of an $n \times n$ square in the traditional tile assembly model has tile complexity $\Theta(\log n / \log \log n)$ [ACGH01, RW00], or $\Theta(\sqrt{\log n})$ in a rather extreme generalization of allowable pairwise glue affinities [ACG$^+$05].

More generally, we show how to self-assemble arbitrary shapes made up of n unit squares in a variety of precise formulations of the problem. Our simplest construction builds the shape using 2 glues, 16 tiles, $O(\text{diameter})$ stages, and $O(1)$ bins, but it only glues tiles together according to a spanning tree, which is what we call the *partial connectivity model* (Section 5.1). All other constructions have *full connectivity*: any two adjacent unit squares are built by tiles with matching glues along their shared edge. In particular, if we scale an arbitrary hole-free shape larger by a factor of 2, then we can self-assemble with full connectivity using 8 glues, $O(1)$ tiles, and $O(n)$ stages and bins (Section 5.2). We also show how to simulate a traditional tile assembly construction with t tiles by a staged assembly using 3 glues, $O(1)$ tiles, $O(\log \log t)$ stages, $O(t)$ bins, and a scale factor of $O(\log t)$ (Section 5.3). If the shape happens to be monotone in one direction, then we can avoid scaling and still obtain full connectivity, using 9 glues, $O(1)$ tiles, $O(\log n)$ stages, and $O(n)$ bins (details omitted in this version).

Table 1 summarizes our results in more detail, in particular elaborating on possible trade-offs between the complexities. The table captures one additional aspect of our

---

[1] Here we view the mixing time required in each stage (and the volume of each bin) as a constant, mainly because it is difficult to analyze precisely from a thermodynamic perspective, as pointed out in [Adl00]. In our constructions, we believe that a suitable design of the relative concentrations of tiles (a feature not captured by the model) leads to reasonable mixing times.

Table 1. Summary of the glue, tile, bin, and stage complexities, the temperature τ, the scale factor, the connectivity, and the planarity of our staged assemblies and the relevant previous work

$n \times n$ square	Glues	Tiles	Bins	Stages	τ	Scale	Conn.	Planar
Previous work [ACGH01, RW00]	$\Theta(\frac{\log n}{\log\log n})$	1	1	1	2	1	full	yes
Jigsaw technique (§4.1)	9	$O(1)$	$O(1)$	$O(\log n)$	1	1	full	yes
Crazy mixing (§4.2)	16	$O(1)$	B	$O(\lceil\frac{\log n}{B^2}\rceil\log B)$	2	1	full	yes
Crazy mixing, $B = \sqrt{\log n}$	16	$O(1)$	$\sqrt{\log n}$	$O(\log\log n)$	2	1	full	yes

General shape with n tiles	Glues	Tiles	Bins	Stages	τ	Scale	Conn.	Planar
Previous work [SW04]	$\Theta(K/\log K)$	1	1	1	2	unbounded	partial	no
Arbitrary shape with n tiles (§5.1)	2	16	$O(\log n)$	$O(\text{diameter})$	1	1	partial	no
Hole-free shape with n tiles (§5.2)	8	$O(1)$	$O(n)$	$O(n)$	1	2	full	no
Simulation of 1-stage tiles T (§5.3)	3	$O(1)$	$O(|T|)$	$O(\log\log|T|)$	1	$O(\log|T|)$	partial	no
Monotone shapes with n tiles (omitted)	9	$O(1)$	$O(n)$	$O(\log n)$	1	1	full	yes

constructions: Planarity. Consider two jigsaw puzzle pieces with complex borders lying on a flat surface. It may not be possible to slide the two pieces together while both remain on the table. Rather, one piece must be lifted off the table and dropped into position. Our current model of assembly intuitively permits supertiles to be placed into position from the third dimension, despite the fact that it may not be possible to assemble within the plane. A *planar* construction guarantees assembly of the final target shape even if we restrict assembly of supertiles to remain completely within the plane. This feature seems desirable, though it may not be essential in two dimensions because reality will always have some thickness in the third dimension (2.5D). However, the planarity constraint (or *spatiality* constraint in 3D) becomes more crucial in 3D assemblies, so this feature gives an indication of which methods should generalize to 3D; see Section 6.

Related Work. There are a handful of existing works in the field of DNA self-assembly that have proposed very basic multiple stage assembly procedures. John Reif introduced a step-wise assembly model for local parallel biomolecular computing [Rei99]. In more recent work Park et. al. have considered a simple hierarchical assembly technique for the assembly of DNA lattices [PPA+06]. Somei et. al. have considered a microfluidic device for stepwise assembly of DNA tiles [SKFM05]. While all of these works use some form of stepwise or staged assembly, they do not study the complexity of staged assembly to the depth that we do here. Further, none consider the concept of bin complexity.

2 The Staged Assembly Model

In this section, we present basic definitions common to most assembly models, then we describe the staged assembly model, and finally we define various metrics to measure the efficiency of a staged assembly system.

Tiles and tile systems. A *(Wang)* tile t is a unit square defined by the ordered quadruple $\langle\text{north}(t), \text{east}(t), \text{south}(t), \text{west}(t)\rangle$ of glues on the four edges of the tile. Each *glue* is taken from a finite alphabet Σ, which includes a special "null" glue denoted null. For simplicity of bounds, we do not count the null glue in the *glue complexity* $g = |\Sigma| - 1$.

A *tile system* is an ordered triple $\langle T, G, \tau \rangle$ consisting of the *tileset* T (a set of distinct tiles), the *glue function* $G : \Sigma^2 \rightarrow \{0, 1, \ldots, \tau\}$, and the *temperature* τ (a positive integer). It is assumed that $G(x, y) = G(y, x)$ for all $x, y \in \Sigma$ and that $G(\text{null}, x) = 0$ for all $x \in \Sigma$. Indeed, in all of our constructions, as in the original model of Adleman [Adl00], $G(x, y) = 0$ for all $x \neq y^2$, and each $G(x, x) \in \{1, 2, \ldots, \tau\}$. The *tile complexity* of the system is $|T|$.

Configurations. Define a *configuration* to be a function $C : \mathbb{Z}^2 \rightarrow T \cup \{\text{empty}\}$, where empty is a special tile that has the null glue on each of its four edges. The *shape* of a configuration C is the set of positions (i, j) that do not map to the empty tile. The shape of a configuration can be disconnected, corresponding to several distinct supertiles.

Adjacency graph and supertiles. Define the *adjacency graph* G_C of a configuration C as follows. The vertices are coordinates (i, j) such that $C(i, j) \neq$ empty. There is an edge between two vertices (x_1, y_1) and (x_2, y_2) if and only if $|x_1 - x_2| + |y_1 - y_2| = 1$. A *supertile* is a maximal connected subset G' of G_C, i.e., $G' \subseteq G_C$ such that, for every connected subset H, if $G' \subseteq H \subseteq G_C$, then $H = G'$. For a supertile S, let $|S|$ denote the number of nonempty positions (tiles) in the supertile. Throughout this paper, we will informally refer to (lone) tiles as a special case of supertiles.

If every two adjacent tiles in a supertile share a positive strength glue type on abutting edges, the supertile is *fully connected.*

Two-handed assembly and bins. Informally, in the two-handed assembly model, any two supertiles may come together (without rotation or flipping) and attach if their strength of attachment, from the glue function, meets or exceeds a given temperature parameter τ.

Formally, for any two supertiles X and Y, the *combination* set $C^\tau_{(X,Y)}$ of X and Y is defined to be the set of all supertiles obtainable by placing X and Y adjacent to each other (without overlapping) such that, if we list each newly coincident edge e_i with edge strength s_i, then $\sum s_i \geq \tau$.

We define the assembly process in terms of bins. Intuitively, a bin consists of an initial collection of supertiles that self-assemble at temperature τ to produce a new set of supertiles P. Formally, with respect to a given set of tile-types T, a *bin* is a pair (S, τ) where S is a set of initial supertiles whose tile-types are contained in T, and τ is a temperature parameter. For a bin (S, τ), the set of *produced* supertiles $P'_{(S,\tau)}$ is defined recursively as follows: (1) $S \subseteq P'_{(S,\tau)}$ and (2) for any $X, Y \in P'_{(S,\tau)}$, $C^\tau_{(X,Y)} \subseteq P'_{(S,\tau)}$. The set of *terminally* produced supertiles of a bin (S, τ) is $P_{(S,\tau)} = \{X \in P' \mid Y \in P', C^\tau_{(X,Y)} = \emptyset\}$. We say the set of supertiles P is *uniquely* produced by bin (S, τ) if each supertile in P' is of finite size. Put another way, unique production implies that every producible supertile can grow into a supertile in P.

Intuitively, P' represents the set of all possible supertiles that can self-assemble from the initial set S, whereas P represents only the set of supertiles that cannot grow any

---

[2] With a typical implementation in DNA, glues actually attach to unique complements rather than to themselves. However, this depiction of the glue function is standard in the literature and does not affect the power of the model.

further. In the case of unique assembly of P, the latter thus represents the eventual, final state of the self-assembly bin. Our goal is therefore to produce bins that yield desired supertiles in the uniquely produced set P.

Given a collection of bins, we model the process of mixing bins together in arbitrarily specified patterns in a sequence of distinct stages. In particular, we permit the following actions: We can *create* a bin of a single tile type $t \in T$, we can *merge* multiple bins together into a single bin, and we can *split* the contents of a given bin into multiple new bins. In particular, when splitting the contents of a bin, we assume the ability to extract only the unique terminally produced set of supertiles P, while filtering out additional partial assemblies in P'. Intuitively, given enough time for assembly and a large enough volume of tiles, a bin that uniquely produces P should consist of almost entirely the terminally produced set P. We formally model the concept of mixing bins in a sequence of stages with the *mix graph*.

Mix graphs. An r-*stage* b-*bin mix graph* M consists of $rb + 1$ vertices, m_* and $m_{i,j}$ for $1 \leq i \leq r$ and $1 \leq j \leq b$, and an arbitrary collection of edges of the form $(m_{r,j}, m_*)$ or $(m_{i,j}, m_{i+1,k})$ for some i, j, k.

Staged assembly systems. A *staged assembly system* is a 3-tuple $\langle M_{r,b}, \{T_{i,j}\}, \{\tau_{i,j}\}\rangle$ where $M_{r,b}$ is an r-stage b-bin mix graph, each $T_{i,j}$ is a set of tile types, and each $\tau_{i,j}$ is an integer temperature parameter. Given a staged assembly system, for each $1 \leq i \leq r$, $1 \leq j \leq b$, we define a corresponding bin $(R_{i,j}, \tau_{i,j})$ where $R_{i,j}$ is defined as follows:

1. $R_{1,j} = T_{1,j}$ (this is a bin in the first stage);

2. For $i \geq 2$, $R_{i,j} = \left(\bigcup_{k:\ (m_{i-1,k}, m_{i,j}) \in M_{r,b}} P_{(R_{(i-1,k)}, \tau_{i-1,k})} \right) \cup T_{i,j}$.

3. $R_* = \bigcup_{k:\ (m_{r,k}, m_*) \in M_{r,b}} P_{(R_{(r,k)}, \tau_{r,k})}$.

Thus, the jth bin in the ith stage takes its initial set of seed supertiles to be the terminally produced supertiles from a collection of bins from the previous stage, the exact collection specified by $M_{r,b}$, in addition to a set of added tile types $T_{i,j}$. Intuitively, the mix graph specifies how each collection of bins should be mixed together when transitioning from one stage to the next. We define the set of terminally produced supertiles for a staged assembly system to be $P_{(R_*, \tau_*)}$. In this paper, we are interested in staged assembly systems for which each bin yields unique assembly of terminal supertiles. In this case we say a staged assembly system uniquely produces the set of supertiles $P_{(R_*, \tau_*)}$.

Throughout this paper, we assume that, for all i, j, $\tau_{i,j} = \tau$ for some fixed global temperature τ, and we denote a staged assembly system as $\langle M_{r,b}, \{T_{i,j}\}, \tau \rangle$.

3   Assembly of $1 \times n$ Lines

As a warmup, we develop a staged assembly for the $1 \times n$ rectangle ("line") using only three glues and $O(\log n)$ stages. The assembly uses a divide-and-conquer approach to split the shape into a constant number of recursive pieces. Before we turn to the simple divide-and-conquer required here, we describe the general case, which will be useful later. This approach requires the pieces to be combinable in a unique way, forcing the creation of the desired shape. We consider the *decomposition tree* formed by

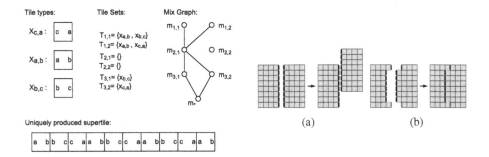

Fig. 1. A sample staged assembly system that uniquely assembles a 1×10 line. The temperature is $\tau = 1$, and each glue a, b, c has strength 1. The tile complexity is 3, the stage complexity is 3, and the bin complexity is 2.

Fig. 2. (a) The shifting problem encountered when combining rectangle supertiles. (b) The jigsaw solution: two supertiles that combine uniquely into a fully connected square supertile.

the recursion, where sibling nodes should uniquely assemble to their parent. The staging proceeds bottom-up in this tree. The height of this tree corresponds to the stage complexity, and the maximum number of distinct nodes at any level corresponds to the bin complexity. The idea is to assign glues to the pieces in the decomposition tree to guarantee unique assemblage while using few glues.

Theorem 1. *There is a planar temperature-1 staged assembly system that uniquely produces a (fully connected) 1×2^k line using 3 glues, 6 tiles, 6 bins, and $O(k)$ stages.*

Proof. The decomposition tree simply splits a 1×2^k line into two $1 \times 2^{k-1}$ lines. All tiles have the null glue on their top and bottom edges. If the 1×2^k line has glue a on its left edge, and glue b on its right edge, then the left and right $1 \times 2^{k-1}$ inherit these glues on their left and right edges, respectively. We label the remaining two inner edges—the right edge of the left piece and the left edge of the right piece—with a third glue c, distinct from a and b. Because $a \neq b$, the left and right piece uniquely attach at the inner edges with common glue c. This recursion also maintains the invariant that $a \neq b$, so three glues suffice overall. Thus there are only $\binom{3}{2} = 6$ possible 1×2^k lines of interest, and we only need to store these six at any time, using six bins. At the base case of $k = 0$, we just create the nine possible single tiles. The number of stages beyond that creation is exactly k.

Corollary 1. *There is a planar temperature-1 staged assembly system that uniquely produces a (fully connected) $1 \times n$ line using 3 glues, 6 tiles, 10 bins, and $O(\log n)$ stages.*

Proof. We augment the construction of Theorem 1 applied to $k = \lfloor \log n \rfloor$. When we build the 1×2^i lines for some i, if the binary representation of n has a 1 bit in the ith position, then we add that line to a new output bin. Thus, in the output bin, we accumulate powers of 2 that sum to n. As in the proof of Theorem 1, three glues suffice to guarantee unique assemblage in the output bin. The number of stages remains $O(\log n)$.

4   Assembly of $n \times n$ Squares

Figure 2(a) illustrates the challenge with generalizing the decomposition-tree technique from $1 \times n$ lines to $n \times n$ squares. Namely, the naïve decomposition of a square into two $n \times n/2$ rectangles cannot lead to a unique assembly using $O(1)$ glues with temperature 1 and full connectivity: by the pigeon-hole principle, some glue must be used more than once along the shared side of length n, and the lower instance of the left piece may glue to the higher instance of the right piece. Even though this incorrect alignment may make two unequal glues adjacent, in the temperature-1 model, a single matching pair of glues is enough for a possible assembly.

4.1   Jigsaw Technique

To overcome this shifting problem, we introduce the *jigsaw technique*, a powerful tool used throughout this paper. This technique ensures that the two supertiles glue together uniquely based on geometry instead of glues. Figure 2(b) shows how to cut a square supertile into two supertiles with three different glues that force unique combination while preserving full connectivity.

Theorem 2. *There is a planar temperature-1 staged assembly of a fully connected $n \times n$ square using 9 glues, $O(1)$ tiles, $O(1)$ bins, and $O(\log n)$ stages.*

Proof. We build a decomposition tree by first decomposing the $n \times n$ square by vertical cuts, until we obtain tall, thin supertiles; then we similarly decompose these tall, thin supertiles by horizontal cuts, until we obtain constant-size supertiles. Table 2 describes the general algorithm. Figure 3 shows the decomposition tree for an 8×8 square. The height of the decomposition tree, and hence the stage complexity, is $O(\log n)$.

We assign glue types to the boundaries of the supertiles to guarantee unique assemblage based on the jigsaw technique. The assignment algorithm is similar to the $1 \times n$ line, but we use three glues on each edge instead of one, for a total of nine glues instead of three.

It remains to show that the bin complexity is $O(1)$. We start by considering the vertical decomposition. At each level of the decomposition tree, there are three types of intermediate products: leftmost supertile, rightmost supertile and middle supertiles. The leftmost and rightmost supertiles are always in different bins. The important thing

Table 2. Algorithm for vertical decomposition. (Horizontal decomposition is symmetric.).

Algorithm DecomposeVertically (supertile S):
    — Here S is a supertile with n rows and m columns; S is not necessarily a rectangle.
 1. **Stop vertical partitioning when width is small enough:**
        If $m \leq 3$, DecomposeHorizontally(S) and return.
 2. **Find the column along which the supertile is to be partitioned:**
        Let $i := \lfloor (m+1)/2 \rfloor$.
        Divide supertile S along the ith column into a left supertile S_1 and right supertile S_2 such that
        tiles at position $(1, i)$ and (n, i) belong to S_1 and the rest of the ith column belongs to S_2.
 3. **Now decompose recursively:**
        DecomposeVertically (S_1)
        DecomposeVertically (S_2)

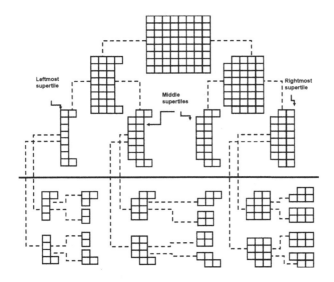

Fig. 3. Decomposition tree for 8×8 square in the jigsaw technique

to observe is that the middle supertiles always have the same shape, though it is possible to have two different sizes—the number of columns can differ by one. In one of these sizes, the number of columns is even and, in the other, the number is odd. Thus we need to separate bins for the even- and odd-columned middle supertiles. For each of the even- or odd-columned supertiles, each of left and right boundaries of the supertile can have three choices for the glue types. Therefore, there is a constant number of different types of middle supertiles at each level of the decomposition tree. Thus, for vertical decomposition, we need $O(1)$ bins. Each of the supertiles at the end of vertical decomposition undergoes horizontal decomposition. A similar argument applies to the horizontal decomposition as well. Therefore, the number of bins required is $O(1)$.

4.2 Crazy Mixing

For each stage of a mix graph on B bins, there are up to $\Theta(B^2)$ edges that can be included in the mix graph. By picking which of these edges are included in each stage, $\Theta(B^2)$ bits of information can be encoded into the mix graph per stage. The large amount of information that can be encoded in the mixing pattern of a stage permits a very efficient trade-off between bin complexity and stage complexity. In this section, we consider the complexity of this trade-off in the context of building $n \times n$ squares.

It is possible to view a tile system as a compressed encoding of the shape it assembles. Thus, information theoretic lower bounds for the descriptional or Kolmogorov complexity of the shape assembled can be applied to aspects of the tile system. From this we obtain the following lower bound:

Theorem 3. *Any staged assembly system with a fixed temperature and bin complexity B that uniquely assembles an $n \times n$ square with $O(1)$ tile complexity must have stage complexity $\Omega(\frac{\log n}{B^2})$ for almost all n.*

Our upper bound achieves a stage complexity that is within a $O(\log B)$ factor of this lower bound:

Theorem 4. *For any n and B, there is a temperature-2 fully connected staged assembly of an $n \times n$ square using 16 glues, $O(1)$ tiles, B bins, and $O(\frac{\log n}{B^2} \log B + \log B)$ stages.*

In the interest of space, the proofs of these two theorems are omitted in this version.

We conjecture that this stage complexity bound can be achieved by a temperature-1 assembly by judicious use of the jigsaw technique.

5 Assembly of General Shapes

In this section, we describe a variety of techniques for manufacturing arbitrary shapes using staged assembly with $O(1)$ glues and tiles.

5.1 Spanning-Tree Technique

The *spanning-tree technique* is a general tool for making an arbitrary shape with the connectivity of a tree. We start with a sequential version of the assembly:

Theorem 5. *Any shape S with n tiles has a partially connected temperature-1 staged assembly using 2 glues, at most 2^4 tiles, $O(\log n)$ bins, and $O(\mathrm{diameter}(S))$ stages.*

Proof. Take a breadth-first spanning tree of the adjacency graph of the shape S. The depth of this tree is $O(\mathrm{diameter}(S))$. Root the tree at an arbitrary leaf. Thus, each vertex in the tree has at most three children. Color the vertices with two colors, black and white, alternating per level. For each edge between a white parent and a black child, we assign a white glue to the corresponding tiles' shared edge. For each edge between a black parent and a white child, we assign a black glue to the corresponding tiles' shared edge. All other tile edges receive the null glue. Now a tile has at most three edges of its color connecting to its children, and at most one edge of the opposite color connecting to its parent.

To obtain the sequential assembly, we perform a particular postorder traversal of the tree: at node v, visit its child subtrees in decreasing order of size. To combine at node v, we mix the recursively computed bins for the child subtrees together with the tile corresponding to node v. The bichromatic labeling ensures unique assemblage. The number of intermediate products we need to store is $O(\log n)$, because when we recurse into a second child, its subtree must have size at most $2/3$ of the parent's subtree.

5.2 Scale Factor 2

Although the spanning-tree technique is general, it probably manufactures structurally unsound assemblies. Next we show how to obtain full connectivity of general shapes, while still using only a constant number of glues and tiles.

Theorem 6. *Any simply connected shape has a staged assembly using a scale factor of 2, 8 glues, $O(1)$ tiles, $O(n)$ stages, and $O(n)$ bins. The construction maintains full connectivity.*

Proof. Slice the target shape with horizontal lines to divide the shape into $1 \times k$ strips for various values of k, which scale to $2 \times 2k$ strips. These strips can overlap along horizontal edges but not vertical edges. Define the *strip graph* to have a vertex for each strip and an edge between two strips that overlap along a horizontal edge. Because the shape is simply connected, the strip graph is a tree. Root this tree at an arbitrary strip, defining a parent relation.

A recursive algorithm builds the subtree of the strip graph rooted at an arbitrary strip s. As shown in Figure 4(a), the strip s may attach to the rest of the shape at zero or more places on its top or bottom edge. One of these connections corresponds to the parent of s (unless s is the overall root). As shown in Figure 4(b), our goal is to form each of these attachments using a jigsaw tab/pocket combination, where bottom edges have tabs and top edges have pockets, extending from the rightmost square up to but not including the leftmost square.

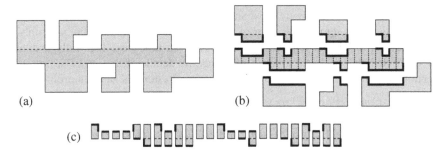

(a)   (b)

(c)

Fig. 4. Constructing a horizontal strip in a factor-2 scaled shape (a), augmented by jigsaw tabs and pockets to attach to adjacent pieces (b), proceeding column-by-column (c)

The horizontal edges of each tab or pocket uses a pair of glues. The unit-length upper horizontal edge uses one glue, and the possibly longer lower horizontal edge uses the other glue. The pockets at the top of strip s use a different glue pair from the tabs at the bottom of strip s. Furthermore, the pocket or tab connecting s to its parent uses a different glue pair from all other pockets and tabs. Thus, there are four different glue pairs (for a total of eight glues). If the depth of s in the rooted tree of the strip graph is even, then we use the first glue pair for the top pockets, the second glue pair for the bottom tabs, except for the connection to the parent which uses either the third or fourth glue pair depending on whether the connection is a top pocket or a bottom tab. If the depth of s is odd, then we reverse the roles of the first two glue pairs with the last two glue pairs. All vertical edges of tabs and pockets use the same glue, 8.

To construct the strip s augmented by tabs and pockets, we proceed sequentially from left to right, as shown in Figure 4(c). The construction uses two bins. At the kth step, the primary bin contains the first $k-1$ columns of the augmented strip. In the secondary bin, we construct the kth column by brute force in one stage using 1–3 tiles and 0–2 distinct internal glues plus the desired glues on the boundary. Because the column specifies only two glues for horizontal edges, at the top and bottom, we can use any two other glues for the internal glues. All of the vertical edges of the column use different glues. If k is odd, the left edges use glues 1–3 and the right edges uses glues 4–6, according to y

coordinate; if k is even, the roles are reversed. (In particular, these glues do not conflict with glue 8 in the tabs and pockets.) The only exception is the first and last columns, which have no glues on their left and right sides, respectively. Now we can add the secondary bin to the primary bin, and the kth column will uniquely attach to the right side of the first $k - 1$ columns. In the end, we obtain the augmented strip.

During the building of the strip, we attach children subproblems. Specifically, once we assemble the rightmost column of an attachment to one or two children strips, we recursively assemble those one or two children subtrees in separate bins, and then mix them into s's primary bin. Because the glues on the top and bottom sides of s differ, as do the glues of s's parent, and because of the jigsaw approach, each child we add has a unique place to attach. Therefore we uniquely assemble s's subtree. Applying this construction to the root of the tree, we obtain a unique assembly of the entire shape.

5.3   Simulation of One-Stage Assembly with Logarithmic Scale Factor

In this section, we show how to use a small number of stages to combine a constant number of tile types into a collection of supertiles that can simulate the assembly of an arbitrary set of tiles at temperature $\tau = 1$, given that these tiles only assemble fully connected shapes. In the interest of space, the details of this proof are omitted. Extending this simulation to temperature-2 one-stage systems is an open problem.

Theorem 7. *Consider an arbitrary single stage, single bin tile system with tile set T, all glues of strength at most 1, and that assembles a class of fully connected shapes. There is a temperature-1 staged assembly system that simulates the one-stage assembly of T up to an $O(\log |T|)$ size scale factor using 3 glues, $O(1)$ tiles, $O(|T|)$ bins, and $O(\log \log |T|)$ stages. At the cost of increasing temperature to $\tau = 2$, the construction achieves full connectivity.*

6   Future Directions

There are several open research questions stemming from this work.

One direction is to relax the assumption that, at each stage, all supertiles self-assemble to completion. In practice, it is likely that at least some tiles will fail to reach their terminal assembly before the start of the next stage. Can a staged assembly be robust against such errors, or at least detect these errors by some filtering, or can we bound the error propagation in some probabilistic model.

Another direction is to develop a model of the assembly time required by a mixing operation involving two bins of tiles. Such models exist for (one-stage) *seeded self-assembly*—which starts with a seed tile and places singleton tiles one at a time—but this model fails to capture the more parallel nature of two-handed assembly in which large supertiles can bond together without a seed. Another interesting direction would be to consider nondeterministic assembly in which a tile system is capable of building a large class of distinct shapes. Is it possible to design the system so that certain shapes are assembled with high probability?

Finally, we have focused on two-dimensional constructions in this paper. This focus provides a more direct comparison with previous models, and it is also a case of practical interest, e.g., for manufacturing sieves. Many of our results also generalize to 3D (or any constant dimension), at the cost of increasing the number of glues and tiles. For example, the spanning-tree model generalizes trivially, and a modification to the jigsaw idea enables many of the other results to carry over. So far, we have not worked out the exact performance measures for these 3D analogs, but we do not expect this to be difficult.

Acknowledgments. We thank M. S. AtKisson and Edward Goldberg for extensive discussions about the bioengineering application.

References

[ACG⁺02]   Adleman, L., Cheng, Q., Goel, A., Huang, M.-D., Kempe, D., de Espanés, P.M., Rothemund, P.W.K.: Combinatorial optimization problems in self-assembly. In: Proceedings of the Thirty-Fourth Annual ACM Symposium on Theory of Computing (electronic), pp. 23–32. ACM Press, New York (2002)

[ACG⁺05]   Aggarwal, G., Cheng, Q., Goldwasser, M.H., Kao, M.-Y., de Espanes, P.M., Schweller, R.T.: Complexities for generalized models of self-assembly. SIAM Journal on Computing 34(6), 1493–1515 (2005)

[ACGH01]   Adleman, L., Cheng, Q., Goel, A., Huang, M.-D.: Running time and program size for self-assembled squares. In: Proceedings of the 33rd Annual ACM Symposium on Theory of Computing, pp. 740–748. ACM Press, New York (2001)

[Adl00]   Adleman, L.M.: Toward a mathematical theory of self-assembly. Technical Report 00-722, Department of Computer Science, University of Southern California (January 2000)

[BRW05]   Barish, R.D., Rothemund, P.W.K., Winfree, E.: Two computational primitives for algorithmic self-assembly: Copying and counting. Nano Letters 5(12), 2586–2592 (2005)

[KS06]   Kao, M.-Y., Schweller, R.: Reducing tile complexity for self-assembly through temperature programming. In: Proceedings of the 17th Annual ACM-SIAM Symposium on Discrete Algorithm, pp. 571–580. ACM Press, New York (2006)

[MLRS00]   Mao, C., LaBean, T.H., Reif, J.H., Seeman, N.C.: Logical computation using algorithmic self-assembly of DNA triple-crossover molecules. Nature 407, 493–496 (2000)

[PPA⁺06]   Park, S.H., Pistol, C., Ahn, S.J., Reif, J.H., Lebeck, A.R., Dwyer, C., LaBean, T.H.: Finite-size, fully addressable DNA tile lattices formed by hierarchical assembly procedures. Angewandte Chemie 45, 735–739 (2006)

[Rei99]   Reif, J.: Local parallel biomolecular computation. In: Proc. DNA-Based Computers, pp. 217–254 (1999)

[Rot06]   Rothemund, P.W.K.: Folding DNA to create nanoscale shapes and patterns. Nature 440, 297–302 (2006)

[RPW04]   Rothemund, P.W.K., Papadakis, N., Winfree, E.: Algorithmic self-assembly of DNA sierpinski triangles. PLoS Biology 2(12), 424 (2004)

[RW00]   Rothemund, P.W.K., Winfree, E.: The program-size complexity of self-assembled squares. In: Proceedings of the 32nd Annual ACM Symposium on Theory of Computing, pp. 459–468. ACM Press, New York (2000)

[See98]   Seeman, N.C.: DNA nanotechnology. In: Siegel, R.W., Hu, E., Roco, M.C. (eds.) WTEC Workshop Report on R&D Status and Trends in Nanoparticles, Nanostructured Materials, and Nanodevices in the United States (January 1998)

[SKFM05]   Somei, K., Kaneda, S., Fujii, T., Murata, S.: A microfluidic device for dna tile self-assembly. In: DNA, pp. 325–335 (2005)

[SQJ04]   Shih, W.M., Quispe, J.D., Joyce, G.F.: A 1.7-kilobase single-stranded DNA that folds into a nanoscale octahedron. Nature 427, 618–621 (2004)

[SW04]   Soloveichik, D., Winfree, E.: Complexity of self-assembled shapes. In: Ferretti, C., Mauri, G., Zandron, C. (eds.) DNA Computing. LNCS, vol. 3384, pp. 344–354. Springer, Heidelberg (2005)

[Wan61]   Wang, H.: Proving theorems by pattern recognition—II. The Bell System Technical Journal 40(1), 1–41 (1961)

[Win98]   Winfree, E.: Algorithmic Self-Assembly of DNA. PhD thesis, California Institute of Technology, Pasadena (1998)

[WLWS98]   Winfree, E., Liu, F., Wenzler, L.A., Seeman, N.C.: Design and self-assembly of two-dimensional DNA crystals. Nature 394, 539–544 (1998)

Activatable Tiles: Compact, Robust Programmable Assembly and Other Applications

Urmi Majumder, Thomas H. LaBean, and John H. Reif

Department of Computer Science,
Duke University, Durham, NC, USA
{urmim,thl,reif}@cs.duke.edu
http://www.cs.duke.edu/~reif

Abstract. While algorithmic DNA self-assembly is, in theory, capable of forming complex patterns, its experimental demonstration has been limited by significant assembly errors. In this paper we describe a novel protection/deprotection strategy to strictly enforce the direction of tiling assembly growth to ensure the robustness of the assembly process. Tiles are initially inactive, meaning that each tile's output pads are protected and cannot bind with other tiles. After other tiles bind to the tile's input pads, the tile transitions to an active state and its output pads are exposed, allowing further growth. We prove that an activatable tile set is an instance of a compact, error-resilient and self-healing tile-set. We also describe a DNA design for activatable tiles and a deprotection mechanism using DNA polymerase enzymes and strand displacement. We conclude with a discussion on some applications of activatable tiles beyond computational tiling.

Keywords: DNA-assembly, error-correction, molecular computation.

1 Introduction

The potential of self-assembling DNA nanostructures is derived from the predictable properties of DNA hybridization as well as from the assembly's theoretical power to instantiate any computable pattern [3]. Winfree [1] formalized this process of tiling assembly growth when he proposed Tile Assembly Model (TAM) which describes how a complex structure can spontaneously form from simple components called "tiles"; this assembly can also perform computation. However, the main problem for a practical implementation of TAM based assemblies is that tile additions are very error-prone. Experiments show that error rates can be as high as 1% to 8% [4,5]. The primary kind of error encountered in DNA tile assembly experiments is known as the *error by insufficient attachment* [7], which occurs when a tile violates the TAM rule stating that a tile may only be added if it binds strongly[1] enough. Thus there is a mismatch between theoretical models of DNA tiles and reality, providing major challenges in applying this model to real experiments.

---

[1] In the TAM for temperature $\tau = 2$, a tile binds strongly either using at least one strong bond or two weak bonds.

M.H. Garzon and H. Yan (Eds.): DNA 13, LNCS 4848, pp. 15–25, 2008.

There have been several designs of error-resilient tile sets [6,7,8] that perform "proofreading" on redundantly encoded information [8] to decrease assembly errors. Recall that the primary kinds of error in assembly experiments are: (i) growth error that occurs when a tile with one weak bond attaches at a location where a tile with two weak bonds should have been attached, (ii) facet nucleation error that occurs when a weakly binding tile attaches to a site where no tile should currently attach and (iii) spontaneous nucleation error that occurs when a large assembly grows without a seed tile. Each of these error-resilient tile sets [6,7,8], however, addresses only certain errors and proposes a construction that works with limited classes of tile sets. Additionally, most constructions result in greatly increased tile set size, hindering practical implementation. This leads to a major open question in error-resilient self-assembly: Is it possible to design a compact tile set that can address all three kinds of errors simultaneously? Our *activatable tile set* is an effort towards achieving this ultimate goal.

Limitations of Previous Approaches towards Robust Assembly: Existing error-resilient tile sets assume directional growth. This is a very strong assumption because experiments show that real tiles do not behave in such a fashion. The assumption, however, underlies the growth model in TAM. Thus, a potential solution to minimizing assembly errors is to enforce this directionality constraint. Observe that if we start with a set of "deactivated" tiles which activate in a desired order, we can enforce a directional assembly at the same scale as the original one. Such a system can be built with minimal modifications of existing DNA nanostructures [9,10,11].

Previous Approaches to direct Tiling Assembly Procedures: The snaked-proofreading technique of Chen et al. [7] provided the main inspiration for *activatable tiles*. This scheme replaces each original tile by a $k \times k$ block of tiles. The assembly process for a block doubles back on itself such that nucleation error cannot propagate without locally forcing another insufficient attachment. Can such a growth order be enforced at the original scale of the assembly? Other motivating work has been from Dirks et al. [2], who designed a system where monomer DNA nanostructures, when mixed together, do not hybridize until an initiator strand is added. Can the idea of triggered self-assembly be used in the context of computational DNA tiling?

The answers to both questions are yes. The key idea is to start with a set of "protected" DNA tiles, which we call *activatable tiles*; these tiles do not assemble until an initiator nanostructure is introduced to the solution. The initiator utilizes strand displacement to "strip" off the protective coating on the input sticky end(s) of the appropriate neighbors [12]. When the input sticky ends are completely hybridized, the output sticky ends are exposed. DNA polymerase enzyme can perform this deprotection, since it can act over long distances (e.g: across tile core) unlike strand displacement. The newly exposed output sticky ends, in turn, strip the protective layer off the next tile along the growing face of the assembly. The use of polymerase in this context is justified because of its successful use in PCR, a biochemistry technique often used for exponentially amplifying DNA. PCR has been so successful that it has several commercial applications including genetic fingerprinting, paternity testing, hereditary disease detection, mutagenesis and more. Further PCR amplification of megabase DNA has also been done [21]. In nature most organisms copy their DNA in the same way making polymerase an excellent choice for reliable deprotection over long distances. Many repeated rounds of primer polymerization are required in conventional PCR. In contrast, we are using only a single round of primer polymerization (similar to a single round of PCR) to expose the desired sticky ends in our activatable tiles. Other proteins, such as helicase which are useful for DNA replication may be used for unwrapping our protection strand, but we have not yet investigated this direction quite thoroughly. Another important observation in this context is that although polymerase and the activatable tile are of comparable sizes, when the polymerase attaches to the primer, which is bound to the protection strand, it is only bound at the concave open face of the assembly (ensured by the sequential assembly growth) and hence there is no possibility of steric hindrance.

Enzyme-free Activated Tiles: The most relevant previous work that has been recently brought to our attention is probably that of Fujibayashi et al. [23,24]: the Protected Tile Mechanism (PTM) and the Layered Tile Mechanism (LTM) which utilize DNA protecting molecules to form kinetic barriers against spurious assembly. Although this is an enzyme-free circuit, in the PTM, the output sticky ends are not protected and thus they can bind to a growing assembly before the inputs are deprotected and hence cause an error. In the LTM, the output sticky ends are protected only by 3 nucleotides each and can be easily displaced causing the above-mentioned error. Error resilience can only be guaranteed if we can ensure a deprotection from input to output end.

Our Results and the Organization of the Paper: Section 1 introduced the notion of deprotection and discussed the need for activatable tiles in computational assemblies. Section 2 describes the abstract and kinetic models for activatable tiles that build on Winfree's original TAMs, with the primary difference being that each tile now has an associated finite state machine. In Section 3, we prove that the activatable tile set is an instance of a compact, error-resilient and self-healing tile set. In Section 4, we describe the DNA design of an example one dimensional activatable tile and its deprotection using both strand displacement and DNA polymerization. In Section 5 we discuss some applications of activatable tiles beyond computational assemblies as a concentration/sensing system and reaction catalyzation. In Section 6 we conclude the paper.

2   The Activatable Tile Assembly Models

An abstract model is a theoretical abstraction from reality that is often easier to work with conceptually as well as mathematically. Since Winfree has already established the framework for tiling assembly models with his TAM, we build our abstract Activatable Tile Assembly Model (aATAM) and the kinetic Activatable Tile Assembly Model (kATAM) discussed in this section on Winfree's abstract and kinetic TAMs respectively [1].

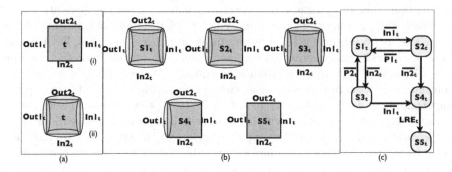

Fig. 1. (a-i) Original Abstract Rule Tile R, (a-ii) Protected version of R, (b) Different states associated with the activatable R (aR), (c) State Transition Diagram for aR. The $In1_t$ and the $In2_t$ denote the sticky ends that displaces the protections $P1_t$ and $P2_t$ from the input ends of the tile t while LRE_t is the long range effector that displaces the protection from the output end.

2.1   The Abstract Activatable Tile Assembly Model (aATAM)

The simplest version of activatable tiles starts with a set of "protected" *rule tiles*[2] that do not assemble until a pre-assembled initiator assembly, consisting of a *seed tile* and multiple *boundary tiles*, is introduced to the mixture. In the more complex version, the initiator is the seed tile alone and the boundary tiles have a protection-deprotection scheme similar to that of the rule tiles.

The aATAM is similar to the original abstract TAM (aTAM) due to Winfree [1] except that each tile type t has an associated finite state machine (FSM) M_t and hence, each tile has a state. The new abstract rule tile is shown in Figure 1(a-ii). Unlike the original tile [Figure 1(a-i)], it has all its sides protected. The states in the FSM M_t arise from the presence or absence of protection on the four sides of the tile type t (as shown in Figure 1(b)). The state transition diagram is shown in Figure 1(c).

2.2   The Kinetic Activatable Tile Assembly Model (kATAM)

The kATAM is based on Winfree's original model kTAM, but due to the the stochastic nature of the protection on all sides of the tile, additional errors need to be modeled. Therefore we need more free parameters than just r_f and $r_{r,b}$ for modeling assembly growth. Figure 2 shows the different states possible in the finite state machine for the kATAM and Figure 1(Right) shows the state transition diagram. In addition to the assumptions of kTAM, the main assumptions of kATAM are: (i) The input protection is only reversible while the output pads are still protected, (ii) Output protection is irreversible, meaning once a tile is

---

[2] The three main types of tiles in TAM are : (i) Rule tiles, responsible for computation in algorithmic self-assembly, (ii) Seed tile that nucleates the assembly and (iii) Boundary tiles that provide two dimensional input for computation.

completely deprotected, it cannot return to the stage where every side of the tile
has a protective cover. Monomers in solution are thus either entirely protected
or entirely deprotected.

The main features of the kinetic model are: (1) a tile can get knocked off
the growth site after output deprotection. These unprotected tiles, however, are
added to the growth site at a different rate, r'_f, that will later be shown to be
much smaller than r_f, (2) with one input match, the tile in S8 (S2) transitions
to S9 (S3) at the rate of r_{dp} (deprotection) and returns to S8 (S2) at the rate
of r_p (protection), (3) When both inputs are matched, the output pads (S5) are
deprotected at the rate r_{dp_out}. Note that r_{dp}, r_p and r_{dp_out} are free parameters
whose value depends on the experimental situation. The kinetic parameters can
be derived for an example deprotection system. The description is omitted due
to space constraints. Interested readers can refer to [13].

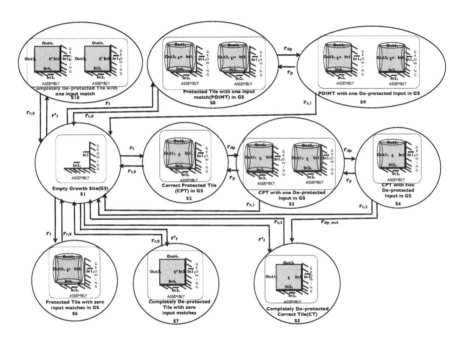

Fig. 2. State transition diagram for kATAM

Forward Rate of Erroneous Tiles: Since there are many free parameters in
the kinetic model, such as $r_f, r_{r,b}, r_p$ and others we decrease the dimensionality
of the parameter space by combining some of the parameters together e.g. r_p, r_{dp}
and r_{dp_out}. This is done by computing the rate at which tiles become completely
deprotected after reaching a growth site, thus neglecting the intermediate states
in Figure 2. This new rate corresponds to the rate at which a tile reaches state
S5 if it is in S1. We call this rate r_{eff} and assume that r_{eff} is a function of G_{se}
such that $r_{eff} = k_f e^{(-2+\epsilon_1)G_{se}}$, where ϵ_1 is a constant between 0 and 1. Note

that r_{eff} is similar to r_f in the original kTAM. Based on the continuous time Markov Chain (CTMC) in Figure 2, we can evaluate r_{eff} as

$$r_{eff} = r_f \frac{r_{dp}}{(r_{dp} + r_{r,0})} \frac{r_{dp}}{(r_p + r_{dp} + r_{r,1})} \frac{r_{dp\_out}}{(r_{dp\_out} + r_{r,2} + r_p)}. \tag{1}$$

One primary assumptions in the model are

$$r_{r,1} > r_f > r_{eff} > r_{r,2} \text{ and}$$
$$r_{r,1} = e^{-G_{se}}, r_{r,2} = e^{-2G_{se}}, r_{eff} = e^{(-2+\epsilon_1)G_{se}}, r_f = e^{(-2+\epsilon_1+\epsilon_2)G_{se}}$$
$$\text{for some } 0 < \epsilon_1, \epsilon_2 < 1. \tag{2}$$

For simplicity of the model, we can ensure that $\epsilon_2 \ll \epsilon_1$ by adjusting the kinetic parameters in the deprotection system (e.g. toehold length in the strand displacement events, nucleotide concentration and template length for polymerization etc). Hence $r_{eff} \gg r_{r,2}$. Another important assumption we make is that DNA polymerization is irreversible and, hence, at equilibrium every tile is completely deprotected.

Based on these assumptions we can first obtain the expected fraction of completely deprotected tiles that leaves $S5$ as $\frac{r_{r,2}}{r_{r,2}+r^*}e^{-G_{mc}}$ and hence derive r'_f, the forward rate of erroneous tiles as $e^{(-2+\epsilon_2)G_{se}}$.

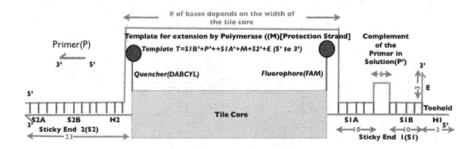

Fig. 3. Protection Strategy for a DNA Tile

3   Compact Proofreading with Activatable Tiles

Activatable tiles provide error-resilience to a growing assembly by enforcing directional growth. Ideally the output ends are never available until the corresponding input ends are completely hybridized, thus preventing both errors by insufficient attachment as well as nucleation errors. There is a small probability, however, of errors by insufficient attachment caused by tiles that leave a growth site after output deprotection. Furthermore, the computation still occurs at the original scale, unlike Chen's snaked proofreading technique [7] which increases the lattice size by a multiplicative factor of k^2. Hence, activatable tiles indeed provide compact error-resilience. Since the seed is the only completely unprotected tile when the assembly begins and the concentration of completely unprotected rule or boundary tiles existing in solution at any given time is very low, activatable tiles can also prevent spontaneous nucleation and enforce "controlled growth".[3] We can formally prove that activatable tiles are indeed an instance of compact proofreading technique. Soloveichik et al. gave a concise definition of compact proofreading [14] and we adapt it to our ATAM:

Definition 1. *Given a small constant $0 < q < 1$, a sequence of deterministic tile systems $\{T_1, T_2, T_3, \ldots\}$ is a compact proofreading scheme for pattern P if (i) T_N produces the full infinite pattern P under the aATAM, (ii) T_N has poly$(\log N)$ tile types (poly(n) denotes $n^{O(1)}$) and (iii) T_N produces the correct $N \times N$ initial portion of the pattern P with probability at least q in time $O(Npoly(logN))$ in the kATAM for some values of the free parameters in the model.*

Theorem 1. *The activatable Tile System A_N is a compact proofreading scheme.*

Proof. Let the tile system in aTAM be T_N and the activatable tile system be A_N. A_N is the same as T_N except that each tile type has an associated FSM. Since in aATAM activatable tiles can bind to a growth site only if they can bind strongly enough (just as in aTAM), A_N can produce the whole system correctly under aATAM so the first condition is satisfied. Moreover, $|A_N| = |T_N|$, the only difference being that we start the assembly with a "protected" version of T_N. Since this work is concerned with only deterministic tile systems, the argument of Soloveichik et al. [14] applies and we need only a constant number of tile types so long the tile set has a locally deterministic assembly sequence.

The argument for the third condition is similar to that of Chen et al. [7]. In this model, errors are only caused by insufficient attachments; these errors are caused by tiles dissociating from growth sites after their output protection has been stripped off. In an insufficient attachment event, first an unprotected monomer (with a single binding site match) attaches at the rate of r'_f. However, before this tile is knocked off at the rate of $r_{r,1}$, a second tile (protected/unprotected) can attach to the first tile at the rate $r'_f + r_{eff}$. Thus, based on the corresponding CTMC we can say that the rate of an insufficient attachment is

$$r_{insuf} = \frac{r'_f(r'_f + r_{eff})}{r_{r,1} + r'_f + r_{eff}} = e^{(-3+\epsilon_1+\epsilon_2)G_{se}} \frac{1 + e^{-(\epsilon_1-\epsilon_2)G_{se}}}{1 + e^{-(1-\epsilon_1)G_{se}} + e^{-(1-\epsilon_2)G_{se}}} \quad (3)$$

Our goal with respect to a particular growth site is to bury the correct tile k levels deep before an insufficient attachment event occurs.[4] In other words, if we have a $k \times k$ square whose left bottom corner location is occupied by this tile, then the $k \times k$ square completes before an insufficient attachment event occurs. This puts the tile under consideration into a "k-frozen" state. The process of

---

[3] Controlled growth is defined to be the growth occurring for parameter values in a certain part of the kinetic parameter space, such that (i) growth does occur, (ii) errors are rare and (iii) growth not seeded by the seed tile is rare [15].

[4] The time taken for single tile attachment is $O(1/r_{eff})$ which is less than $1/r_{insuf}$.

tile attaching or detaching in a 2D assembly can be modeled as a random walk.[5] Note that the forward growth (tile association at the output ends of the current tile) happens at the rate of $r_{eff} + r'_f$ while the backward growth (dissociation of the current tile) has a rate of $r_{r,2}$. Thus, the average rate of growth (the mean of forward and backward rates) $\bar{r}$ is $\frac{1}{2}(r_{eff} + r'_f + r_{r,2})$ and the expected time taken for this $k \times k$ square to grow is $O(k^4/\bar{r})$ since in a 2D random walk, we have to take k^4 steps in expectation in order to cover k^2 locations.

Thus, for any small ϵ_{insuf}, one can find a constant c_{insuf} such that, with probability $1 - \epsilon_{insuf}$, no insufficient attachment happens at this specific location but a correct tile becomes k-frozen within time $O(k^4/\bar{r})$. In other words, $(k^4/\bar{r}) < (c_{insuf}/r_{insuf})$. Hence, for a given k, such that with high probability a given growth site is filled correctly and buried k levels deep in $O(k^4/\bar{r})$ time. For constant kinetic parameters and k, this time is also constant. Hence we can use the same argument as Adleman et al. [19] and show that the $N \times N$ square is completed in expected $O(N)$ time. $\qquad \square$

Compact Self-healing with Activatable Tiles: The impact of activatable tiles goes beyond the compact error-resilience which is a primary concern for fault tolerant self-assembly. In case of gross external damage, e.g. a hole created in a growing tiling assembly, activatable tiles can repair the damage with minimal error by enforcing directional growth. Since the original, self-assembled lattice was formed by algorithmic accretion in the forward direction, only forward re-growth is capable of rebuilding the correct structure. The protected monomers in the solution ensure a forward directional accretion. There is a small probability, however, of backward growth from the unprotected monomers that were once part of the original tiling assembly and dissociated after outputs are deprotected. The likelihood is comparatively small since the forward reaction rate depends on concentration of the monomers and the protected tiles are much more abundant than their unprotected counterparts. Defining size in terms of number of tiles, we conclude the following theorem:

Theorem 2. *With high probability, a damaged hole of size S (small compared to the assembly size) is repaired in time $O(S^2)$, for suitable kATAM parameters.*

Proof. Observe that the maximum rate of error due to backward growth is bounded by r'_f while the forward rate of growth is $r_{eff} + r'_f$. Observe that $\bar{r} > r'_f$. Using the same technique as in Theorem 1, we can prove that the hole can be repaired in $O(S^2/\bar{r})$ by a 2D random walk on the set of S tile positions on the 2D plane. We can further argue that for any small ϵ_{heal} ($0 < \epsilon_{heal} < 1$), one can find a constant c_{heal} such that with probability $1 - \epsilon_{heal}$, $(S^2/\bar{r}) < (c_{heal}/r'_f)$. For a given S, we can compute G_{se} so that there is no backward growth when a hole of size at most S gets repaired in $O(S^2)$ time assuming constant parameters. $\qquad \square$

---

[5] The stochastic process of tile attachment and detachment in self-assembly has often been modeled as a random walk [7]. Further this is similar to the lattice gas model where modeling interactions as random walks is quite well established.

4   DNA Design of One Dimensional Activatable Tiles

The DNA design of one dimensional (1D) activatable tiles is very helpful in understanding the more complex DNA design of two dimensional (2D) activatable tiles. It is also motivated by the need for a protection strategy for tiles that self-assemble into a 1D lattice, such as the boundary of the computational tiling. Hence we first describe the DNA design of a 1D activatable tile. Figure 3 gives the sequence design of a 1D activated tile. Some of the key features of the tile design are: (i) The sticky ends are protected by the protection strand M, (ii) For adjacent tiles, the protection strand needs to be arranged in a different manner so as to satisfy both constraints on the direction for sticky end matching as well as the template for polymerization (not shown here), resulting in two kinds of tile types, (iii) The 3 base portion (E) at the $3'$ end of the protection strand in the tile design prevents polymerization of the toehold $H1$, (iv) The portion of the protection strand which hybridizes to the primer P is held tightly in a hairpin loop of six bases between two subportions of the input sticky end, (v) The fluorophore and the quencher are positioned such that the flourophore is quenched only when correct tiles hybridize.

How does an activated tile deprotect its neighbor? The idea is quite simple: the toehold H on the input sticky end $S1$ of the protected tile (say Tile 1) is used to displace the protection strand M on it; after the input sticky end of the Tile 1 and the output sticky end of the deprotected tile (say Tile 2) are completely hybridized, the protection strand M is freed from the input end of Tile 1; the primer P can now attach to the complementary portion P' on the protection strand M that was earlier held tightly in a hairpin loop. Polymerase next binds to the $3'$ end of the primer and extends it to the output end of Tile 1. Eventually, the output sticky end of Tile 1 is exposed.

Our DNA design for 2D activatable tiles is a direct extension of our 1D activatable tiles. Interested readers can refer to [13].

5   Other Applications of Activatable Tiles

Beyond the applications to computational tiling, activatable tiles can also be used as a novel system for sensing and concentration. For example, one can design a modified activatable tile to include a docking site for a specific target molecule. Initially, the tiles are in the inactive state; they are neither bound to a target molecule nor they are assembled together. When a target molecule binds to the tile's docking site, the tile transitions from an inactive to an active state. Tiles in the active state can assemble. As the activated tiles assemble, the target molecules are concentrated making an excellent concentration system.

Activatable tiles can also be used for reaction catalyzation. Suppose that for some small k, the goal is to gather k distinct types of target molecules to initiate or catalyze a chemical reaction. Just as with the sensing system, one can design k distinct activated tiles, each with a docking site for a different target molecule. These tiles become active only when they are carrying their target molecules.

Once activated, these k distinct tiles assemble into a small tiling lattice, putting the target molecules in close proximity, and allowing them to react. Additionally, the reaction products can be used to disassemble the lattices and deactivate the tiles, allowing them to be reused. Observe that the binding site on the same face of each tile type is so designed that after assembly, the molecules bound to the tiles will be close to each other. They are never bound inside the lattice and therefore, the reaction can never become slower. Although this is quite a novel idea, the concept of DNA directed chemistry has been explored quite extensively in the recent years (See [22]).

6   Conclusion

In spite of the fact that it may be impossible to eliminate errors completely from the assembly process, activatable tiles appear to be quite promising. Thus, as a part of future work, we not only intend to have an experimental validation, but also evaluate our deprotection strategy with computer simulation, particularly compare it with the simulation results from Fujibayashi et al.'s enzyme-free activated tile model [23,24]. We conclude with one interesting open question: Can combining overlay redundancy techniques [16] with the idea of activatable tiles further improve the compact error-resilience of self-assembly experiments?

Acknowledgments. This work is supported by NSF ITR Grants EIA-0086015 and CCR-0326157, NSF QuBIC Grants EIA-0218376 and EIA-0218359, and DARPA/AFSOR Contract F30602-01-2-0561. The authors thank Erik Winfree for bringing the work of Satoshi Murata to our attention. Majumder also thanks Erik Halvorson for useful edits and discussion.

References

1. Winfree, E.: Algorithmic Self-Assembly of DNA Caltech (1998)
2. Dirks, R.M., Pierce, N.A.: PNAS 101(43), 15275–15278 (2004)
3. Wang, H.: Bell System Tech Journal 40(1), 1–41 (1961)
4. Barish, R.D., Rothemund, P.W.K., Winfree, E.: Nano Letters 5(12), 2586–2592
5. Rothemund, P.W.K., Papadakis, N., Winfree, E.: PLoS Biology 2 (12), 424 (2004)
6. Reif, J.H., Sahu, S., Yin, P.: DNA10. In: Ferretti, C., Mauri, G., Zandron, C. (eds.) DNA Computing. LNCS, vol. 3384, pp. 293–307. Springer, Heidelberg (2005)
7. Chen, H.L., Goel, A.: DNA 10. In: Ferretti, C., Mauri, G., Zandron, C. (eds.) DNA Computing. LNCS, vol. 3384, pp. 62–75. Springer, Heidelberg (2005)
8. Winfree, E., Bekbolatov, R.: DNA 9. In: Chen, J., Reif, J.H. (eds.) DNA Computing. LNCS, vol. 2943, pp. 126–144. Springer, Heidelberg (2004)
9. Yan, H., Park, S.H., Finkelstein, G., Reif, J.H., LaBean, T.H.: Science 301(5641), 1882–1884 (2003)
10. LaBean, T.H., Yan, H., Kopatsch, J., Liu, F., Winfree, E., Reif, J.H., Seeman, N.C.: J. Am. Chem. Soc. 122, 1848–1860 (2000)
11. Winfree, E., Liu, F., Wenzler, L.A., Seeman, N.C.: Nature 394 (1998)
12. Thompson, B.J., Camien, M.N., Warner, R.C.: PNAS 73(7), 2299–2303 (July 1976)

13. http://www.cs.duke.edu/~reif/paper/urmi/activatable/activatable.pdf
14. Soloveichik, D., Winfree, E.: SICOMP 36(6), 1544–1569 (2007)
15. Winfree, E.: Simulations of Computing by Self-Assembly Caltech CS Report, 22 (1998)
16. Sahu, S., Reif, J.H.: DNA 12. In: Mao, C., Yokomori, T. (eds.) DNA Computing. LNCS, vol. 4287, pp. 223–238. Springer, Heidelberg (2006)
17. Saturno, J., Blanco, L., Salas, M., Esterban, J.A: J. Bio. Chem. 270(52), 31235–31243 (1995)
18. Thompson, B.J., Escarmis, C., Parker, B., Slater, W.C., Doniger, J., Tessman, I., Warner, R.C.: J. Mol. Biol. 91, 409–419 (1975)
19. Adleman, L., Cheng, Q., Goel, A., Huang, M.D.: Proceedings of STOC. pp. 740–748 (2001)
20. Schulman, R., Winfree, E.: DNA 10. In: Ferretti, C., Mauri, G., Zandron, C. (eds.) DNA Computing. LNCS, vol. 3384, pp. 319–328. Springer, Heidelberg (2005)
21. Grothues, D., Cantor, C.R., Smith, C.L.: Nuc. Acid. Res. 21(5), 1321–1322 (1993)
22. Rosenbaum, D.M., Liu, D.R.: J. Am. Chem. Soc. 125, 13924–13925 (2003)
23. Fujibayashi, K., Murata, S.: DNA 10. In: Ferretti, C., Mauri, G., Zandron, C. (eds.) DNA Computing. LNCS, vol. 3384, pp. 113–127. Springer, Heidelberg (2005)
24. Fujibayashi, K., Zhang, D., Winfree, E., Murata, S.: In submission

Constant-Size Tileset for Solving an NP-Complete Problem in Nondeterministic Linear Time

Yuriy Brun

Department of Computer Science
University of Southern California
Los Angeles, CA 90089
ybrun@usc.edu

Abstract. The tile assembly model, a formal model of crystal growth, is of special interest to computer scientists and mathematicians because it is universal [1]. Therefore, tile assembly model systems can compute all the functions that computers compute. In this paper, I formally define what it means for a system to nondeterministically decide a set, and present a system that solves an NP-complete problem called SubsetSum. Because of the nature of NP-complete problems, this system can be used to solve all NP problems in polynomial time, with high probability. While the proof that the tile assembly model is universal [2] implies the construction of such systems, those systems are in some sense "large" and "slow." The system presented here uses $49 = \Theta(1)$ different tiles and computes in time linear in the input size. I also propose how such systems can be leveraged to program large distributed software systems.

1   Introduction

Self-assembly is a process that is ubiquitous in nature. Systems form on all scales via self-assembly: atoms self-assemble to form molecules, molecules to form complexes, and stars and planets to form galaxies. One manifestation of self-assembly is crystal growth: molecules self-assembling to form crystals. Crystal growth is an interesting area of research for computer scientists because it has been shown that, in theory, under careful control, crystals can compute [2]. The field of DNA computation demonstrated that DNA can be used to compute [3], solving NP-complete problems such as the satisfiability problem [4,5]. This idea of using molecules to compute nondeterministically is the driving motivation behind my work.

Winfree showed that DNA computation is Turing-universal [6]. While DNA computation suffers from relatively high error rates, the study of self-assembly shows how to utilize redundancy to design systems with built-in error correction [7,8,9,10,11]. Researchers have used DNA to assemble crystals with patterns of binary counters [12] and Sierpinski triangles [13], but while those crystals are deterministic, generating nondeterministic crystals may hold the power to solving complex problems quickly.

M.H. Garzon and H. Yan (Eds.): DNA 13, LNCS 4848, pp. 26–35, 2008.
© Springer-Verlag Berlin Heidelberg 2008

Two important questions about self-assembling systems that create shapes or compute functions are: "what is a minimal tile set that can accomplish this goal?" and "what is the minimum assembly time for this system?" For nondeterministic computation, the following question is also important: "what is the probability of assembling the crystal that encodes the solution?" Researchers have answered these questions for n-long linear polymers [14] and $n \times n$ squares (minimum tileset of size $\Theta(\frac{\log n}{\log \log n})$ and optimal assembly time of $\Theta(n)$) [15,16,17]. A key issue related to assembling squares is the assembly of small binary counters, which theoretically can have as few as 7 tile types [18].

Other early attempts at nondeterministic computation include a proposal by Lagoudakis et al. to solve the satisfiability problem [19]. They informally define a system that nondeterministically computes whether or not an n-variable boolean formula is satisfiable using $\Theta(n^2)$ distinct tiles. In contrast, all the systems I present in this paper use $\Theta(1)$ distinct tiles.

Barish et al. have demonstrated a DNA implementation of tile systems, one that copies an input and another that counts in binary [12]. Similarly, Rothemund et al. have demonstrated a DNA implementation of a tile system that computes the *xor* function, resulting in a Sierpinski triangle [13]. These systems grow crystals using double-crossover complexes [20] as tiles. The theoretical underpinnings of these systems are closely related to the work presented here because these systems compute functions.

1.1   Tile Assembly Model

The tile assembly model [15,1,2] is a formal model of crystal growth. It was designed to model self-assembly of molecules such as DNA. It is an extension of a model proposed by Wang [21]. The model was fully defined by Rothemund and Winfree [15], and the definitions I use are similar to those. Full formal definitions can be found in [22].

Intuitively, the model has *tiles*, or squares, that stick or do not stick together based on various *binding domains* on their four sides. Each tile has a binding domain on its north, east, south, and west side. The four binding domains, elements of a finite alphabet Σ, define the type of the tile. The strength of the binding domains are defined by the *strength function* g. The placement of some tiles on a 2-D grid is called a *configuration*, and a tile may *attach* in empty positions on the grid if the total strength of all the binding domains on that tile that match its neighbors exceeds the current *temperature* (a natural number). Finally, a *tile system* $\mathbb{S}$ is a triple $\langle T, g, \tau \rangle$, where T is a finite set of tiles, g is a strength function, and $\tau \in \mathbb{N}$ is the temperature, where $\mathbb{N} = \mathbb{Z}_{\geq 0}$.

Starting from a *seed configuration* S, tiles may attach to form new configurations. If that process terminates, the resulting configuration is said to be *final*. At some times, it may be possible for more than one tile to attach at a given position, or there may be more than one position where a tile can attach. If for all sequences of tile attachments, all possible final configurations are identical, then $\mathbb{S}$ is said to produce a *unique* final configuration on S. The *assembly time* of the system is the minimal number of steps it takes to build a final configuration, assuming maximum parallelism.

In [22] and [23], I give formal definitions of what it means for a tile system to compute functions, both deterministically and nondeterministically. Here, I am interested in computing a particular subset of functions, the characteristic functions of subsets of the natural numbers. A characteristic function of a set has value 1 on arguments that are elements of that set and value 0 on arguments that are not elements of that set. Typically, in computer science, programs and systems that compute such functions are said to decide the set. Since for all constants $m \in \mathbb{N}$, the cardinalities of $\mathbb{N}^m$ and $\mathbb{N}$ are the same, it makes sense to talk about deciding subsets of $\mathbb{N}^m$.

Let $\Omega \subseteq \mathbb{N}^m$ be a set. A tile system $\mathbb{S} = \langle T, g, \tau \rangle$ *nondeterministically decides* Ω with identifier tile $r \in T$ iff for all $\boldsymbol{a} \in \mathbb{N}^m$, there exists a seed configuration S that encodes $\boldsymbol{a}$ and for all final configurations F that $\mathbb{S}$ produces on S, $r \in F(\mathbb{Z}^2)$ iff $\boldsymbol{a} \in \Omega$, and there exists at least one final configuration F with r attached. In other words, the *identifier* tile r attaches to one or more of the nondeterministic executions iff the seed encodes an element of Ω.

This paper provides the definitions necessary for understanding the below constructions and theorems. More complete versions of the definitions and formal proofs of the theorems presented below can be found in [24]. In the remainder of this paper, I require systems to encode their inputs in binary, and call the set of tiles used to encode the input Γ.

2   Solving SubsetSum

SubsetSum is a well known NP-complete problem. The set *SubsetSum* is a set of pairs: a finite sequence $\boldsymbol{B} = \langle B_1, B_2, \cdots, B_n \rangle \in \mathbb{N}^n$, and a target number $v \in \mathbb{N}$, such that $\langle \boldsymbol{B}, v \rangle \in SubsetSum$ iff $\exists \boldsymbol{c} = \langle c_1, c_2, \cdots, c_n \rangle \in \{0, 1\}^n$ such that $\sum_{i=1}^{n} c_i B_i = v$. In other words, the sum of some subset of numbers of $\boldsymbol{B}$ equals exactly v.

In order to explain the system that nondeterministically decides *SubsetSum*, I will first define three smaller systems that perform pieces of the necessary computation. The first system subtracts numbers, and given the right conditions, will subtract a B_i from v. The second system computes the identity function and just copies information (this system will be used when a B_i should not be subtracted from v). The third system nondeterministically guesses whether the next B_i should or should not be subtracted. Finally, I will add a few other tiles that ensure that the computations went as planned and attach an identifier tile if the execution found that $\langle \boldsymbol{B}, v \rangle \in SubsetSum$. The system works by nondeterministically choosing a subset of $\boldsymbol{B}$ to subtract from v and comparing the result to 0.

2.1   Subtraction

In this section, I will describe a system that subtracts positive integers. It is similar to one of the addition systems from [22], contains 16 tiles, and will subtract one bit per row of computation.

Figure 1(a) shows the 16 tiles of T_-. The value in the middle of each tile t represents that tile's $v(t)$ value. Intuitively, the system will subtract the ith bit on the ith row. The tiles to the right of the ith location will be blue; the tile in the ith location will be yellow; the next tile, the one in the $(i+1)^{st}$ location, will be magenta; and the rest of the tiles will be green. The purpose of the yellow and magenta tiles is to compute the diagonal line, marking the ith position on the ith row.

$\mathbb{S}_-$ is a system that is capable of subtracting numbers, and it does so in time linear in the input. Full proofs of these statements are available in [24]. In Figure 1(b), the system computes $221 - 214 = 7$ In Figure 1(c), the system attempts to compute $221 - 246$, but because $246 > 221$, the computation fails.

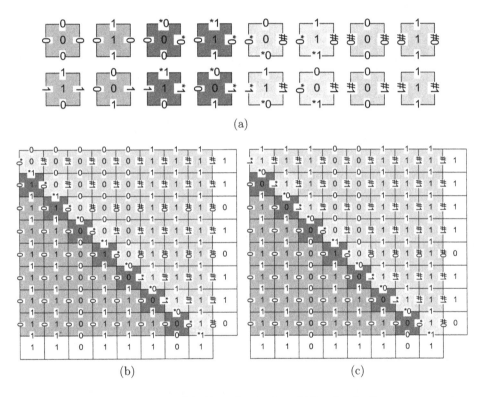

(a)

(b)   (c)

Fig. 1. There are 16 tiles in T_- (a). The value in the middle of each tile t represents that tile's $v(t)$ value. In (b), the system subtracts $214 = 11010110_2$ from $221 = 110111101_2$ to get $7 = 111_2$. The inputs are encoded along the bottom row ($221 = 110111101_2$) and right-most column ($214 = 11010110_2$). The output is on the top row ($7 = 00000111_2$). Because $214 \leq 221$, all the west binding domains of the left-most column contain a 0. In (c), the system attempts to subtract $246 = 11110110$ from $221 = 110111101_2$, but because $246 > 221$, the computation fails and indicates its failure with the top- and left-most west binding domain containing a 1.

This system is very similar to an adding system from [22], but not the smallest adding system from [22]. While this system has 16 tiles, it is possible to design a subtracting system with 8 tiles, that is similar to the 8-tile adding system from [22].

2.2   Identity

I now describe a system that ignores the input on the right-most column, and simply copies upwards the input from the bottom row. This is a fairly straight-forward system that will not need much explanation. Figure 2(a) shows the 4 tiles in T_x and Figure 2(b) shows a sample execution of the $\mathbb{S}_x$ system.

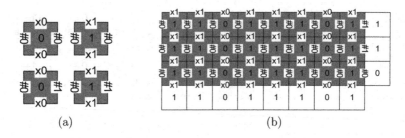

(a)   (b)

Fig. 2. There are 4 tiles in T_x (a); the value in the middle of each tile t represents that tile's $v(t)$ value. In an example of an $\mathbb{S}_x$ execution (b), the system simply copies the input on the bottom row upwards, to the top column.

$\mathbb{S}_x$ is a system that is capable of computing the identify function, and it does so in time linear in the input. Again, full proofs of these statements are available in [24].

2.3   Nondeterministic Guess

In this section, I describe a system that nondeterministically decides whether or not the next B_i should be subtracted from v. It does so by encoding the input for either the $\mathbb{S}_-$ system or the $\mathbb{S}_x$ system.

$\mathbb{S}_?$ is a system that is capable of nondeterministically preparing a valid seed configuration for either $\mathbb{S}_-$ or $\mathbb{S}_x$, and it does so in time linear in the input. Full proofs of these statements are available in [24].

Figure 3 shows two possible executions of $\mathbb{S}_?$. In Figure 3(b), the system attaches tiles with ! east-west binding domains, preparing a valid seed for $\mathbb{S}_-$, and in Figure 3(c), the system attaches tiles with x east-west binding domains, preparing a valid seed for $\mathbb{S}_x$. Only one tile, the orange tile, attaches nondeterministically, determining which tiles attach to its west.

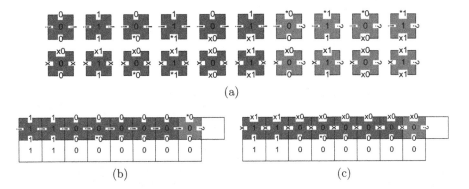

(a)

(b)   (c)

Fig. 3. There are 20 tiles in $T_?$ (a). The value in the middle of each tile t represents that tile's $v(t)$ value. Unlike the red tiles, the orange tiles do not have unique east-south binding domain pairs, and thus will attach nondeterministically. In (b), the system attaches tiles with ! east-west binding domains, preparing a valid seed for $\mathbb{S}_-$, and in (c), the system attaches tiles with x east-west binding domains, preparing a valid seed for $\mathbb{S}_x$.

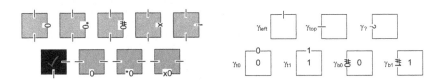

Fig. 4. There are 9 tiles in $T_\checkmark$ (a); the black tile with a $\checkmark$ in the middle will serve as the identifier tile. There are 7 tiles in Γ_{SS} (b); the value in the middle of each tile t represents that tile's $v(t)$ value and each tile's name is written on its left.

2.4   Deciding *SubsetSum*

I have described three systems that I will now use to design a system to decide *SubsetSum*. Intuitively, I plan to write out the elements of $\boldsymbol{B}$ on a column and v on a row, and the system will nondeterministically choose some of the elements from $\boldsymbol{B}$ to subtract from v. The system will then check to make sure that no subtracted element was larger than the number it was being subtracted from, and whether the result is 0. If the result is 0, then a special identifier tile will attach to signify that $\langle \boldsymbol{B}, v \rangle \in SubsetSum$.

Theorem 1. *Let $\Sigma_{SS} = \{0, 1, {}^*0, {}^*1, \#0, \#1, x0, x1, \#0, \#1, ?, !, 0, 1, x0, x1, {}^*0, {}^*1\}$. Let $T_{SS} = T_- \cup T_x \cup T_? \cup T_\checkmark$, where $T_\checkmark$ is defined by Figure 4(a). Let $g_{SS} = 1$ and $\tau_{SS} = 2$. Let $\mathbb{S}_{SS} = \langle T_{SS}, g_{SS}, \tau_{SS} \rangle$. Then $\mathbb{S}_{SS}$ nondeterministically decides SubsetSum with the black $\checkmark$ tile from $T_\checkmark$ as the identifier tile.*

I refer the reader to [24] for a full proof of theorem 1.

Figure 5 shows an example execution of $\mathbb{S}_{SS}$. Figure 5(a) encodes a seed configuration with $v = 75 = 1001011_2$ along the bottom row and $\boldsymbol{B} = \langle 11 = 1011_2,$

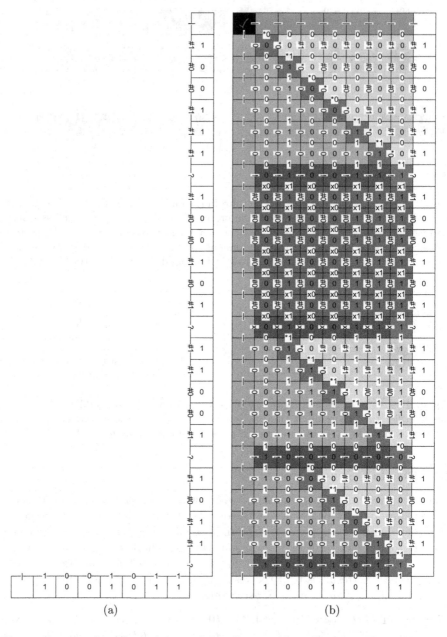

Fig. 5. An example of $\mathbb{S}_{SS}$ solving a *SubsetSum* problem. Here, $v = 75 = 1001011_2$, and $\boldsymbol{B} = \langle 11 = 1011_2, 25 = 11001_2, 37 = 100101_2, 39 = 100111_2 \rangle$. The seed configuration encodes v on the bottom row and $\boldsymbol{B}$ on the right-most column (a). The fact that $75 = 11 + 25 + 39$ implies that $\langle \boldsymbol{B}, t \rangle \in SubsetSum$, thus at least one final configuration (b) contains the ✓ tile.

$25 = 11001_2$, $37 = 100101_2$, $39 = 100111_2$⟩ along the right-most column. Note that the seed is encoded using the tiles shown in Figure 4(b). Tiles from T_{SS} attach to the seed configuration, nondeterministically testing all possible values of $c \in \{0,1\}^4$. Figure 5(b) shows one such possible execution, the one that corresponds to $c = \langle 1,1,0,1 \rangle$. Because $11 + 25 + 39 = 75$, the ✓ tile attaches in the top left corner.

The assembly time of $\mathbb{S}_{SS}$ is linear in the size of the input (number of bits in $\langle B, v \rangle$), and assuming each tile that may attach to a configuration at a certain position attaches there with a uniform probability distribution, the probability that a single nondeterministic execution of $\mathbb{S}_{SS}$ succeeds in attaching a ✓ tile if $\langle B, v \rangle \in SubsetSum$ is at least $\left(\frac{1}{2}\right)^n$. The proofs of both these statements can be found in [24].

Therefore, a parallel implementation of $\mathbb{S}_{SS}$, such as a DNA implementation like those in [12,13], with 2^n seeds has at least a $1 - \frac{1}{e} \geq 0.5$ chance of correctly deciding whether a $\langle B, v \rangle \in SubsetSum$. An implementation with 100 times as many seeds has at least a $1 - \left(\frac{1}{e}\right)^{100}$ chance.

Note that T_{SS} has 49 computational tile types and uses 7 tile types to encode the input.

3   Software Systems

Fault and adversary tolerance have become not only desirable but required properties of software systems because mission-critical systems are commonly distributed on large networks of insecure nodes. Further, users of such distributed systems may desire their private data to remain private. It is possible for computers on a large network to act as tiles to compute. For example, one can solve NP-complete problems by reducing them to $SubsetSum$ and then using $\mathbb{S}_{SS}$ to solve them, as illustrated in Figure 6. Such a software system can leverage the error-correction work in tile assembly [7,8,9,10,11] to automate fault and adversary tolerance, and distribute computation over network in a way that no small group of nodes nodes the private inputs to the computation [25,26,27].

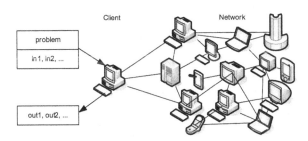

Fig. 6. A schematic of a system implementing a tile-style architecture

4   Contributions

The tile assembly model is a formal model of self-assembly and crystal growth. Here, I defined what it means for a tile system to decide a set and designed and explored a system that decides an NP-complete problem *SubsetSum*. The system computes at temperature two and uses 49 computational tile types and 7 tile types to encode the input. The system computes in time linear in the input size and each nondeterministic assembly has a probability of success of at least $\left(\frac{1}{2}\right)^n$, and that probability can be brought exponentially close to 1 at a linear cost in the number of seeds.

On the way to defining a system that decides *SubsetSum*, I also defined a system that deterministically subtracts numbers. This system uses 16 computational tile types and executes in time linear in the input size. I stated without proof that there exists an 8-tile subtracting system based on the 8-tile adding system from [22].

Finally, I described some preliminary work on using theoretical self-assembly to design complex computational software systems.

References

1. Winfree, E.: Simulations of computing by self-assembly of DNA. Technical Report CS-TR:1998:22, California Insitute of Technology, Pasadena, CA, USA (1998)
2. Winfree, E.: Algorithmic Self-Assembly of DNA. PhD thesis, California Insitute of Technology, Pasadena, CA, USA (June 1998)
3. Adleman, L.: Molecular computation of solutions to combinatorial problems. Science 266, 1021–1024 (1994)
4. Braich, R., Johnson, C.R., Rothemund, P.W.K., Hwang, D., Chelyapov, N., Adleman, L.: Solution of a satisfiability problem on a gel-based DNA computer. In: Proceedings of DNA Computing: 6th International Workshop on DNA-Based Computers (DNA 2000), Leiden, The Netherlands, pp. 27–38 (June 2000)
5. Braich, R., Chelyapov, N., Johnson, C.R., Rothemund, P.W.K., Adleman, L.: Solution of a 20-variable 3-SAT problem on a DNA computer. Science 296(5567), 499–502 (2002)
6. Winfree, E.: On the computational power of DNA annealing and ligation. DNA Based Computers 199–221 (1996)
7. Winfree, E., Bekbolatov, R.: Proofreading tile sets: Error correction for algorithmic self-assembly. In: Proceedings of the 43rd Annual IEEE Symposium on Foundations of Computer Science (FOCS 2002), Madison, WI, USA, June 2003, vol. 2943, pp. 126–144. IEEE Computer Society Press, Los Alamitos (2003)
8. Baryshnikov, Y., Coffman, E.G., Seeman, N., Yimwadsana, T.: Self correcting self assembly: Growth models and the hammersley process. In: Carbone, A., Pierce, N.A. (eds.) DNA Computing. LNCS, vol. 3892, Springer, Heidelberg (2006)
9. Chen, H.L., Goel, A.: Error free self-assembly with error prone tiles. In: Ferretti, C., Mauri, G., Zandron, C. (eds.) DNA Computing. LNCS, vol. 3384, Springer, Heidelberg (2005)
10. Reif, J.H., Sahu, S., Yin, P.: Compact error-resilient computational DNA tiling assemblies. In: Ferretti, C., Mauri, G., Zandron, C. (eds.) DNA Computing. LNCS, vol. 3384, Springer, Heidelberg (2005)

11. Winfree, E.: Self-healing tile sets. Nanotechnology: Science and Computation, 55–78 (2006)
12. Barish, R., Rothemund, P.W.K., Winfree, E.: Two computational primitives for algorithmic self-assembly: Copying and counting. Nano Letters 5(12), 2586–2592 (2005)
13. Rothemund, P.W.K., Papadakis, N., Winfree, E.: Algorithmic self-assembly of DNA Sierpinski triangles. PLoS Biology 2(12), 424 (2004)
14. Adleman, L., Cheng, Q., Goel, A., Huang, M.-D., Wasserman, H.: Linear self-assemblies: Equilibria, entropy, and convergence rates. In: Proceedings of the 6th International Conference on Difference Equations and Applications (ICDEA 2001), Augsburg, Germany (June 2001)
15. Rothemund, P.W.K., Winfree, E.: The program-size complexity of self-assembled squares. In: Proceedings of the ACM Symposium on Theory of Computing (STOC 2000, Portland, OR, USA, pp. 459–468. ACM Press, New York (2000)
16. Adleman, L., Cheng, Q., Goel, A., Huang, M.-D., Kempe, D., de Espanes, P.M., Rothemund, P.W.K.: Combinatorial optimization problems in self-assembly. In: Proceedings of the ACM Symposium on Theory of Computing (STOC 2002), Montreal, Quebec, Canada, pp. 23–32. ACM Press, New York (2002)
17. Adleman, L., Goel, A., Huang, M.-D., de Espanes, P.M.: Running time and program size for self-assembled squares. In: Proceedings of the ACM Symposium on Theory of Computing (STOC 2002), Montreal, Quebec, Canada, pp. 740–748. ACM Press, New York (2001)
18. de Espanes, P.M.: Computerized exhaustive search for optimal self-assembly counters. In: Proceedings of the 2nd Foundations of Nanoscience: Self-Assembled Architectures and Devices (FNANO 2005), Snowbird, UT, USA, pp. 24–25 (April 2005)
19. Lagoudakis, M.G., LaBean, T.H.: 2D DNA self-assembly for satisfiability. DIMACS Series in Discrete Mathematics and Theoretical Computer Science 54, 141–154 (1999)
20. Fu, T.J., Seeman, N.C.: DNA double-crossover molecules. Biochemistry 32(13), 3211–3220 (1993)
21. Wang, H.: Proving theorems by pattern recognition. II. Bell System Technical Journal 40, 1–42 (1961)
22. Brun, Y.: Arithmetic computation in the tile assembly model: Addition and multiplication. Theoretical Computer Science 378, 17–31 (2007)
23. Brun, Y.: Nondeterministic polynomial time factoring in the tile assembly model. Theoretical Computer Science (2007), doi:10.1016/j.tcs.2007.07.051
24. Brun, Y.: Solving NP-complete problems in the tile assembly model. Theoretical Computer Science (2007), doi:10.1016/j.tcs.2007.07.052
25. Brun, Y., Medvidovic, N.: An architectural style for solving computationally intensive problems on large networks. In: Proceedings of Software Engineering for Adaptive and Self-Managing Systems (SEAMS 2007), Minneapolis, MN, USA (May 2007)
26. Brun, Y., Medvidovic, N.: Fault and adversary tolerance as an emergent property of distributed systems' software architectures. In: Proceedings of the 2nd International Workshop on Engineering Fault Tolerant Systems (EFTS 2007), Dubrovnik, Croatia (September 2007)
27. Brun, Y.: Discreetly distributing computation via self-assembly. Technical Report USC-CSSE-2007-714, Center for Software Engineering, University of Southern California (2007)

Solutions to Computational Problems Through Gene Assembly[*]

Artiom Alhazov[2], Ion Petre[1,2], and Vladimir Rogojin[2]

[1] Academy of Finland
[2] Computational Biomodelling Laboratory
Turku Center for Computer Science, FIN-20520 Turku, Finland
aalhazov@abo.fi, ipetre@abo.fi, vrogojin@abo.fi

Abstract. Gene assembly in ciliates is an impressive computational process. Ciliates have a unique way of storing their genetic information in two fundamentally different forms within their two types of nuclei. Micronuclear genes are broken into blocks (called MDSs), with MDSs shuffled and separated by non-coding material; some of the MDSs may even be inverted. During gene assembly, all MDSs are sorted in the correct order to yield the transcription-able macronuclear gene. Based on the intramolecular model for gene assembly, we prove in this paper that gene assembly may be used in principle to solve computational problems. We prove that any given instance of the hamiltonian path problem may be encoded in a suitable way in the form of an 'artificial' gene so that gene assembly is successful on that gene-like pattern if and only if the given problem has an affirmative answer.

1 Introduction

Ciliates are unicellular organisms existing for over a billion years, forming a group of thousands of species. A common feature they share is that their cell contains two kinds of nuclei that have different functionality - micronuclei act as germline nuclei and macronuclei act as the somatic nuclei.

During the sexual reproduction, the macronuclei are destroyed and one haploid micronucleus is transformed into a macronucleus. The gene operations have a definite computational flavor: some DNA segments (internally eliminated sequences, IES) are eliminated, others (macronuclear destined sequences, MDS) are reordered; some MDSs are also inverted. The process is driven by splicing on specific sequences on the ends of MDSs, called pointers: the end of each MDS matches the beginning of the MDS that should follow it in the assembled gene. Two main models exist for the gene assembly, one intermolecular, see [13,14] and one intramolecular, see [8,18]. In this article we consider the latter one.

In 1994 a famous experiment of L. Adleman took place giving an example how biological processes can be interpreted as computing (a small instance of

---

[*] A. Alhazov (artiom@math.md) and V. Rogojin are on leave of absence from Institute of Mathematics and Computer Science of Academy of Sciences of Moldova, Chisinau MD-2028 Moldova.

M.H. Garzon and H. Yan (Eds.): DNA 13, LNCS 4848, pp. 36–45, 2008.

hamiltonian path problem, HPP was represented by DNA molecules and solved by molecular biology tools). The current paper considers replacing such DNA operations as annealing by ciliate operations, therefore, we speak about ciliate-based computing.

Indeed, what ciliates do in gene assembly is sorting, inversion and excision of DNA sequences. Therefore, our strategy is to encode an arbitrary instance of HPP into a *hypothetical* micronuclear gene, assemble the gene using the intramolecular model, and filter the result of the assembly to get the answer to HPP, if there is any.

This is a novel approach to DNA computing, using a model for gene assembly in ciliates. Although the computational flavor of ciliates has been shown previously in[13,14,15] where the Turing universality of various assembly models was proved, this is the first attempt at using (in principle) gene assembly for solving mathematical problems. If ever implemented in living cells, the solution potentially has the advantage that the cell itself implements many steps of the procedure, including selecting the resulting substring and its replication. It is important to underline that we only propose here a conceptual (theoretical) approach to ciliate-based computing. We only briefly discuss some issues related to potential experimental implementations of our approach in Section 8.

2   Definitions

For an alphabet Σ we denote by Σ^* the set of all finite strings over Σ. We denote the empty string by Λ. For strings u, v over Σ we say that u is a *substring* of v, denoted $u \leq v$, if $v = xuy$, for some strings x, y. Let $\overline{\Sigma} = \{\overline{a} \mid a \in \Sigma\}$ be complement symbols of Σ; we call $u \in (\Sigma \cup \overline{\Sigma})^*$ a *signed string*. The complement operation is extended to signed strings by $\overline{\overline{a}} = a$, $a \in \Sigma$ and $\overline{a_1 a_2 \cdots a_k} = \overline{a_k} \cdots \overline{a_2}\, \overline{a_1}$, $a_i \in \Sigma \cup \overline{\Sigma}$, $1 \leq i \leq k$.

We call a (directed) *graph* a tuple $G = (V, E)$, where V is a finite set of nodes, and $E \subseteq \{(p, q) \mid p, q \in V\}$ is a set of edges. A sequence $q_1 q_2 \cdots q_k$ of nodes $q_i \in V$, $1 \leq i \leq k$ is called a path if $(q_i, q_{i+1}) \in E$, $1 \leq i \leq k - 1$.

The *hamiltonian Path Problem* for a directed graph $G = (V, E)$, given the initial node p and a final node q is the problem of deciding whether G has an acyclic path from p to q containing all nodes of the graph (it is implicit in this definition that all nodes are visited only once). Such a path is called *hamiltonian*. The hamiltonian path problem is a known NP-complete problem, see [16].

For some results we need also the following graph construction, which we call *bipartite transformation*: given a graph $G = (V, E)$ we construct a graph $bi(G) = G' = (V', E')$ where $V' = \{p, p' \mid p \in V\}$ and $E' = \{(p', q) \mid (p, q) \in E\} \cup \{(p, p') \mid p \in V\}$. In other words, we split each node p in two nodes p and p' connected by an edge, and replace the edges (p, q) of the original graph by (p', q). This gives us a bipartite graph where every edge (p, q) in G corresponds to a path $pp'q$ in G'.

Consider a graph $G = (V, E)$ with $V = \{p_1, \cdots, p_n\}$ and the hamiltonian path problem from p_1 to p_n. Due to the technical reasons, throughout the paper we use the following transformation: add two nodes $b, e \notin V$ to G and look for paths from b to e in graph $ext(G, p_1, p_n) = G'' = (V \cup \{b, e\}, E \cup \{(b, p_1), (p_n, e)\})$. Notice that the edge from b is unique and so is the edge to e, and there are no edges to b and no edges from e. Clearly, u is a path in G from p_1 to p_n if and only if bue is a path in G' from b to e. Therefore, this HPP is equivalent to the original one.

Example 1. For the graph $G_1 = (V_1 = \{1, 2, 3\}, E_1 = \{(2, 1), (3, 1), (3, 2)\})$ we illustrate in Figure 1 the graph $ext(G_1, 3, 1)$.

Fig. 1. A hamiltonian path in $ext(G_1, 3, 1)$ is $b321e$

3   Gene Assembly

The following three molecular operations are postulated in the intramolecular model to explain the gene assembly process, see [8] and [18]:

- *Loop, Direct Repeat* (ld) is applied on a pair of directly repeating pointers in the molecule. The molecule folds on itself to form a loop so that recombination is facilitated on the two occurrences of that pointer. As a result, the part of the molecule between repeating pointers is excised from the molecule in the form of circular molecule, while the parts from both sides of the excised molecule splice together;
- *Hairpin, Inverted Repeat* (hi) is applied on a pair of pointers, where one is an inverted repeat of the other one. The molecule is folded as a hairpin to facilitate recombination on those pointers. As a result of the operation, the part of the molecule flanked by the repeating pointers is inverted;
- *Double Loop, Alternating Direct Repeat* (dlad) is applicable to the overlapping direct repetitions of pointers, i.e., if we have a molecule of the form $\cdots p \cdots q \cdots p \cdots q \cdots$. The molecule folds to form a double loop so that a double recombination on p and q is facilitated. As a result, the parts of the molecule between the first and the second occurrences of p and q are exchanged.

A sequence of nucleotides is considered to act as a pointer only when placed at the border of an MDS and an IES. Note that after applying an operation on a certain pointer, that pointer remains as a sequence of nucleotides in the molecule, but ceases to participate in other operations, because it does not reside anymore on the border between an IES and an MDS.

We represent each molecule through its sequence of MDSs. In turn, we represent each MDS through its incoming and outgoing pointers, as well as through the sequence of pointers incorporated in the MDS as a result of applying previous operations. To formalize this definition, let $\Sigma_P = \{p_1, p_2, \cdots, p_n\}$ be the set of pointers. Then we represent an MDS by a triple $M = (p, u, q)$ where $p, q \in \Sigma_P$ are called *incoming* and *outgoing* pointers, respectively, and $u \in \Sigma_P^*$ is the *content*. We say that the length of M is $|M| = |puq|$. Let us denote by Σ_M the set $\{(p, u, q) \mid p, q \in \Sigma_P, \ u \in \Sigma_P^*\}$ of all MDSs. The complement of an MDS (p, u, q) is $(\overline{q}, \overline{u}, \overline{p})$ and $\overline{\Sigma_M} = \{\overline{M} \mid M \in \Sigma_M\}$. Finally, we call *descriptors* the strings from the set $S = (\Sigma_P \cup \overline{\Sigma_P} \cup \Sigma_M \cup \overline{\Sigma_M})^*$.

It is important to note that we consider in this paper descriptors in which pointers may have an arbitrary number of occurrences. Although in any successful assembly only two such occurrences are actually used, this multiplicity is the foundation of our ciliate-based search algorithm for a solution to HPP: choosing non-deterministically various occurrences of a given pointer in the assembly yield the detection of various paths in the given graph.

We formalize the ld, hi, and dlad operations as rewriting rules on descriptors as shown bellow. Note that all rewriting rules are non-deterministic: in general, for a given input, a rule may be applied in several ways, leading to different results. We assume here a non-deterministic computing paradigm: a descriptor may be assembled successfully if there exists a sequence of rules leading to its assembly, see bellow for formal details.

1 $\psi_1(q, u, p)\psi_2(p, v, r)\psi_3 \Rightarrow^{\mathsf{ld}_p} \psi_1(q, upv, r)\psi_3$;

2.1 $\psi_1(p, u, q)\psi_2(\overline{p}, \overline{v}, \overline{r})\psi_3 \Rightarrow^{\mathsf{hi}_p} \psi_1 p\overline{\psi_2}(\overline{q}, \overline{u}\ \overline{p}\ \overline{v}, \overline{r})\psi_3$;

2.2 $\psi_1(q, u, p)\psi_2(\overline{r}, \overline{v}, \overline{p})\psi_3 \Rightarrow^{\mathsf{hi}_p} \psi_1(q, upv, r)\overline{\psi_2}\ \overline{p}\psi_3$;

3.1 $\psi_1(p, u_1, r_1)\psi_2(q, u_2, r_2)\psi_3(r_3, u_3, p)\psi_4(r_4, u_4, q)\psi_5 \Rightarrow^{\mathsf{dlad}_{p,q}}$
$\quad \psi_1 p\psi_4(r_4, u_4 q u_2, r_2)\psi_3(r_3, u_3 p u_1, r_1)\psi_2 q\psi_5$;

3.2 $\psi_1(p, u_1, r_1)\psi_2(r_2, u_2, q)\psi_3(r_3, u_3, p)\psi_4(q, u_4, r_4)\psi_5 \Rightarrow^{\mathsf{dlad}_{p,q}}$
$\quad \psi_1 p\psi_4 q\psi_3(r_3, u_3 p u_1, r_1)\psi_2(r_2, u_2 q u_4, r_4)\psi_5$;

3.3 $\psi_1(r_1, u_1, p)\psi_2(q, u_2, r_2)\psi_3(p, u_3, r_3)\psi_4(r_4, u_4, q)\psi_5 \Rightarrow^{\mathsf{dlad}_{p,q}}$
$\quad \psi_1(r_1, u_1 p u_3, r_3)\psi_4(r_4, u_4 q u_2, r_2)\psi_3 p\psi_2 q\psi_5$;

3.4 $\psi_1(r_1, u_1, p)\psi_2(r_2, u_2, q)\psi_3(p, u_3, r_3)\psi_4(q, u_4, r_4)\psi_5 \Rightarrow^{\mathsf{dlad}_{p,q}}$
$\quad \psi_1(r_1, u_1 p u_3, r_3)\psi_4 q\psi_3 p\psi_2(r_2, u_2 q u_4, r_4)\psi_5$;

3.1′ $\psi_1(p, u_1, r_1)\psi_2(q, u_2, p)\psi_4(r_4, u_4, q)\psi_5 \Rightarrow^{\mathsf{dlad}_{p,q}}$
$\quad \psi_1 p\psi_4(r_4, u_4 q u_2 p u_1, r_1)\psi_2 q\psi_5$;

3.2′ $\psi_1(p, u_1, q)\psi_3(r_3, u_3, p)\psi_4(q, u_4, r_4)\psi_5 \Rightarrow^{\mathsf{dlad}_{p,q}}$
$\quad \psi_1 p\psi_4 q\psi_3(r_3, u_3 p u_1 q u_4, r_4)\psi_5$;

3.3′ $\psi_1(r_1, u_1, p)\psi_2(q, u_2, r_2)\psi_3(p, u_3, q)\psi_5 \Rightarrow^{\mathsf{dlad}_{p,q}}$
$\quad \psi_1(r_1, u_1 p u_3 q u_2, r_2)\psi_3 p\psi_2 q\psi_5$;

where ψ_i are descriptors, q, r, r_i are pointers, u, v, u_i are sequences of pointers.

Consider also the operation cut on descriptors defined in the following way: $\mathsf{cut}(\psi_1(b, u, e)\psi_2) = (b, u, e)$. We call an MDS (b, u, e) a *successful assembly* of the descriptor ψ if $(b, u, e) = \mathsf{cut}(\phi_1(\phi_2(\cdots \phi_k(\psi)\cdots)))$, with $\phi_1, \cdots, \phi_k$ being some ld, hi, or dlad rules.

For a descriptor ψ we denote by $\mathcal{L}_{ld}(\psi)$ ($\mathcal{L}_{hi}(\psi)$, $\mathcal{L}_{dlad}(\psi)$) the set of all MDSs assembled successfully from ψ using only ld-rules (hi, dlad, respectively).

For more details about the formalization of the gene structure and the intramolecular operations we refer to [4], [5], [6], [7], [9], [10], [11], [20], [21], as well as to the monograph [3].

4 Computing Through Gene Assembly

Our principle of computing through gene assembly is the following: given a (mathematical) problem, we encode its input into a descriptor as defined in the previous section in such a way that the problem has a solution if and only if the associated descriptor has a successful assembly with certain properties. Moreover, the result of the assembly encodes the solution to the problem.

As our computational problem of choice we consider in this paper the hamiltonian path problem (HPP): given a directed graph $G = (V, E)$ and two nodes $p, q \in V$ one needs to decide whether or not G has a hamiltonian path from p to q. To solve the problem through the gene assembly by ld only, hi only or dlad only, we encode the set of edges of graph $G' = ext(G, p, q)$ into certain descriptors ψ_G^{ld}, ψ_G^{hi} or ψ_G^{dlad} respectively. Also, we encode any path v in G' through an MDS M_v using a construction described bellow. We prove then in each case that the graph G contains a hamiltonian path u if and only if descriptors ψ_G^{ld}, ψ_G^{hi} and ψ_G^{dlad} can be successfully assembled to descriptors containing MDS M_{bue}.

Let $G = (V, E)$ be a directed graph and $f = (p', q') \in E$ an edge of G. We then associate to f the MDS $M_f = (p', \Lambda, q')$. In general, for a set of edges $\{(q_1, q_2), (q_2, q_3), \cdots (q_{k-1}, q_k)\}$ of G, we encode the path $u = q_1 q_2 \cdots q_{k-1} q_k$ of G through the MDS $M_u = (q_1, q_2 \cdots q_{k-1}, q_k)$.

We say that *a node r appears in an MDS (p, u, q)* if symbol r appears in the string puq.

5 Computing Using ld Only

In this section we consider a (theoretical) solution to the hamiltonian path problem through gene assembly with ld only to be used throughout the assembly.

Let $G = (V, E)$ be a directed graph with $V = \{p_1, p_2, \cdots, p_n\}$, $n > 0$, and consider the hamiltonian path problem with p_1 as the starting node and p_n as the ending node. We reduce it to the same problem for graph $G' = ext(G, p_1, p_n)$, starting node b and ending node e.

We say that *a descriptor ψ_G is associated to G* if it is of the form

$$\psi_G^{ld} = (b, \Lambda, p_1)\alpha_G^{n-1}(p_n, \Lambda, e), \text{ where } \alpha_G = \prod_{(p,q)\in E} (p, \Lambda, q) \text{ is a descriptor}$$

encoding all edges of G. Note that in general there are many descriptors associated to G, depending on the order in which the edges are encoded in α_G. As far as our solution to HPP is concerned, we may freely choose any of them.

Example 2. Consider the graph G_1 from Example 1. Then $\psi_{G_1}^{\mathsf{ld}} = (b, \Lambda, 3)(2, \Lambda, 1)$ $(3, \Lambda, 1)(3, \Lambda, 2)(2, \Lambda, 1)(3, \Lambda, 1)(3, \Lambda, 2)(1, \Lambda, e)$ is associated to G_1, and $\mathcal{L}_{\mathsf{ld}}(\psi_{G_1}^{\mathsf{ld}})$ $= \{(b, 321, e), (b, 31, e)\}$.

Using the encoding presented above, we can now prove that gene assembly solves the HPP problem. Let G be a directed graph and consider the hamiltonian path problem from b to e for $G' = ext(G, p_1, p_n)$. Let ψ_G^{ld} be an arbitrary descriptor associated to G. Then the following results hold.

Lemma 1. *Any successfully assembled MDS $M \in \mathcal{L}_{\mathsf{ld}}(\psi_G^{\mathsf{ld}})$ is associated to a path from b to e in G'.*

Lemma 2. *For every acyclic path u from b to e in G', $M_u \in \mathcal{L}_{\mathsf{ld}}(\psi_G^{\mathsf{ld}})$.*

Theorem 1. *The hamiltonian path problem for graph $G = (V, E)$ and nodes p_1, p_n has an affirmative answer if and only if there exists $M \in \mathcal{L}_{\mathsf{ld}}(\psi_G^{\mathsf{ld}})$ where all nodes appear and $|M| = |V| + 2$. In this case, M is an encoding of a hamiltonian path of $G' = ext(G, p_1, p_n)$ from b to e.*

6   Computing Using **hi** Only

In this section we consider a (theoretical) solution to the HPP problem through the gene assembly by hi operation only.

Consider a directed graph $G = (V, E)$ with $V = \{p_1, \cdots, p_n\}$, $n > 0$ and the hamiltonian path problem with p_1 as the starting node and p_n as the ending node. We solve an equivalent HPP for $G' = ext(G, p_1, p_n)$ from b to e instead.

We say that *a descriptor ψ_G^{hi} is associated to G* if it is of the form

$$\psi_G^{\mathsf{hi}} = (b, \Lambda, p_1) \prod_{(p,q) \in E \cup \{(p_n, e)\}} g_{p,q}, \text{ where } g_{p,q} = (x, \Lambda, y)(p, \Lambda, q)(\overline{z}, \Lambda, \overline{y})$$

is a descriptor encoding an edge (p, q). The order of the descriptors $g_{p,q}$ in ψ_G^{hi} is not important.

Example 3. Consider the graph G_1 from Example 1. Then $\psi_{G_1}^{\mathsf{hi}} = (b, \Lambda, 3)$ $(x, \Lambda, y)(2, \Lambda, 1)(\overline{z}, \Lambda, \overline{y})$ $(x, \Lambda, y)(3, \Lambda, 1)(\overline{z}, \Lambda, \overline{y})(x, \Lambda, y)(3, \Lambda, 2)(\overline{z}, \Lambda, \overline{y})(x, \Lambda, y)$ $(1, \Lambda, e)(\overline{z}, \Lambda, \overline{y})$ is associated to G_1. The corresponding successful assemblies are $\mathcal{L}_{\mathsf{hi}}(\psi_{G_1}^{\mathsf{hi}}) = \{(b, 321, e), (b, 31, e)\}$.

Using the encoding presented above, we can now prove that hi-operations solve the HPP problem. Consider a directed graph G and the hamiltonian path problem from b to e for $G' = ext(G, p_1, p_n)$. Then the following results hold.

Lemma 3. *Any successfully assembled MDS $M \in \mathcal{L}_{\mathsf{hi}}(\psi_G^{\mathsf{hi}})$ is associated to a path from b to e in G'.*

We omit the proof, since its idea is similar to that from Lemma 1.

Lemma 4. *For every path u from b to e in G' without repeating edges, there exists an MDS $M_u \in \mathcal{L}_{\text{hi}}(\psi_G^{\text{hi}})$.*

Theorem 2. *The hamiltonian path problem for graph $G = (V, E)$ and nodes p_1, p_n has an affirmative answer if and only if there exists MDS $M \in \mathcal{L}_{\text{hi}}(\psi_G^{\text{hi}})$ where all nodes appear and $|M| = |V| + 2$. In this case, M is an encoding of a hamiltonian path of $G' = ext(G, p_1, p_n)$ from b to e.*

Consider the bipartite transformation $bi(G)$ applied to graph G. In this case, for any node p the edge (p, p') is encoded only once, so only acyclic paths are assembled. Therefore, the following corollary holds.

Corollary 1. *The following statements are equivalent: (a) The hamiltonian path problem for graph $G = (V, E)$ and nodes p_1, p_n has an affirmative answer; (b) there exists $M \in \mathcal{L}_{\text{hi}}(\psi_{bi(G)}^{\text{hi}})$ where all nodes appear (c) there exists $M \in \mathcal{L}_{\text{hi}}(\psi_{bi(G)}^{\text{hi}})$ with $|M| = 2|V| + 2$. Moreover, the MDS M from (b) and (c) is an encoding of a hamiltonian path of $ext(bi(G), p_1, p'_n)$ from b to e.*

7   Computing Using **dlad** Only

We now consider a (theoretical) solution to the HPP problem through the gene assembly using only **dlad** operation.

Consider a directed graph $G = (V, E)$ with $V = \{p_1, \cdots, p_n\}$, $n > 0$ and the hamiltonian path problem in $G' = ext(G, p_1, p_n)$ from b to e.

We say that *a descriptor ψ_G^{dlad} is associated to G* if it is of the form

$$\psi_G^{\text{dlad}} = g_{b,p_1} \left(\prod_{(p,q) \in E \cup \{(p_n, e)\}} g_{p,q} \right) (r, x)^{|E|}, \text{ where } g_{p,q} = (p, q)(x, y)$$

is a descriptor encoding an edge (p, q); we may choose any order of encoding edges (with the exception g_{b,p_1} must be the first) in ψ_G^{dlad}.

Example 4. Consider the graph G_1 from Example 1. Then $\psi_{G_1}^{\text{dlad}} = (b, \Lambda, 3)$ $(x, \Lambda, y)(2, \Lambda, 1)(x, \Lambda, y)$ $(3, \Lambda, 1)(x, \Lambda, y)(3, \Lambda, 2)(x, \Lambda, y)(1, \Lambda, e)(x, \Lambda, y)(r, \Lambda, x)$ $(r, \Lambda, x)(r, \Lambda, x)(r, \Lambda, x)(r, \Lambda, x)$ is associated to G_1. The successful assemblies are $\mathcal{L}_{\text{dlad}}(\psi_{G_1}^{\text{dlad}}) = \{(b, 321, e), (b, 31, e)\}$.

Equipped with this encoding, we now prove that **dlad** solves the HPP problem. Consider a directed graph G and the hamiltonian path problem in G' from b to e. The following results hold.

Lemma 5. *Any successfully assembled MDS $M \in \mathcal{L}_{\text{dlad}}(\psi_G^{\text{dlad}})$ is associated to a path from b to e.*

We omit the proof, since its idea is again similar to that in Lemma 1.

Lemma 6. *For any path u without repeating edges, exists MDS $M_u \in \mathcal{L}_{\text{dlad}}(\psi_G^{\text{dlad}})$.*

Theorem 3. *The hamiltonian path problem for graph $G = (V, E)$ and nodes p_1, p_n has an affirmative answer if and only if there exists MDS $M \in \mathcal{L}_{\mathsf{dlad}}(\psi_G^{\mathsf{dlad}})$ where all nodes appear and $|M| = |V| + 2$. In this case, M is an encoding of a hamiltonian path of $G' = ext(G, p_1, p_n)$ from b to e.*

Consider the bipartite transformation $bi(G)$ applied to graph G. In this case, for any node p the edge (p, p') is encoded only once, so only acyclic paths are assembled. Therefore, the following corollary holds.

Corollary 2. *The following statements are equivalent: (a) The hamiltonian path problem for graph $G = (V, E)$ and nodes p_1, p_n has an affirmative answer; (b) there exists $M \in \mathcal{L}_{\mathsf{dlad}}(\psi_{bi(G)}^{\mathsf{dlad}})$ where all nodes appear; (c) there exists $M \in \mathcal{L}_{\mathsf{dlad}}(\psi_{bi(G)}^{\mathsf{dlad}})$ with $|M| = 2|V| + 2$. Moreover, the MDS M from (b) and (c) is an encoding of a hamiltonian path of $ext(bi(G), p_1, p_n')$ from b to e.*

8   Discussion

It has been observed many times in the literature that gene assembly in ciliates has a definite computational flavor. Two mathematical models were proposed to model gene assembly as a computational process transforming one structure into another one. Moreover, it has been shown that both models are Turing universal: assuming that a Turing machine may be encoded in the form of an artificial gene of high enough length and present in a high enough number of copies, then the Turing machine may be simulated through gene assembly, see [13,14,15]. The approach that we take in this paper is different. Given a mathematical problem such as HPP, we ask the question how to encode the problem into a gene pattern such that solving the problem is equivalent with assembling the gene. Using each of the three operations ld, hi, and dlad, we show that the construction is indeed possible, at least theoretically. It is important to underline here the connection with the computational principle in the celebrated experiment of Adleman [1]. While in [1], one encodes the given graph into a set of molecules that recombine among themselves to yield in principle the encodings of all paths through the graph, we encode our graph into a set of sequences that are placed in an arbitrary order on a chromosome-like molecule. This molecule may be assembled in many possible ways; in fact, the encodings of all paths of a certain length may be assembled in this way. Although a micronuclear gene is presented in several copies in a ciliate, it remains to be tested experimentally if a ciliate would assemble two or more identical copies of our artificial gene into several different forms. Answering this question would clarify the scale of a prototype experiment to test our approach, in terms of the number of ciliates required.

Some recent results of [2] and [22] suggest that RNA-template could be used to control and direct gene assembly. Based on this, one may attempt to implement our ciliate-based solutions to HPP. For example, one may inject templates to indicate all possibilities in which two MDSs may recombine. The amount of such templates would thus be at most quadratic in the number of MDSs used by our encoding. Clearly, this can only be validated through laboratory experiments.

Acknowledgments

I. Petre gratefully acknowledges support by Academy of Finland, project 108421, A. Alhazov and V. Rogojin gratefully acknowledge support by Academy of Finland, project 203667.

References

1. Adleman, L.M.: Molecular computation of solutions to combinatorial problems. Science 226, 1021–1024 (1994)
2. Angeleska, A., Jonoska, N., Saito, M., Landweber, L.F.: RNA-Template Guided DNA Assembly. In: Garzon, M., Yan, H. (eds.) Preliminary Proceedings on DNA13 meeting, University of Memphis, Memphis, p. 364 (2007)
3. Ehrenfeucht, A., Harju, T., Petre, I., Prescott, D.M., Rozenberg, G.: Computation in Living Cells: Gene Assembly in Ciliates. Springer, Heidelberg (2003)
4. Ehrenfeucht, A., Harju, T., Petre, I., Prescott, D.M., Rozenberg, G.: Formal systems for gene assembly in ciliates. Theoret. Comput. Sci. 292, 199–219 (2003)
5. Ehrenfeucht, A., Harju, T., Petre, I., Rozenberg, G.: Characterizing the micronuclear gene patterns in ciliates. Theory of Comput. Syst. 35, 501–519 (2002)
6. Ehrenfeucht, A., Petre, I., Prescott, D.M., Rozenberg, G.: String and graph reduction systems for gene assembly in ciliates. Math. Structures Comput. Sci. 12, 113–134 (2001)
7. Ehrenfeucht, A., Petre, I., Prescott, D.M., Rozenberg, G.: Circularity and other invariants of gene assembly in ciliates. In: Ito, M., Păun, G., Yu, S. (eds.) Words, semigroups, and transductions, pp. 81–97. World Scientific, Singapore (2001)
8. Ehrenfeucht, A., Prescott, D.M., Rozenberg, G.: Computational aspects of gene (un)scrambling in ciliates. In: Landweber, L.F., Winfree, E. (eds.) Evolution as Computation, pp. 216–256. Springer, Heidelberg (2001)
9. Harju, T., Petre, I., Li, C., Rozenberg, G.: Parallelism in gene assembly. In: Ferretti, C., Mauri, G., Zandron, C. (eds.) DNA Computing. LNCS, vol. 3384, pp. 138–148. Springer, Heidelberg (2005)
10. Harju, T., Petre, I., Rozenberg, G.: Gene assembly in ciliates: Molecular operations. In: Păun, G., Rozenberg, G., Salomaa, A. (eds.) Current Trends in Theoretical Computer Science (2004)
11. Harju, T., Petre, I., Rozenberg, G.: Gene assembly in ciliates: formal frameworks. In: Păun, G., Rozenberg, G., Salomaa, A. (eds.) Current Trends in Theoretical Computer Science (2004)
12. Kari, L., Landweber, L.F.: Computational power of gene rearrangement. In: Winfree, E., Gifford, D.K. (eds.) Proceedings of DNA Bases Computers, V. American Mathematical Society, pp. 207–216 (1999)
13. Landweber, L.F., Kari, L.: The evolution of cellular computing: Nature's solution to a computational problem. In: Proceedings of the 4th DIMACS Meeting on DNA-Based Computers, Philadelphia, PA, pp. 3–15 (1998)
14. Landweber, L.F., Kari, L.: Universal molecular computation in ciliates. In: Landweber, L.F., Winfree, E. (eds.) Evolution as Computation, Springer, New York (2002)
15. Onolt-Ishdorj, T., Petre, I., Rogojin, V.: Computational Power of Intramolecular Gene Assembly. Computability in Europe (submitted 2007)
16. Papadimitriou, C.H.: Computational Complexity. Addison-Wesley, Reading (1994)

17. Prescott, D.M.: The DNA of ciliated protozoa. Microbiol. Rev. 58(2), 233–267 (1994)
18. Prescott, D.M., Ehrenfeucht, A., Rozenberg, G.: Molecular operations for DNA processing in hypotrichous ciliates. Europ. J. Protistology 37, 241–260 (2001)
19. Petre, I.: Invariants of gene assembly in stichotrichous ciliates. IT, Oldenbourg Wissenschftsverlag 3, 161–167 (2006)
20. Prescott, D.M., Rozenberg, G.: How ciliates manipulate their own DNA – A splendid example of natural computing. Natural Computing 1, 165–183 (2002)
21. Prescott, D.M., Rozenberg, G.: Encrypted genes and their reassembly in ciliates. In: Amos, M. (ed.) Cellular Computing, Oxford University Press, Oxford (2003)
22. Vijayan, V., Nowacki, M., Zhou, Y., Doak, T., Landweber, L.: Programming a Ciliate Computer: Template-Guided In Vivo DNA Rearrangements in Oxytricha. In: Garzon, M., Yan, H. (eds.) Preliminary Proceedings on DNA13 meeting, University of Memphis, Memphis, p. 172 (2007)

Toward Minimum Size Self-Assembled Counters

Ashish Goel and Pablo Moisset de Espanés

Department of Management Science and Engineering and (by courtesy) Computer Science,
Stanford University, Terman 311, Stanford CA 94305
ashishg@stanford.edu
pmoisset@usc.edu

Abstract. DNA self-assembly is a promising paradigm for nanotechnology. In this paper we study the problem of finding tile systems of minimum size that assemble a given shape in the Tile Assembly Model, defined by Rothemund and Winfree [14]. We present a tile system that assembles an $N \times \lceil \log_2 N \rceil$ rectangle in asymptotically optimal $\Theta(N)$ time. This tile system has only 7 tiles. Earlier constructions need at least 8 tiles [7]. We managed to reduce the number of tiles without increasing the assembly time. The new tile system works at temperature 3.

The new construction was found by the combination of exhaustive computerized search of the design space and manual adjustment of the search output.

1 Introduction

Self-Assembly (SA) is the process by which autonomous components assemble into complexes following rules of local interaction only. SA is ubiquitous in Nature. Chemistry and Biology provide many examples, such as the formation of crystals and the growth of some organisms. SA is a promising paradigm for assembling shapes and patterns at molecular scale. The ability to construct many objects of intricate design may be useful in the fields of nano-electronics [9] and Material Sciences. The Watson-Crick law of pairing, together with the small size of bases, make DNA an attractive material to build self-assembled systems. There are numerous experimental results that support this approach [11,12,13,16,17,18,20].

Rothemund and Winfree proposed a theoretical model for DNA SA, the Tile Assembly Model (TAM. In the TAM, the DNA compounds are modeled as square tiles with glues on their sides. The individual tiles can stick to a growing assembly, as long as the glues on their sides provide enough sticking strength. Adleman *et al.* [1] added the notion of time complexity to the model. Some variants of the TAM have been explored in [2,3,8].

In [14], Rothemund and Winfree studied the problem of assembling an $N \times N$ square starting from a single tile. In their construction, they first built a rectangle from a base row by simulating a binary counter. Then, they completed the square by other means. Their counter construction required 12 tiles and needed $\Theta(N \cdot \log N)$ time to finish the assembly. Adleman *et al.* [1] presented a new counter that assembles in asymptotically optimal $\Theta(N)$ time, but requires 15 different tiles. Chen, Cheng, Goel and Moisset [7] improved the result by finding a counter that uses only 8 tiles and also achieves $\Theta(N)$ assembly time.

M.H. Garzon and H. Yan (Eds.): DNA 13, LNCS 4848, pp. 46–53, 2008.
© Springer-Verlag Berlin Heidelberg 2008

Reducing the number of tiles to assemble a given shape has a practical motivation. The cost of materials, i.e. DNA, and the time to carry out an experiment is closely related to the number of tiles in the design. Also, finding a smaller, or the smallest number of tiles to accomplish a given task is a theoretical problem of independent interest. Tiles that assemble a given shape are analogous to a computer program that outputs that shape. Minimizing the number of required tiles is similar to minimizing the program size.

In some computational problems, there is a trade-off between program size and running time. The natural question to ask is if reducing the number of tiles to build a counter forces to increase the assembly time.

T he main results of the paper are: In Section 3, we show a set of 7 tiles that assembles an $N \times \lceil \log_2 N \rceil$ rectangle from an initial base row in asymptotically optimal $\Theta(N)$ time. It is the smallest counter known so far, and it does not incur an increased assembly time. In spite of the small number of tiles, the construction is more involved than those found in [1,14]. The proof of correctness of the new counter is non-trivial and it is outlined in Section 4. For a complete, formal proof, see [4]. The counter with 7 tiles was originally published in [5], but it was described informally and no proof of correctness was given.

The process of finding a working design with only 7 tiles is of independent interest. It relied an exhaustive computerized search. This search was not guaranteed to output a correct design. It was only meant to suggest a candidate set of tiles which had to be verified manually. In fact, the search program produced a set that was flawed, and had to be corrected by hand. Interestingly, the manual modification of the candidate set fell outside the design space our program searched. In any case, the resulting design is so involved that it is unlikely that it could have been found without using the computer-aided approach. Details of the search process will appear in [6].

2 Definitions

The Tile Assembly Model (TAM): The tile assembly model [14,1] extends the theoretical model of tiling by Wang [15] to include a mechanism for growth based on the physics of molecular SA. We will present a succinct definition, with minor modifications for ease of explanation.

A tile is an oriented unit square with the north, east, south and west edges labeled from some alphabet Σ of glues. For each tile $t \in T$, the labels of its four edges are denoted $\sigma_N(t)$, $\sigma_E(t)$, $\sigma_S(t)$, and $\sigma_W(t)$. Sometimes we will describe a tile t as the quadruple $(\sigma_N(t), \sigma_E(t), \sigma_S(t), \sigma_W(t))$. Consider the triple $\langle T, G, \tau \rangle$ where T is a finite set of tiles, $\tau \in \mathbb{Z}_{>0}$ is the *temperature*, and G is the *glue strength* function from Σ to $\mathbb{Z}_{\geq 0}$, where Σ is the set of glues.

Given $p = (x, y), p' = (x', y') \in \mathbb{Z}^2$, we say p and p' are *position adjacent* iff $|x - x'| + |y - y'| = 1$. A *shape* is a finite, connected (under the adjacency relation defined above) subset of $\mathbb{Z}^2$. Let $Dom(f)$ denote the domain of a function f. A *supertile* S of T is a partial function from $\mathbb{Z}^2$ to T such that $Dom(S)$ is a shape. For a supertile S, we will write $[S]$ to represent $Dom(S)$.

Let C and D be two supertiles. Suppose there exist some $t \in T$ and some $(x, y) \in \mathbb{Z}^2$ such that $(x, y) \notin Dom(C)$, $D(x, y) = t$ and $D = C$ except at (x, y). If $(x, y + 1) \in Dom(C)$ and $\sigma_N(t) = \sigma_S(C(x, y + 1))$, let $f_{N,C,t}(x, y) = G(\sigma_N(t))$ and let $f_{N,C,t}(x, y) = 0$ otherwise. Informally $f_{N,C,t}(x, y)$ is the strength of the bond between C and the north side of t. Define $f_{S,C,t}(x, y)$, $f_{E,C,t}(x, y)$ and $f_{W,C,t}(x, y)$ similarly. Then we say that tile t is *attachable* to C at position (x, y) iff $f_{N,C,t}(x, y) + f_{S,C,t}(x, y) + f_{E,C,t}(x, y) + f_{W,C,t}(x, y) \geq \tau$, and we write $C \rightarrow_T D$ to denote the transition from C to D in attaching a tile to C at position (x, y). Informally, $C \rightarrow_T D$ iff D can be obtained from C by adding a tile t such that the total strength of interaction between t and C is at least τ.

A *tile system* is a quadruple $\mathbf{T} = \langle T, s, G, \tau \rangle$, where T, G, τ are as above and s is a special supertile called the "seed". The notion of a *derived supertile* of a tile system $\mathbf{T} = \langle T, s, G, \tau \rangle$ is defined recursively:

1. The seed s is a derived supertile of $\mathbf{T}$, and
2. if $C \rightarrow_T D$ and C is a derived supertile of $\mathbf{T}$, then D is also a derived supertile of $\mathbf{T}$.

Informally, a derived supertile is either just the seed (condition 1 above), or obtained by legal addition of a single tile to another derived supertile (condition 2).

A *terminal supertile* of the tile system $\mathbf{T}$ is a derived supertile A such that there is no supertile B for which $A \rightarrow_T B$. Let $\rightarrow_T^*$ denote the reflexive transitive closure of $\rightarrow_T$. If there is a terminal supertile A such that for any derived supertile B, $B \rightarrow_T^* A$, we say that the tile system *uniquely produces* A. A tile system $\mathbf{T}$ *uniquely produces a shape* W iff it uniquely produces some supertile Γ and $[\Gamma]$ is identical (up to translation) to W.

We will now add the notion of running time to this model. We associate with each tile $t \in T$ a non-negative probability $P(t)$, such that $\sum_{t \in T} P(t) = 1$. We assume that the tile system has an infinite supply of each tile, and $P(t)$ models the concentration of tile t in the system. Now SA of the tile system corresponds to a continuous time Markov process where the states are in a one to one correspondence with derived supertiles, and the initial state corresponds to the seed s. Suppose a single tile t can be added to a derived supertile C to produce supertile D. Then there is a transition from state C to D in the Markov chain, and the rate of the transition is $P(t)$. Suppose the tile system produces a unique terminal supertile $A_{\mathbf{T}}$. In the Markov chain, the time for reaching $A_{\mathbf{T}}$ from s is a random variable. The "running time" of the SA process is defined as the expected value of this random variable. Note that the Markov process modeling the SA process is inherently parallel. For details, see [1].

A supertile Γ is *full* iff for all $p, p' \in [\Gamma]$, if $p' = p + (1, 0)$ then $\sigma_E(\Gamma(p)) = \sigma_W(\Gamma(p'))$ and if $p' = p - (1, 0)$ then $\sigma_W(\Gamma(p)) = \sigma_E(\Gamma(p'))$ and if $p' = p + (0, 1)$ then $\sigma_N(\Gamma(p)) = \sigma_S(\Gamma(p'))$ and if $p' = p - (0, 1)$ then $\sigma_S(\Gamma(p)) = \sigma_N(\Gamma(p'))$. Intuitively, a supertile is full if there are no glue mismatches in the abutting edges of adjacent tiles.

General Purpose Counter (GPC): A quadruple $\langle T, T_s, G, \tau \rangle$, where T and T_s are finite sets of tiles with glues from some alphabet Σ, $G : \Sigma \rightarrow \mathbb{Z}_{\geq 0}$, and τ is a temperature, is a *general purpose counter* iff for all integers $h > 1$, for all integers $w \geq \lceil \log_2 h \rceil$, there exists a supertile $s_{h,w}$ of T_s such that:

1. $\langle T \cup T_s, s_{h,w}, G, \tau \rangle$ uniquely produces a supertile, denote it $\Gamma_{h,w}$, such that $[\Gamma_{h,w}] = \{0, -1, \ldots, -w+1\} \times \{0, 1, \ldots, h-1\}$.
2. $[s_{h,w}] = \{0, -1, \ldots, -w+1\} \times \{0\}$.
3. For all $(x,y) \in \{0, -1, \cdots, -w+1\} \times \{1, 2, \ldots, h-1\}$, $\Gamma_{h,w}(x,y) \in T$.

Informally, the seed row s_w has width w and is made out of tiles in T_s. The tiles in T will grow the rest of the $h \times w$ rectangle on top of s_w. The *size of a GPC* $\langle T, T_s, G, \tau \rangle$ is $|T|$.

The General Purpose Counter problem: Given a temperature τ, find the least positive integer m such that there exist an alphabet Σ and sets of tiles T and T_s with glues from Σ, and there exists $G : \Sigma \to \mathbb{Z}_{\geq 0}$, such that $\langle T, T_s, G, \tau \rangle$ is a GPC and $|T| = m$.

Informally, we would like to find the smallest set of tiles that assembles a rectangle whose size is determined solely by the initial supertile, i.e. the seed, of the SA process. We would like the size of this set of tiles to be independent of the size of the desired rectangle. We will also assume the shape of the seed has to be a horizontal line. Constructions with these properties were used by Rothemund and Winfree [14] and by Adleman *et al.* [1] as a "subroutine" to assemble squares. Since the techniques to assemble rectangles in [14,1,7] are based on repeated addition of binary numbers, we refer to these constructions as *counters*. Information theory imposes a logarithmic lower bound on the width of the counter. Hence, we impose the $w \geq \lceil \log_2 h \rceil$ constraint. Our choice of 2 as base of the logarithm is somewhat arbitrary.

3   A Counter of Size 7

In this section we present a GPC of size 7 that works at temperature 3. We also outline the proof of correctness. Before describing the counter, we introduce some notation.

A supertile Γ is said to be *rectangular* iff there are positive integers w and h such that $[\Gamma] = \{0, -1, \ldots, -w+1\} \times \{0, 1, \ldots, h-1\}$. We will call w and h the *width* and *height* of Γ, respectively.

Let Γ be a rectangular supertile, and let w and h be the width and height of Γ, respectively. For all $k \in \{0, 1, \ldots, w-1\}$, let $\mathcal{C}_{\Gamma,k} = (\Gamma(-k,0), \Gamma(-k,1), \ldots, \Gamma(-k, h-1))$. For all $k \in \{0, 1, \ldots, w-1\}$, we will refer to the restriction of Γ to $\{-k\} \times \{0, 1, \ldots, h\}$ as *the k-th column* of Γ. Similarly, for all $k \in \{0, 1, \ldots, h-1\}$, let $\mathcal{R}_{\Gamma,k} = (\Gamma(-w+1, k), \Gamma(-w+2, k), \ldots, \Gamma(0, k))$. For all $k \in \{0, 1, \ldots, h-1\}$, we will refer to the restriction of Γ to $\{0, -1, \ldots, -w+1\} \times \{k\}$ as the *k-th row* of Γ.

The set of tiles: We begin by giving a pictorial representation of the counter in Figure 1. Define $\mathcal{T} = \{T_1, T_2, \cdots, T_7\}$, and define the glue-strength function $G : \{a, b, c, d, e, f, g\}^2 \to \{0, 1, 2, 3\}$ so that $G(a) = G(b) = 3$, $G(c) = G(d) = G(e) = 2$ and $G(f) = G(g) = 1$.

The supertile $\mathcal{B}_w$: Given a sequence S, and a positive integer k, we will write S_k to denote the k-th element of S. For all positive integers k and l, define $S_{k,l}$ as the subsequence of S comprising all elements from S_k through S_l. Define the sequence concatenation operator $\bullet$ in the usual way. For all positive integers k, for all finite sequences S, we will write $k \times S$ to denote $S \bullet S \bullet \cdots \bullet S$, where S is concatenated k times.

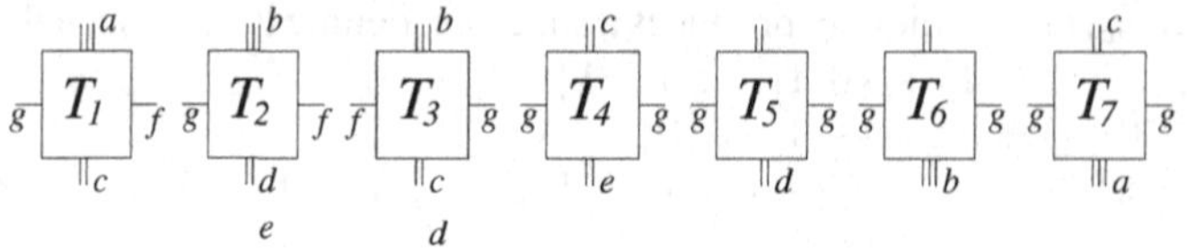

Fig. 1. Counter with 7 tiles

Define the following infinite sequences of tiles with period 6.

$$\bar{D} = (T_4, T_3, T_6, T_2, T_6, T_5, T_4, T_3, T_6, T_2, T_6, T_5, \ldots)$$
$$D = (T_6, T_5, T_4, T_1, T_7, T_3, T_6, T_5, T_4, T_1, T_7, T_3, \ldots)$$
$$\bar{E} = (T_4, T_1, T_7, T_3, T_6, T_5, T_4, T_1, T_7, T_3, T_6, T_5, \ldots)$$
$$E = (T_6, T_2, T_6, T_5, T_4, T_3, T_6, T_2, T_6, T_5, T_4, T_3, \ldots)$$

For all positive integers w, define the following w sequences of length 2^w:

1. $C^{(0,w)} = 2^{w-1} \times (T_1, T_7)$
2. For all odd and positive $k \le w - 1$, $C^{(k,w)} = 2^{w-k-1} \times (\bar{D}_{1,2^k} \bullet D_{1,2^k})$
3. For all even and positive $k \le w - 1$, $C^{(k,w)} = 2^{w-k-1} \times (\bar{E}_{1,2^k} \bullet E_{1,2^k})$

For all positive integers w, define the rectangular supertile $\mathcal{B}_w$ in such a way that the width of $\mathcal{B}_w$ is w, the height of $\mathcal{B}_w$ is 2^w and for all $(k, i) \in \{0, 1, \ldots, w - 1\} \times \{0, 1, \ldots, h - 1\}$, $\mathcal{B}_w(-k, i) = C_{i+1}^{(k,w)}$. Note that $C_{\mathcal{B}_w,k} = C^{(k,w)}$. Define s_w as the 0-th row of $\mathcal{B}_w$. Figure 3 shows $\mathcal{B}_3$, as an example.

4   Results and Proof Outlines

We state now the main result of the paper:

Theorem 1. *For all positive integers w, the tile system $\mathbf{T}_w = \langle T, G, s_w, 3 \rangle$ uniquely produces $\mathcal{B}_w$.*

Note that the height of $\mathcal{B}_w$ is exactly 2^w. Minor modifications to s_w allow the assembly of rectangles of all heights up to 2^w. The details about the modifications are omitted.

For reasons of space, we present only an informal outline of the proof of correctness here. For details see [4].

The proof is constructive, showing that if s_w is the seed row and the temperature is 3, the tiles in T uniquely assemble $\mathcal{B}_w$. The first step is to show that $\mathcal{B}_w$ is a full supertile, i.e. there are no glue mismatches between adjacent tiles. This fact follows from the definition of $\mathcal{B}_w$. Then we prove that $\mathcal{B}_w$ can be derived from s_w. The proof of this fact is constructive, showing a particular derivation of $\mathcal{B}_w$ from s_w. The process is sketched in Figure 2.

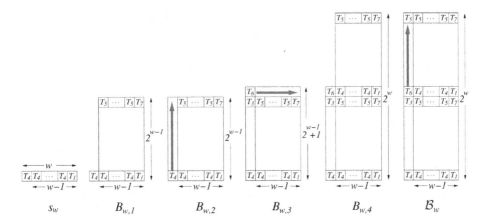

Fig. 2. The inductive step

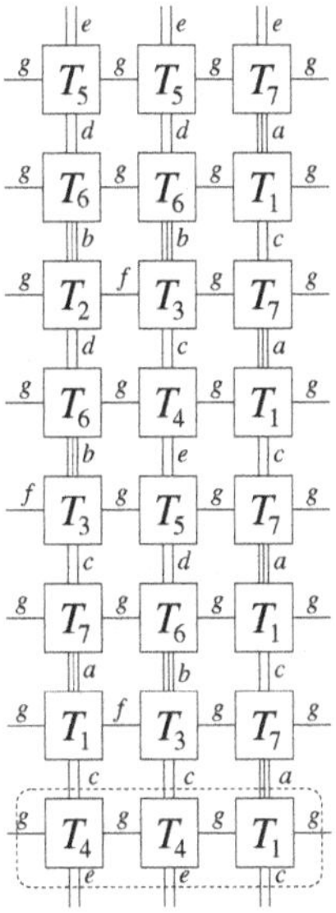

Fig. 3. The supertile $\mathcal{B}_3$. The dashed line encloses s_3.

We prove $\mathcal{B}_w$ can be derived from s_w by induction on w, exploiting the recursive structure of $\mathcal{B}_w$. Roughly speaking, $\mathcal{B}_w$ contains two copies of $\mathcal{B}_{w-1}$. Therefore, we use the inductive hypothesis to prove that we can derive the supertile $B_{w,1}$ from s_w. This follows from $B_{w,1}$ being $\mathcal{B}_{w-1}$ with an extra tile attached to its west side. It follows form $\mathcal{B}_w$ being full that we can start growing the westmost column of $\mathcal{B}_w$, deriving $B_{w,2}$ from $B_{w,1}$. Using a similar argument, we add one row to the northmost side of $B_{w,2}$ to obtain $B_{w,3}$. Now we use the inductive hypothesis, and grow another copy of $\mathcal{B}_{w-1}$ on top of $B_{w,3}$, yielding $B_{w,4}$. Finally, we use appeal to $\mathcal{B}_w$ being full to prove we can finish assembly the westmost column of $\mathcal{B}_w$.

We know that $\mathcal{B}_w$ is produced from s_w. Therefore, we just need to prove that production is unique, which is done through a case analysis.

We conclude by stating the time complexity of $\mathbf{T}_w$ is $\Theta(2^w)$, which follows from the derivation of $\mathcal{B}_w$ used to prove that $\mathcal{B}_w$ derives from s_w, and from results in [10]. The proof of the next theorem relies on Lemmas 3.3, 3.4, and 4.1, and Theorem 4.4 in [10].

Theorem 2. *There exists a concentration function* $P : \mathcal{T} \to (0, 1)$ *such that for all positive integers* w, *the time complexity of* $\mathbf{T}_w = \langle \mathcal{T}, G, s_w, 3 \rangle$ *is* $\Theta(2^w)$.

Proof outline: Define P as a constant valued function with value $1/7$. The bound $\Omega(2^w)$ is trivial. Call E_w the equivalent acyclic graph induced by the derivation of $\mathcal{B}_w$ described in Figure 2. E_w is identical to the DAG G_N defined in the proof of Lemma 3.3 in [10], if $N = 2^w$. The length of the longest path in E_w is $O(2^w)$. By Theorem 4.4 in [10], the time complexity of $\mathbf{T}_w$ is $O(2^w)$. $\square$

5   Open Problems

Although the counter presented here uses fewer tiles than any other known counter, there are still some unanswered questions.

1. Our counter works at temperature 3, which is undesirable for lab implementations. Experience shows [12,19] that it is possible to obtain a reasonable approximation to the TAM at temperature 2 using DNA tiles. The question whether or not there exists a GPC of size 7 at temperature 2 remains open.
2. Our counter produces full supertiles. Dropping that constraint could potentially result in smaller counters that work at lower temperature. We are currently pursuing that goal.
3. The exhaustive exploration techniques we used to find the tile system do not scale up well past 7 or 8 tiles. This is consequence of the combinatorial explosion of the design space as the number of tiles grow. Perhaps it is possible to find an efficient algorithm that yields sub-optimal results.

Acknowledgments

We would like to thank Len Adleman, Ming-Deh Huang, Yuri Brun and Manoj Gopalkrishnan for useful discussion, and especially Dustin Reishus for his comments on the first manuscript.

References

1. Adleman, L., Cheng, Q., Goel, A., Huang, M.: Running time and program size for self-assembled squares. In: Proceedings of the thirty-third annual ACM symposium on Theory of computing, pp. 740–748. ACM Press, New York (2001)
2. Aggarwal, G., Goldwasser, M., Kao, M., Schweller, R.T.: Complexities for generalized models of self-assembly. In: Proceedings of symposium on discrete algorithms, ACM Press, New York (2004)

3. Cheng, Q., Moisset de Espanés, P.: Resolving two open problems in the self-assembly of squares. Technical Report 03-793, University of Southern California (2003)
4. Moisset de Espanés, P., Goel, A.: Toward minimum size self-assembled counters. In: 13th International Meeting on DNA Computing (2007)
5. Moisset de Espanés, P.: Computerized exhaustive search for optimal self-assembly counters. In: FNANO 2005: Proccedings of the 2nd Annual Foundations of Nanoscience Conference, pp. 24–25 (2005)
6. Moisset de Espanés, P.: Systems self-assembly: Multidisciplinary snapshots, N. Krasnogor, S. Gustafson, D. Pelta, J.L. Verdegay (eds.) Elsevier (2007)
7. Goel, A., Chen, H., Cheng, Q., Moisset de Espanés, P.: Invadable self-assembly, combining robustness with efficiency. In: Proceedings of symposium on discrete algorithms, ACM Press, New York (2004)
8. Kao, M.-Y., Schweller, R.: Reducing tile complexity for self-assembly through temperature programming (2006)
9. Rothemund, P., Cook, M., Winfree, E.: Self assembled circuit patterns. In: Proceedings of DNA Computing, Springer, Heidelberg (2003)
10. Goel, A., Cheng, Q., Moisset de Espanés, P.: Optimal self-assembly of counters at temperature two. In: Foundation of Nanoscience (2004)
11. Rothemund, P.: Theory and Experiments in Algorithmic Self-Assembly. PhD thesis, University of Southern California (2001)
12. Rothemund, P.W., Papadakis, N., Winfree, E.: Algorithmic self-assembly of dna sierpinski triangles. PLoS Biol. 2(12) (December 2004)
13. Rothemund, P.W.K.: Design of dna origami. In: ICCAD 2005: Proceedings of the 2005 IEEE/ACM International conference on Computer-aided design, pp. 471–478. IEEE Computer Society, Washington (2005)
14. Rothemund, P.W.K., Winfree, E.: The program-size complexity of self-assembled squares (extended abstract). In: Proceedings of the thirty-second annual ACM symposium on Theory of computing, pp. 459–468. ACM Press, New York (2000)
15. Wang, H.: Proving theorems by pattern recognition ii. Bell Systems Technical Journal 40, 1–42 (1961)
16. Winfree, E.: Algorithmic Self-Assembly of DNA. PhD thesis, California Institute of Technology, Pasadena (1998)
17. Winfree, E., Liu, F., Wenzler, L., Seeman, N.: Design and self-assembly of two-dimensional dna crystals (6 pages). Nature 394, 539–544 (1998)
18. Winfree, E., Yang, X., Seeman, N.: Universal computation via self-assembly of dna: Some theory and experiments. In: Proceedings of the Second Annual Meeting on DNA Based Computers, Princeton University (June 1996)
19. Winfree, E., Bekbolatov, R.: Proofreading tile sets: Error correction for algorithmic self-assembly. In: DNA, pp. 126–144 (2003)
20. Yurke, B., Turberfield, A., Mills Jr, A., Simmel, F., Neumann, J.: A dna-fuelled molecular machine made of DNA. Nature 406, 605–608 (2000)

A Realization of DNA Molecular Machine That Walks Autonomously by Using a Restriction Enzyme

Hiroyuki Sekiguchi, Ken Komiya, Daisuke Kiga, and Masayuki Yamamura

Interdisciplinary Graduate School of Science and Engineering,
Tokyo Institute of Technology, 4259 Nagatsuta, Yokohama 226-8502, Japan
sekiguchi@es.dis.titech.ac.jp, {komiya,kiga,my}@dis.titech.ac.jp

Abstract. In this paper, we propose an autonomous molecular walking machine using DNA. This molecular machine follows a track of DNA equipped with many single-strand DNA stators arranged in a certain pattern. The molecular machine achieves autonomous walk by using a restriction enzyme as source of power. With a proposed machine we can control its moving direction and we can easily extend walking patterns in two or three dimensions. Combination of multiple legs and ssDNA stators can control the walking pattern. We designed and performed a series of feasibility study with molecular biology experiments.

1 Introduction

Several molecular machines, which can walk along a DNA track, have been proposed [1-6]. Two of six molecular machines [1-2] need fuel DNA and cannot walk autonomously. The others have achieved autonomous walk by using DNA cleaving activity of a restriction enzyme or DNAzyme. "Walking DNAzyme"[3] and "free-running DNA motor"[4] can walk along only one dimensional track. "Spider molecules"[5] has multiple legs and can walk along patterns in two or three dimensions, but it can only walk to random direction. There are few molecular machines that can walk autonomously on two or three dimensions along the designed route. "Unidirectional DNA walker"[6] satisfies these features but it requires to design an appropriate ground pattern to program target patterns when we extend it to two or more dimension.

In this paper, we propose new molecular walking machine that can achieve these requirements by using a restriction enzyme and a track of DNA equipped with many single-strand DNA stators arranged in a certain pattern. In the following, we first propose the conceptual scheme and show feasibility experiments with a restriction enzyme.

2 A Molecular Walking Machine

2.1 Mechanism How the Molecular Machine Walks

The molecular machine presented in this paper, walks a track of DNA embedded with many single-strand DNA stators. We must prepare a set of ssDNA stators and arrange

M.H. Garzon and H. Yan (Eds.): DNA 13, LNCS 4848, pp. 54–65, 2007.

them in a certain pattern to form the track which the molecular machine follows. The molecular machine has more than three legs. Each of legs anneals one ssDNA stator, and ssDNA stators are equipped with enough distance so that one leg cannot anneal more than two ssDNA stators at once. We can design a route which the molecular machine walks, by patterning the ssDNA stators and legs.

If nothing has occurred, a molecular machine walks randomly like "spidar molecules". We introduced the idea for the molecular machine to cleave the ssDNA stators in the correct order to walk along the route we designed.

Fig. 1 shows the reaction of three legs molecular walking machine. Fig. 1a shows a three legs molecular machine anneals the ssDNA stators. Green leg and red leg anneal green and red ssDNA stators. This is the basic state that the molecular machine binds the track of DNA. Blue leg cannot reach to blue ssDNA stators and is free-floating. The molecular machine must keep more than one legs free-floating to cleave ssDNA stators.

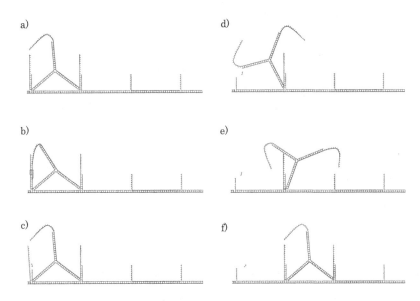

Fig. 1. The structure and reaction of the molecular walking machine

In this stage, blue leg plays an important role to cleave green ssDNA stator. Fig. 1b shows the free-floating blue leg anneals green leg. In the later experiments, we selected a nicking enzyme "*N.Alw* I" to cleave the ssDNA stators. *N.Alw* I can bind and cleave green ssDNA stators when blue leg anneals green leg on the green ssDNA stator.

After cleaving green ssDNA stator, green leg cannot keep annealing there. And blue leg is denatured from green leg, because that double-strand is designed unstable (Fig. 1c). After denatured from green ssDNA stator, blue leg gets to reach blue ssDNA stator. The molecular machine moves ahead to anneal blue ssDNA stator (Fig. 1d, e).

Finally, blue leg anneals blue ssDNA stator. Red leg and blue leg anneal ssDNA stators (Fig. 1f). This state is the same as the state Fig. 1a. Green leg acts to cleave red ssDNA stator in the next step. Red leg will act to cleave blue ssDNA stator after the next step. A molecular machine can cleaves ssDNA stators in the order of green, red, blue, green, and so on, to repeat these motions. This is the mechanism that the track of ssDNA stators is cleaved in the correct order.

2.2 Function of Cleaving ssDNA Stators

It is important for this molecular walking machine to cleave the ssDNA stators. A nicking enzyme like *N.Alw* I binds to double-strand DNA at the recognition site and cleaves only one strand at the cut point to introduce a nick. So the molecular machine can cleave only ssDNA stators.

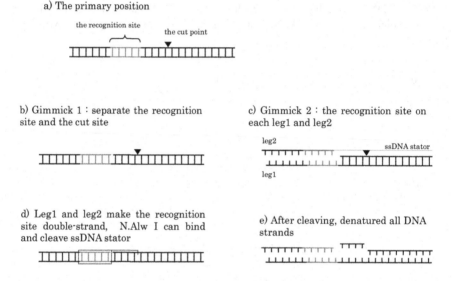

Fig. 2. Function of cleaving DNA stators

N.AlwI has another characteristic that is the cut point isn't included in the recognition site (Fig. 2a). A molecular machine uses this characteristic to control the activity of *N.Alw* I. We designed the molecular machine with following two gimmicks for the control.

First gimmick is that the cut point and the recognition site of *N.Alw* I are separated into the different two DNA strands (Fig. 2b). All ssDNA stators have the cut point and all legs of the molecular machine have the recognition site. The way of control a restriction enzyme is similar to that of using at "Programmable and autonomous computing machine"[7].

"Programmable and autonomous computing machine" uses a restriction enzyme "*Fok* I". As the first choice, we considered using *Fok* I for the molecular machine. *Fok* I isn't the nicking enzyme, so *Fok* I cleaves the molecular machine's leg unless we protect that strand. We first examined to make the activity of *Fok* I the same as that of a nicking enzyme by using of phosrhorothioate-modified DNA [8-9]. Since, we couldn't achieve expected activity, we determined to use *N.Alw* I. See the appendix for detail.

Second gimmick is that the two legs of the molecular machine make the recognition site double-strand (Fig. 2c). When only one leg anneals ssDNA stator, the recognition site is still single strand and *N.Alw* I cannot bind to cleave ssDNA stator. The other leg can anneal to make the recognition site double-stranded. Then *N.Alw* I can cleave ssDNA stator (Fig. 2d). To achieve this function, a leg has two recognition sites; one is for the binding ssDNA stators and another is for the neighborhood ssDNA stators. A combination of the two recognition site leads to the cleaving reaction of Fig. 1.

3   Experiment

3.1   Activity of *N.Alw* I

We first designed an experiment to know the activity of *N.Alw* I. We confirmed the DNA cleaving activity for the case that the recognition site and the cut point are in the different DNA strands.

3.1.1   Materials and Methos

For this experiment, we prepared four sets of DNA strands with different separation point, and one set of DNA strands without separate point for a control (Table. 1) (Fig. 3). We named each strands N-X-Y, where X denotes the number of the separate point and Y denotes the length of each strands. "N-none" means a strand of no separate point and "N-c" means a complement of N-none.

Each five sets of DNA strands were mixed at 0.5 μM in hybridization buffer. NEBuffer2 from New England Biolabs was used as the hybridization buffer. 3 units of *N.Alw* I from New England Biolabs were added to each five sets 20 μl solution. The five sets were incubated at 37°C by 24hours.

Table 1. DNA strands. The recognition site of *N.Alw* I is shown in bold type. The cut point is indicated by /.

Name	Separate Point	sequence(5'...3')
N-1-10	1	GATACAT**GGA**
N-2-11	2	GATACAT**GGA**T
N-3-12	3	GATACAT**GGA**TC
N-4-13	4	GATACAT**GGA**TCA
N-1-18	1	T**CACGG**/CTGAGACACTCT
N-2-16	2	**CACGG**/CTGAGACACTCT
N-3-17	3	A**CGG**/CTGAGACACTCT
N-4-15	4	**CGG**/CTGAGACACTCT
N-none	None	GATACAT**GGA**TCACGG/CTGAGACACTCT
N-c		AGAGTGTCTCAGCCGT**GATCC**ATGTATC

Fig. 3. Separate points

We ran the resulting solutions in 16% PAGE (non-denaturing gel and denaturing gel).

3.1.2 Results
Fig. 4 shows the results. We confirmed the activity of *N.Alw* I at the all sets. We found in the denaturing gel that *N.Alw* I at lane 8 cleaved the DNA strand "N-c"

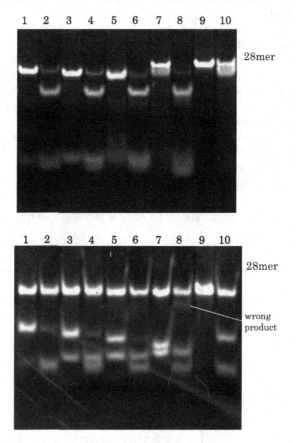

Fig. 4. Result of cleaving activity. 1)2) Separate point 1, 3)4) Separate point 2, 5)6) Separate point 3, 7)8) Separate point 4, 9)10) No separate, Odd number of lane : without *N.Alw* I , and even number of lane : with *N.Alw* I.

which didn't have the cut point. There were the set of DNA with separation point 4 in this solution. We supposed that double-strand DNA twists the recognition site and the cut point at the separation point, and *N.Alw* I cleaved wrong DNA strand. If we used separation point 4 at this molecular machine, *N.Alw* I would cleave not only ssDNA stators but also the legs of the molecular machine. Except for the result of lane 8, *N.Alw* I had the cleaving activity which this molecular machine needs.

We determined to use separation point 1 for later experiments about the molecular machine. The molecular machine needs to anneal the ssDNA stators and to be denatured after cleavage. We need a significant difference in melting temperature before and after cleaving to lead to this behavior. For this reason, the molecular machine should cleave ssDNA stator into as large fragments as possible.

3.2 Half Step of the Molecular Walking Machine

As the next stage, we confirmed the behavior of this molecular machine's legs. Each leg has two recognition sites which are for cleaving ssDNA stators and for cleaving the neighborhood ssDNA stators. The molecular machine cleaves ssDNA stators using two legs. We experimented to confirm that the two legs can make the recognition site double-stranded and cleave ssDNA stators.

3.2.1 Materials and Methods

We prepared three DNA strands for two legs molecular machine and one DNA strand for ssDNA stator (Table. 2). In two legs, leg1 is a strand "N-L1", leg2 is a strand "N-L2". Strand "N-L3" is to be a leg3, when the molecular machine has third leg. "N-S1" is ssDNA stator, annealed and cleaved by leg1. Fig. 5 shows N-L1, N-L2 and N-L3 build two legs molecular machine and the molecular machine anneals to N-S1.Reaction condition is the same as the experiment in chapter 3.1.

Table 2. DNA strands. The recognition site of N.AlwI is shown in bold type. The cut point is indicated by /.

Name	Sequence(5'...3')
N-L1	TATAGATATCAAGTAGTCGATATGCTTCACAGTCTGATCTGAGGTGTGGAGACGTC ATGTGCATGCCAGTGTA**CGATCC**TGCAACG
N-L2	GTGAAGCATATCGACTACTTGATATCTATAGACCTAGTAGTGCC**GATCC**AGATCT ACGTTGCA**GGA**
N-L3	ACATGACGTCTCCACACCTCAGATCAGACT
N-S1	CGTA/CACTGGCATGC

3.2.2 Results

Fig. 6 shows the result that the two legs molecular machine achieved to cleave ssDNA stator. By analtsis of denaturing gel electrophoresis, the band of 10 mer in lane 2 shows N-S1 was cleaved by *N.Alw* I. The band of 10 mer corresponds to the cleaving products of N-S1. From non-denaturing gel, we found the two legs molecular machine annealed N-S1 at lane1. And lane 2 shows that the two legs molecular machine couldn't keep annealing after cleaving, because the band of the cleaving products appeared on the different position of a band which denotes the two legs molecular machine.

The result shows that the two legs cooperated to cleave N-S1 and the unconcerned parts for cleaving of the legs don't have a critical effect on cleaving. We think that the molecular machine presented in this paper, can cleave and be denatured from ssDNA stators. Thus we achieved the motion that the molecular machine lifts up its legs. The molecular machine can walk ahead a half step.

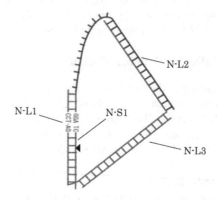

Fig. 5. Two Legs Molecular Machine

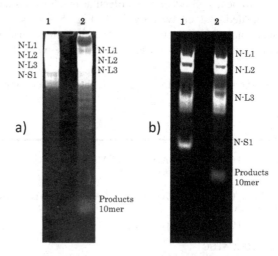

Fig. 6. Activity of two leg machine. a) non-denaturing gel, b) denaturing gel. 1) without *N.Alw* I, 2) with *N.Alw* I.

3 Discussion

There are several further issues to discuss. Firstly, we need the motion to drop off the lifted legs for other half step. The motion is decomposed into two moves. One is a move to anneal the next ssDNA stator, the other is a move to anneal the leg which anneals the neighborhood ssDNA stator. To achieve other half step, we plan to examine the move to the leg annealing the neighborhood ssDNA stator at first. The move can be confirmed by an experiment using two legs machine and a track with two ssDNA stators. In this experiment, we should make the condition that leg1 and leg2 anneal each ssDNA stators. We are going to add a strand which should anneal leg1 and make the recognition site double-stranded, and then *N.Alw* I will be able to bind the recognition site and cleave the ssDNA stator. After cleaving, leg1 is going to move to anneal leg2 for cleaving the neighborhood stator. The move is similar to another move to the next ssDNA stator. If the two moves are realized, the molecular machine will achieve the motion to drop off the legs. To combine this motion and a motion to lift up the legs, the molecular machine can walk one step. If the molecular machine can repeat the one step, it should walk ahead.

Secondly, remark that these anneal—cut—denature—anneal cycle is similar to "free running DNA motor" [5] if the walker has only two legs. They have already achieved multiple cycles under isothermal environment which means autonomous behavior. So we believe we can also achieve these cycles autonomously. The difference is using more than three legs leads to control the moving direction. We are now considering more about the combination of phosphorothioate-modified DNA and restriction enzymes. The results shown in appendix indicate the half of phosphorothioate-modified DNA is cut by *Fok* I. There are two alternative oxygen sites to be phosphorothioated in DNA backbone. We guess *Fok* I was only blocked by the specific phosphorothioation site. The recognition site and the cut site of *Fok* I have longer distance than any known nicking enzymes. We expect this provides good amount of freedom for sequence design for walkers with more than three legs.

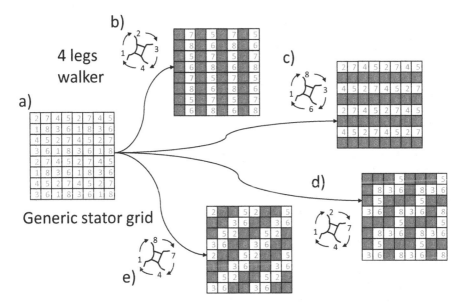

Fig. 7. Programming Two Dimmensional Patterns with Four Legs Walker

Finally, we can program the behavior of our walking machine to make two or three dimmensional patterns when we achieve complete walking steps on multiple legs. Fig.7 shows a portion of such "programmable patterns". Consider a general purpose stator grid and four legs walker. Fig.7a shows an example of a general purpose stator grid with eight kinds of sequences arranged with some regularity. If we program four legs walker with execution order "1-2-3-4" to move "go up straightly", then a solution of walkers make a vertical stripe on the stator grid by cutting stators in the same order "1-2-3-4" as shown in Fig.7b. We can easily program other patterns like horizontal stripe, checker board, and so on, only by changing four legs layout of the walker even on only one general purpose stators grid. Moreover, we need only one kind of restriction enzyme even for any number of legs. This feature can reduce experimental complexity on tuning reaction condition for many kinds of restriction enzymes working in the same efficiency.

4 Conclusion

We proposed an autonomous walking machine that can follow a track with many ssDNA stators along the certain route on two or three dimensions by cleaving ssDNA stators. From a series of feasibility studies, we confirmed the molecular machine can cleave and be denatured from ssDNA stators. So it is able to walk ahead a half step. Although our experiments are still in quite preliminary level, but we believe it is an alternative to construct programmable autonomous walking machines.

References

1. Sherman, W.B., Seeman, N.C.: A Precisely Controlled DNA Biped Walking Device. Nano Lett. 4, 1203–1207 (2004)
2. Shin, J.-S., Pierce, N.A.: A Synthetic DNA Walker for Molecular Transport. J. Am. Chem. Soc. 126, 10834–10835 (2004)
3. Tian, Y., He, Y., Chen, Y., Yin, P., Mao, C.: A DNAzyme That Walks Processively and Autonomously along a One-Dimensional Track. Angew. Chem. Int. Ed. 44, 4355–4358 (2005)
4. Bath, J., Green, S.J., Turberfield, A.J.: A free-running DNA motor powered by a nicking enzyme. Angew. Chem. Int. Ed. 44, 4358–4361 (2005)
5. Pei, R., Taylor, S.K., Stefanovic, D., Rudchenko, S., Mitchell, T.E., Stojanovic, M.N.: Behavior of Polycatalytic Assemblies in a Substrate- Displaying Matrix. J. Am. Chem. Soc. 128, 12693–12699 (2006)
6. Yin, P., Yan, H., Daniell, X.G., Turberfield, A.J., Reif, J.H.: A Unidirectional DNA Walker That Moves Autonomously along a Track. Angew. Chem. Int. Ed. 43, 4906–4911 (2004)
7. Benenson, Y., Paz-Elizur, T., Adar, R., Keinan, E., Livneh, Z., Shapiro, E.: Programmable and autonomous computing machine made of biomolecules. Nature 414, 430–434 (2001)
8. Verma, S., Eckstein, F.: Modified Oligonucleotides:Synthesis and Strategy for Users. Annu. Rev. Biochem. 67, 99–134 (1998)
9. Taylor, J.W., Schmidt, W., Cosstick, R., Okruszek, A., Eckstein, F.: The use of phosphorothioate-modified DNA in restriction enzyme reactions to prepare nicked DNA. Nucleic Acids Res. 13, 8749–8764 (1985)

Appendix: *Fok* I as a Nicking Enzyme

We examined to achieve a function which is the same as a nicking enzyme by using *Fok* I and phosphorothioate-modified DNA. *Fok* I cleaves both of double-strand, this cleaving activity isn't suitable for the molecular walking machine. We expected to block the cleaving activity at the cut point of one strand by phosphorothioatemodified DNA (Fig. 7). Remark that for our walking machine, only legs need phosphorothioate modification.

We experimented to confirm the effect of phosphorothioate-modified DNA. We prepared five DNA strands for the experiment (Table. 3). The five DNA strands compose the following four sets;

1) unmodified and separated the recognition site and the cut poin
2) phosphorothioate-modified and separated the recognition site and the cut point
3) unmodified and unseparated
4) phosphorothioate-modified and unseparated

Each four sets of DNA strands were mixed at 0.3 µM in hybridization buffer.
NEBuffer4 from New England Biolabs was used as the hybridization buffer. 4 units of *Fok* I from New England Biolabs were added to each five sets 20 µl solution. The four sets were incubated at 37°C by 2hours.

We ran the resulting solutions in 16% PAGE (non-denaturing gel).

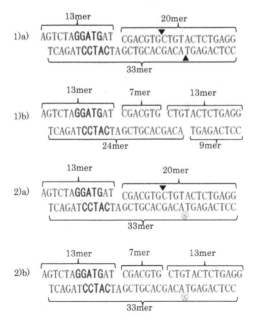

Fig. 8. Phosphorothioate-modified point.1) Unmodified (F-13, F-20, F-33), a) before cleaving b) after cleaving. 2) Phosphorothioate-modified(F-13, F-20, F-33s), a) before cleaving b) after cleaving.

Table 3. DNA strands. The recognition site of FokI is shown in bold type. The cut point is indicated by / , and s indicates the position of phosphorothioate-modified.

Name	Sequence(5'...3')
F-13	AGTCTA**GGATG**AT
F-20	CGACGTG/CTGTACTCTGAGG
F-13+20	AGTCTA**GGATG**ATCGACGTG/CTGTACTCTGAGG
F-33	CCTCAGAGT/ACAGCACGTCGAT**CATCC**TAGACT
F-33s	CCTCAGAGTsACAGCACGTCGAT**CATCC**TAGACT

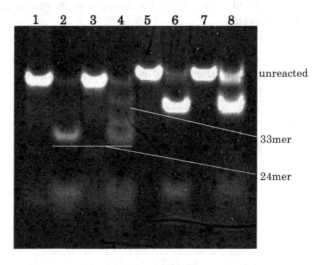

Fig. 9. Result of experiment of phosphorothioate-modified. 1)2) Unmodified and separated (F-13, F-20, F-33). 3)4) Phosphorothioate-modified and separated (F-13, F-20, F-33s). 5)6) Unmodified and not separated (F-13+20, F-33). 7)8) Phosphorothioate-modified and not separated (F-13+20, F-33s). Odd number of lane : without *Fok* I , Even number of lane : with *Fok* I.

For this result, we found phosphorothioate-modified DNA could block the cleving activity a little (Fig. 8). There were phosphorothioate-modified DNA with separate point in lane 4, and no-modified DNA with separate point in lane 2. To compare the two lanes, we confirmed that phosphorothioate-modified DNA was cleaved. The length 24 mer band is the products of cleaving F-33 at the cut point, and there is the same band in lane 4. So this result shows phosphorothioate-modified DNA was cleaved.

In this regard, however there is the length 33 mer band in lane 4. The band shows a few F-33s remained not to be cleaved. And there is the band which shows unreacted

DNA strand. This band isn't observed in lane 2. Unmodified DNA couldn't block the cleaving activity.

The effort to block the cleaving activity by phosphorothioate-modified DNA has failed. But it was observed phosphorothioate-modified DNA had the weak inhibition of the activity. We expect that using phosphorodithioate-modified DNA leads to stronger inhibition of the activity. The molecular machine may be able to get the cleaving function which it needs by phosphorodithioate-modified DNA.

Autonomous Programmable Nanorobotic Devices Using DNAzymes

John H. Reif and Sudheer Sahu

Department of Computer Science, Duke University
Box 90129, Durham, NC 27708-0129, USA
{reif,sudheer}@cs.duke.edu

Abstract. A major challenge in nanoscience is the design of synthetic molecular devices that run *autonomously* and are *programmable*. DNA-based synthetic molecular devices have the advantage of being relatively simple to design and engineer, due to the predictable secondary structure of DNA nanostructures and the well-established biochemistry used to manipulate DNA nanostructures. We present the design of a class of DNAzyme based molecular devices that are autonomous, programmable, and further require no protein enzymes. The basic principle involved is inspired by a simple but ingenious molecular device due to Mao et al [25]. Our DNAzyme based designs include (1) a finite state automata device, *DNAzyme FSA* that executes finite state transitions using DNAzymes, (2) extensions to it including probabilistic automata and non-deterministic automata, (3) its application as a *DNAzyme router* for programmable routing of nanostructures on a 2D DNA addressable lattice, and (4) a medical-related application, *DNAzyme doctor* that provide transduction of nucleic acid expression: it can be programmed to respond to the underexpression or overexpression of various strands of RNA, with a response by release of an RNA.

1 Introduction

1.1 Prior Autonomous Molecular Computing Devices

In the last few years the idea of constructing complex devices at the molecular scale using synthetic materials such as DNA has gone from theoretical conception to experimental reality.

(a) **DNA Tiling Assemblies.** One theoretical concept that had considerable impact on experimental demonstrations was that of Wang Tiling. This is an abstract model that allows for a finite set of 2D rectangles with labeled sides to assemble 2D lattices by appending together tiles at their matching sides. Winfree first proposed the use of DNA nanostructures known as *DNA tiles* to achieve universal computations. DNA tiles self-assemble into 2D lattices as determined by the tiles' pads (ssDNA on the sides of the tiles that can hybridize to other tiles' pads). The last decade has seen major successes in experimental demonstrations of the use of such DNA tiling assemblies to construct patterned lattices and tiling computations. DNA tiling assemblies have been used effectively in construction of periodic two-dimensional lattices, such as those made from double-crossover (DX) DNA tiles [29], rhombus tiles [12], triple-crossover (TX)

M.H. Garzon and H. Yan (Eds.): DNA 13, LNCS 4848, pp. 66–78, 2008.

tiles [9], and "4x4" tiles [31], as well as triangle lattices [11] and hexagonal lattices [5]. They have also been used for the construction of patterned lattices [30] by designing the DNA tile pads to program computations. The use of DNA tiling assembly has two major advantages over most other methods for molecular computation, since it: (i) operates entirely autonomously, without outside mediated changes, and (ii) does not require the use of protein enzymes.

DNA tiling assemblies do have limitations: in particular, in general as currently conceived, they do not allow for the molecular devices (the tiles in their case) to transition between multiple states (except of course for their free or assembled states). In contrast, many complex molecular mechanisms found in the cell can transition into multiple states, allowing far more flexibility of application.

(b) **Autonomous Molecular Computing Devices that Execute Multiple State Transitions.** There are only two other known methods for DNA computation that operate autonomously. Both use ingenious constructions, but require the use of enzymes.

(i) The *whiplash PCR machines* of [14,15,19,28]. These however, can only execute a small number of steps before they require changes in the environment to execute further steps. Also, they require the use of polymerase enzyme.

(ii) The autonomous DNA machines of Shapiro[4,2,3], which execute finite transitions using restriction enzymes. The autonomous DNA machine [3] demonstrated molecular sensing and finite state response capabilities for that could be used for medical applications (though the demonstrations were made in test tubes only, rather than in natural biological environments as would be required for their medical applications). Their paper was important motivational factor in the work described here.

1.2   Our Main Contribution

This paper provides the first known design for a DNA-RNA based devices that (a) operates autonomously, (b) do not require the use of protein enzymes, and (c) allow for the execution of multiple state transitions. Our designs make use of certain prior DNA nanomechanical devices, which will be discussed below.

1.3   DNA Nanomechanical Devices

Prior Nonautonoumous Nanomechanical DNA Devices. A variety of DNA nanomechanical devices have been constructed that exhibit motions such as open/close [23,24,34], extension/contraction [1,8,10], and rotation [13,26,32]. The motion of these devices is mediated by external environmental changes such as the addition and removal of DNA fuel strands [1,8,10,23,24,26,32,34] or the change of ionic strength of the solution [13]. For example, non-autonomous progressive walking devices, mediated by the addition and removal of DNA strands, were constructed both by Seeman [21] and Pierce [22]. Although in many cases ingeniously designed, these devices need external (human or automation-based) intervention for each step of their motions. These synthetic DNA devices are in sharp contrast with cellular protein motors and machines on macroscale that operate autonomously, without requiring any interference.

Recent times have seen significant progress in construction of DNA nanomechanical devices that execute autonomous, progressive motions. Reif [17] gave two designs for autonomous DNA nanomechanical devices that traverse bidirectionally along a DNA nanostructure. Turberfield et al proposed using DNA hybridization energy to fuel autonomous free-running DNA machines [27]. Peng et al [33] was the first to experimentally demonstrate an autonomous DNA walker, which is an autonomous DNA device in which a DNA fragment translocates unidirectionally along a DNA nanostructure. It used DNA ligase and restriction enzymes.

Recently Mao demonstrated two autonomous DNA nanomechanical devices driven by DNA enzymes

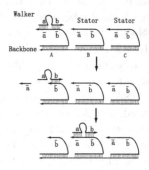

Fig. 1. Overview of Mao's crawler [25] constructed using DNA enzyme

(non-protein), namely (a) a tweezer [7,6] which is a DNA nanostructure that open and closes autonomously and (b) a DNA crawler [25] using DNA enzyme (DNAzyme), which traverses across a DNA nanostructure.

Their crawler device contains a DNAzyme that constantly extracts chemical energy from its substrate molecules (RNA) and uses this energy to fuel the motion of the DNA device. This DNAzyme-based crawler integrates DNAzyme activity and strand-displacement reaction. They use 10-23 DNAzyme, which is a DNA molecule that can cleave RNA with sequence specificity. The 10-23 DNAzyme contains a catalytic core and two recognition arms that can bind to a RNA substrate. When the RNA substrate is cleaved, the short fragment dissociate from the DNAzyme and that provides a toehold for another RNA substrate to pair with short recognition arm of the DNAzyme. The crawler device traverses on a series of RNA stators implanted on a nanostructure as shown in Figure 1. Their crawler is the primary inspiration to our designs. While an ingenious device, there are a number of limitations of Mao's DNAzyme-based crawler: (1) it did not demonstrate the loading and unloading of nanoparticles (2) it only traverses along a one dimensional sequence of ssRNA strands (stators) dangling from a DNA nanostructure, and its route is not programmable (3) it does not execute finite state transitions beyond what are required to move (that is, it does not execute computations).

1.4   Overview of This Paper and Results

The goal of this paper is to address the above limitations, providing DNAzyme based devices with substantially enhanced functionalities. We present the design of *DNAzyme FSA*: a finite state machine based on the activity of DNAzyme and strand displacements in Section 2. DNAzyme FSA can be easily extended to non-deterministic finite state automata and probabilistic automata as described in Section 2.6. In Section 3 we present a medical related application of DNAzyme FSA referred to as *DNAzyme doctor*. DNAzyme doctor is a molecular computer for logical control of RNA expression using DNAzyme. Another application of DNAzyme FSA, *DNAzyme router*: a DNAzyme based system for programmable routing of the walker on a 2D lattice is described in Section 4. All the devices described in this paper are based on selective cleaving activity of DNAzyme and strand displacement processes.

2 DNAzyme FSA: DNAzyme Based Finite State Automata

A *finite state automata* can be described as a 5-tuple $(\Sigma, S, s_0, \delta, F)$, where Σ is a finite non-empty set of symbols called input alphabet, S is a finite non-empty set of states, $s_0 \in S$ is an initial state, δ is the state transition function ($\delta : S \times \Sigma \rightarrow S$), and $F \subset S$ is the set of final states.

In this section, we describe a DNAzyme based finite state automata, referred to as DNAzyme FSA. At any time an RNA sequence encoding an input symbol is examined by the DNAzyme FSA, then an appropriate state transition takes place, and then the RNA sequence encoding the next input symbol is examined. This process continues till all the input symbols are scanned and the output of the DNAzyme FSA is its state at the end of process.

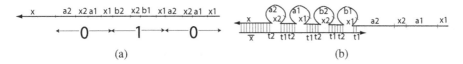

(a)   (b)

Fig. 2. (a) Encoding of 0 and 1 in DNAzyme FSA. (b) Protector strand partially hybridizes with the input strand to form bulge loops. The sticky end formed at the end of the input strand outside of the bulge loops represents the active input symbol. This scheme protects the input symbols other than the currently active symbol from becoming active.

2.1 Encoding the Input Symbols

First of all, we describe the way the input is encoded for the DNAzyme FSA. Input symbols 0 and 1 are encoded as the RNA sequences $x_1 \cdot a_1 \cdot x_2 \cdot a_2$ and $x_1 \cdot b_1 \cdot x_2 \cdot b_2$, respectively, where a_1, a_2, b_1, b_2, x_1, and x_2 are RNA sequences, and $\cdot$ represents concatenation. Figure 2 (a) illustrates this encoding of the input symbols. It should be noted that 0 and 1 share common subsequences x_1 and x_2. Also, there is a special subsequence x at the end of the input subsequence. This is central to the working of the DNAzyme FSA as will be explained later.

2.2 Active Input Symbol

While encoding the input for DNAzyme FSA, it is essential to have a mechanism to detect the current input symbol that is being scanned by DNAzyme FSA. We will refer to this symbol as *active input symbol*. In order to implement this feature in DNAzyme FSA only a small segment of the RNA strand encoding the input symbols is kept active. Most part of it is kept protected by hybridization with a partially complementary sequence, referred to as *protecting sequence*. It has not been shown in the figure but the protecting sequence should not be one continuous strand. Instead it should contain nicks at various positions. This is necessary for the working of device and will be explained later. The active input symbol is represented by the sticky end of the RNA sequence encoding the input. We refer to this nanostructure as *input nanostructure*. Figure 2 (b) illustrates the idea. The input nanostructure encodes the input 010. The active input symbol is rightmost 0 (in 010), and it is encoded by the sticky end of the input nanostructure, and hence is active. However, the leftmost 0 and the 1 are encoded in the protected portion of the input nanostructure. They have been protected by hybridization with a protecting sequence. Since the protecting sequence is partially complementary to the sequence encoding the input symbols, it results in the formation of bulge loops. In the Figure 2 b) a_2, a_1, b_2, and b_1 contain a subsequence complementary to t_2, while x_2 and x_1 contain subsequence complementary to t_1. Since the RNA sequence encoding input is partially complementary to the protecting sequence $t_2.t_1.t_2.t_1...$ it forms the bulge loop structure as shown in the Figure 2 (b). Each input symbol is hence represented by two bulge loops. It should be noted that the special sequence x at the end of the input sequence and $\bar{x}$ at the end of protecting sequence ensure that only the desired alignment of protecting sequence with input sequence is favored. As a result, only the desired input nanostructure as shown in Figure 2 (b) is formed.

2.3   States and Transitions

After the description of the input, next we describe the design of states and transitions in finite state machine. In DNAzyme FSA, a network of DNAzymes is embedded on a two-dimensional plane, and the input nanostructure is routed over it. The state of the DNAzyme FSA at any time is indicated by the DNAzyme that holds the input nanostructure at that time. During each state transition of DNAzyme FSA, the segment of input nanostructure encoding the active input symbol is cleaved, the next bulge loop opens up exposing the segment encoding next input symbol, thereby making it new active input symbol, and the input nanostructure jumps to another DNAzyme that indicates the new state of DNAzyme FSA. In subsequent paragraphs, we will explain in details the complete process of state transition in DNAzyme FSA. As shown in Figure 3 (a), a state transition from one state to another is implemented as two evenly spaced DNAzymes, referred to as *transition machinery* for that state transition. Each of these DNAzymes is tethered to another DNA nanostructure, which forms part of the backbone of the DNAzyme FSA. DNAzyme D_{0,s_1} and D'_{0,s_2} form the transition machinery for state transition from state s_1 to state s_2 for input 0. Similarly, DNAzyme D_{1,s_1} and D'_{1,s_2} form the transition machinery for state transition from state s_1 to state s_2 for input 1. It should be noted that in our nomenclature the first subscript of the DNAzyme specifies the active input symbol and the second subscript specifies the states for a transition machinery.

The foremost thing to ensure in DNAzyme FSA is that if the active input symbol is 0, then the state transition for input 0 should be taken. Similarly, if the active input symbol is 1, then the state transition for input 1 should be taken.

In the transition machinery for state transition for input 0, the DNAzymes D_{0,s_1} and D'_{0,s_2} contain DNA subsequences $\overline{x_2} \cdot \overline{a_1} \cdot \overline{x_1}$ and $\overline{x_1} \cdot \overline{a_2} \cdot \overline{x_2}$ respectively, at their free ends. The DNA subsequences of D_{0,s_1} is partially complementary to the RNA sequence

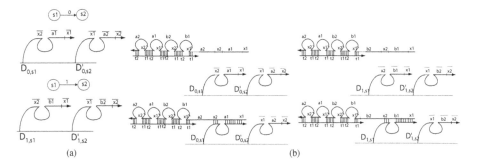

Fig. 3. (a) Figure illustrates the implementation of a state transition through DNAzymes. (b) D_{0,s_1} in the transition machinery for state transition at 0 combines with input nanostructure when active input symbol encoded by the sticky end is 0. When the active input symbol encoded by the sticky end is 1, D_{1,s_1} in the transition machinery for state transition at 1 combines with the input nanostructure.

that encode the symbol 0 ($x_1 \cdot a_1 \cdot x_2 \cdot a_2$). This ensures that only when the sticky end of input nanostructure is $x_1 \cdot a_1 \cdot x_2 \cdot a_2$, it can hybridize with the DNAzyme D_{0,s_1}. Thus a state transition for 0 is not taken in DNAzyme FSA, unless the active input symbol is 0.

Similarly, in the transition machinery for state transition for input 1, the DNAzymes D_{1,s_1} and D'_{1,s_2} contain DNA subsequences $\overline{x_2} \cdot \overline{b_1} \cdot \overline{x_1}$ and $\overline{x_1} \cdot \overline{b_2} \cdot \overline{x_2}$ respectively, at their free ends. These subsequences are partially complementary to the RNA sequence that encode the symbol 1 ($x_1 \cdot b_1 \cdot x_2 \cdot b_2$). As explained earlier, this ensures that a state transition for 1 is not taken in the DNAzyme FSA, unless the active input symbol is 1. Figure 3 (b) further illustrates the idea.

2.4 Description of State Transition

In this section, we will describe the movement of the input nanostructure over the DNAzymes in a transition machinery to carry out the state transition in DNAzyme FSA. Figure 4 (a) shows a transition machinery for input 0. Initially, the input nanostructure is hybridized with the DNAzyme D_{0,s_1}. The sticky end of the input nanostructure represents the active input symbol 0, and therefore, the transition at input 0 is to be performed. First, the DNAzyme D_{0,s_1} cleaves the input nanostructure as shown in Figure 4 (a). Now the sticky end of input nanostructure has only x_2 as complementary subsequence to the subsequence $\overline{x_2} \cdot \overline{a_1} \cdot \overline{x_1}$ at the free end of DNAzyme D_{0,s_1}. However, the longer subsequence $x_2 \cdot a_2$ in its sticky end is complementary with the subsequence $\overline{a_2} \cdot \overline{x_2}$ of DNAzyme D'_{0,s_2}. Therefore, a strand displacement process takes place with the free ends of DNAzymes D_{0,s_1} and D'_{0,s_2} competing against each other to hybridize with sticky end ($x_2 \cdot a_2$) of the input nanostructure. Since D'_{0,s_2} provides a longer complementary subsequence, ultimately D_{0,s_1} is displaced and the input nanostructure is now hybridized with D'_{0,s_2} as shown in Figure 4 (a). It should be noted that the next bulge loop gets opened in this process. An input symbol is encoded across two bulge

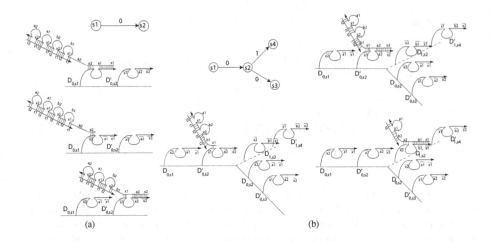

Fig. 4. (a) First half of a state transition by DNAzyme FSA from s_1 to s_2 at input 0 is illustrated. Sequence encoding active input symbol 0 gets cleaved by DNAzyme D_{0,s_1}, input nanostructure moves to next DNAzyme D'_{0,s_2} by strand displacement, and the next bulge loop in the input nanostructure opens up in the process. (b) Second half of a state transition by DNAzyme FSA from s_1 to s_2 at input 0 is shown. The mechanism is similar to the first half. However, in this part the next input symbol and next state transition of DNAzyme FSA is determined, and the input nanostructure lands up on the appropriate transition machinery for the next state transition to begin correctly.

loops in the input nanostructure. As the first half of the sticky end $(x_1 \cdot a_1)$ encoding the half of the active input symbol 0 got cleaved, the current sticky end is $x_2 \cdot a_2 \cdot x_1 \cdot b_1$, that contains half of the sequence encoding symbol 0 and half of the sequence encoding the symbol 1. This completes the first half of the state transition by DNAzyme FSA.

The second half of the transition in DNAzyme FSA takes place in exactly similar manner. Half of the sticky end $(x_2 \cdot a_2)$ of the input nanostructure that encodes the remaining half of the active input symbol 0 gets cleaved, thus leaving only x_1 as complementary to free end of DNAzyme D'_{0,s_2} $(\overline{x_1} \cdot \overline{a_2} \cdot \overline{x_2})$. At this point the sticky end of the input nanostructure is $x_1 \cdot b_1$ which is half of the sequence that encodes the input symbol 1. It indicates that the next active input symbol is 1 and therefore, the next state transition should be from state s_2 at input 1. This is ensured by the DNAzyme FSA in the following way. Since the sticky end of the input nanostructure is $(x_1 \cdot b_1)$, the DNAzyme D_{1,s_2} that has the sequence $\overline{x_2} \cdot \overline{b_1} \cdot \overline{x_1}$ at its free end gets involved in strand displacement with D'_{0,s_2} to hybridize with the sticky end $(x_1 \cdot b_1)$ of input nanostructure. Because of the longer complementary sequence D_{1,s_2} ultimately displaces D'_{0,s_2} and hybridizes with the sticky end of nanostructure. This results in the opening of next bulge loop in input nanostructure as shown in Figure 4 (b).

It should be noted that D_{0,s_2} (with sequence $\overline{x_1} \cdot \overline{b_2} \cdot \overline{x_2}$ at its free end) does not have sequences complementary to the sticky end $(x_1 \cdot b_1)$ of input nanostructure, so it can not get involved in any strand displacement. Therefore, the input nanostructure is guaranteed to move to the DNAzyme D_{1,s_2}. After the opening of the next bulge loop, the new sticky end ($x_1 \cdot b_1 \cdot x_2 \cdot b_2$) of input nanostructure encodes the input symbol 1. Thus, the input nanostructure lands up in the appropriate transition machinery for the next state transition, and the next state transition at input 1 can begin correctly.

It can be argued in a similar manner that during the second half of the transition, if the next active input symbol was to be 0, the input structure would have moved from DNAzyme D'_{0,s_2} to D_{0,s_2} instead of moving to D_{1,s_2}. We omit the explanation here for the sake of brevity.

Figure 4 (b) illustrates the second half of the state transition of DNAzyme FSA.

It should be noted that the strand displacement of the protector strand also takes place during the process. But since it contains nicks, its fragments just wash away in the solution when they get completely displaced.

2.5   Complete State Machine

The components described above can be integrated to implement the complete finite state automata. Any state transition in the DNAzyme FSA can be implemented by two DNAzymes as described earlier. These DNAzymes are embedded on a nanostructure that forms the backbone of the DNAzyme FSA. The addressable nanostructures formed by DNA origami [20] or fully-addressable DNA tile lattices [16] might provide useful nanostructures for this backbone. Hence, the state machine can be laid out on this nanostructure by implanting a network of DNAzymes on it. The input nanostructure traverses over them in a programmable way and keeps getting cleaved in the process.

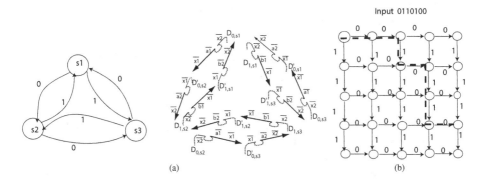

Fig. 5. (a) The DNAzyme implementation of the finite state machine shown on left. (b) Illustration of programmable routing in two dimensions.

Figure 5 (a) shows an implementation of a DNAzyme FSA (at the right) for the finite state automata (at the left). It should be noted that the DNAzymes shown in the Figure 5 (a) are actually implanted on a backbone nanostructure. The dashed lines represent the sides of these DNAzymes that are embedded in the backbone nanostructure.

The output of the DNAzyme FSA is detected using insitu hybridization techniques. The details of the protocol are described in [18].

2.6 Non-deterministic and Probabilistic DNAzyme FSA

A *nondeterministic finite state automata* is a 5-tuple $(\Sigma, S, s_0, \delta, F)$, where Σ is a finite set of input symbols, S is a finite set of states, δ is a state transition function (δ : $S \times (\Sigma \bigcup \{\epsilon\}) \rightarrow P(S)$ where $P(S)$ is the power set of S), ϵ is the empty string, $s_0 \subset S$ is a set of initial states, and $F \subset S$ is a set of final states.

A *probabilistic finite state automata* is a finite state automata in which the state transitions are probabilistic in nature. It can be described as a 5-tuple $(\Sigma, S, s_0, \delta, F)$, where Σ is a finite set of input symbols, S is a finite set of states, δ is a state transition function ($\delta : S \times \Sigma \times S \rightarrow [0, 1]$), $s_0 \subset S$ is a set of initial states, and $F \subset S$ is a set of final states.

The idea extends to the non-deterministic automata directly. Different DNAzyme-FSA described above will work in parallel inside a test-tube. Therefore, the above described scheme will work for non-deterministic automata as well. In case there are more than one transitions possible for one input from one state, each of them will be taken in one DNAzyme-FSA or the other inside the solution, and thus exhibiting non-deterministic nature of the automata. Regarding the output, if the output state in any of the DNAzyme-FSA in solution is an accepting state (or final state), it implies the acceptance of the input by the overall non-deterministic finite state automata.

In case the sequences of all the DNAzymes are identical, then the DNAzyme-FSA described above becomes a probabilistic automata having equal probabilities of transitions from any state to any other state. However, to construct an arbitrary probabilistic finite state automata, the probabilistic transitions can be implemented by using partially complementary sequences in the designs. The sequences of the DNAzymes for transition are chosen in a way so that the ratios of probability of hybridization are in accordance with the transition probabilities.

3 DNAzyme Doctor: A Molecular Computer for Logical Control of RNA Expression Using DNAzyme

The finite state automaton described in Section 2 can be used in various computational and routing applications. In this section we describe DNAzyme doctor, an application related to medical field. It is an autonomous molecular computer for control of RNA expression based on the overexpression and underexpression of other RNAs. Earlier Shapiro[3] had constructed a molecular computer using protein enzymes for logical control of RNA expression. DNAzyme doctor performs the same function, while completely eliminating the use of protein enzymes in the design. For the ease of illustration let us consider a similar example as given in [3]. Suppose a disease is diagnosed positive if RNAs R_1 is underexpressed, R_2 is underexpressed, R_3 is overexpressed, and R_4 is overexpressed. Thus, the detection of the disease can be done by computing logical AND of the above mentioned four RNA expression tests. In case it is established that the disease exists, a curing drug should be released. While in any other case, the drug should not be released. Figure 6 (a) illustrates the aforementioned logic in the form of a state diagram.

The sequences y_1, y_2, y_3 and y_4 are characteristic sequences of RNAs R_1, R_2, R_3, and R_4 respectively. If R_1 is overexpressed then y_1 is in excess, and if R_2 is overexpressed

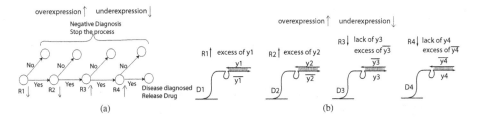

Fig. 6. (a)A state diagram for DNAzyme doctor that controls the release of a drug RNA on the basis of the RNA expression tests for the a disease (b) The figure shows the consequences of overexpression and underexpression of different RNAs on the concentrations of the respective characteristic sequences. The overexpression of R_1 and R_2 results in excess of y_1 and y_2 respectively, and they block the path of input nanostructure by hybridizing with D_1 and D_2. Similarly underexpression of R_3 and R_4 results in excess of $\overline{y_3}$ and $\overline{y_4}$ respectively, to block the path of input nanostructure.

then y_2 is in excess. However, if R_3 is underexpressed, then lack of y_3 and if R_4 is underexpressed, then lack of y_4. But a threshold concentration of $\overline{y_1}, \overline{y_2}, \overline{y_3}, \overline{y_4}$ is thrown into the solution, therefore lack of y_3 causes excess of $\overline{y_3}$, and lack of y_4 causes excess of $\overline{y_4}$.

Since the DNAzyme doctor only needs to perform a logical AND, it can be implemented in a simple way. We make the input nanostructure walk over four DNAzyme stators implanted on a nanostructure in a straight path (more details in [18]). Each DNAzyme stator represents one of the RNA expression test. In case the test is positive, the input nanostructure moves to next DNAzyme stator, otherwise it gets stuck and ultimately floats away in the solution. Therefore, the successful traversal of input nanostructure over all these DNAzyme stators implies that all tests are positive, and hence positive diagnosis of the disease.

In case the first test is negative (ie. overexpression of R_1), then excessively floating y_1 can bind to $\overline{y_1}$ part of the DNAzyme D_1. Similarly if second, third, or fourth tests are negative (ie.. overexpression of R_2, underexpression of R_3 or underexpression of R_4), then excessively floating y_2, y_3, or y_4 can bind to $\overline{y_2}, \overline{y_3}, \overline{y_4}$ portions of DNAzyme D_2, D_3, or D_4, respectively. The principle idea is illustrated in Figure 6. More details of DNAzyme doctor are presented in [18].

4   DNAzyme Router

For any arbitrary path along the network of DNAzymes in a given DNAzyme FSA, an input nanostructure can be designed to traverse along that path. This principle can be used for the design of a programmable routing system. The input nanostructure that moves over the DNAzyme FSA is referred to as *walker* and the complete system as DNAzyme router. The path of the walker is programmed through the state transitions of the automata and the input symbols encoded in the walker. As an example, we can create a state machine on a rectangular grid (Figure 5 (b)), in which you move right if the input is 0, and towards bottom if the input is 1. Then an input nanostructure that represents the input 0110100 can be made to walk through the path shown by dashed lines in Figure 5 (b).

It should be noted that in a DNAzyme router the path does not get destroyed as a result of the motion of the walker. It is the input nanostructure (walker) that gets cleaved in the process, which is equivalent to exhaustion of fuel as a result of motion. Most remarkable feature of DNAzyme router is that we can have multiple walkers moving on the grid independently, each having its own programmed path.

5   Conclusion

We have described the construction of various devices based on the DNAzymes. DNAzymes evolve through invitro selection procedures, and these processes can be designed to generate DNAzymes that cut distinct sequences. In the DNAzyme FSA, the number of DNAzymes required is proportional to the number of transitions in the automata. For binary-coded inputs the number of transitions is proportional to number of states. It should be noted that each of the devices described in the paper need the DNAzymes to be mounted on an addressable two-dimensional nanostructure such as the ones constructed by Rothemund [20] or Park et al [16], which themselves are floating in the solution. The molecular computer for logical control of RNA expression can be useful in medical field if it can be used inside a cell, and the programmable walkers can be a really useful tool in nanopartical transportation systems at nanoscale. In conclusion, the designs provided in this paper might provide useful insight for research into many interesting problems in nanotechnology.

Acknowledgement. The work is supported by NSF EMT Grants CCF-0523555 and CCF-0432038.

References

1. Alberti, P., Mergny, J.: DNA duplex-quadruplex exchange as the basis for a nanomolecular machine. Proc. Natl. Acad. Sci. USA 100, 1569–1573 (2003)
2. Benenson, Y., Adar, R., Paz-Elizur, T., Livneh, Z., Shapiro, E.: DNA molecule provides a computing machine with both data and fuel. Proc. Natl. Acad. Sci. USA 100, 2191–2196 (2003)
3. Benenson, Y., Gil, B., Ben-Dor, U., Adar, R., Shapiro, E.: An autonomous molecular computer for logical control of gene expression. Nature 429, 423–429 (2004)
4. Benenson, Y., Paz-Elizur, T., Adar, R., Keinan, E., Livneh, Z., Shapiro, E.: Programmable and autonomous computing machine made of biomolecules. Nature 414, 430–434 (2001)
5. Chelyapov, N., Brun, Y., Gopalkrishnan, M., Reishus, D., Shaw, B., Adleman, L.: DNA triangles and self-assembled hexagonal tilings. J. Am. Chem. Soc. 126, 13924–13925 (2004)
6. Chen, Y., Mao, C.: Putting a brake on an autonomous DNA nanomotor. J. Am. Chem. Soc. 126, 8626–8627 (2004)
7. Chen, Y., Wang, M., Mao, C.: An autonomous DNA nanomotor powered by a DNA enzyme. Angew. Chem. Int. Ed. 43, 3554–3557 (2004)
8. Feng, L., Park, S., Reif, J., Yan, H.: A two-state DNA lattice switched by DNA nanoactuator. Angew. Chem. Int. Ed. 42, 4342–4346 (2003)

9. LaBean, T., Yan, H., Kopatsch, J., Liu, F., Winfree, E., Reif, J., Seeman, N.: The construction, analysis, ligation and self-assembly of DNA triple crossover complexes. J. Am. Chem. Soc. 122, 1848–1860 (2000)
10. Li, J., Tan, W.: A single DNA molecule nanomotor. Nano Lett. 2, 315–318 (2002)
11. Liu, D., Wang, M., Deng, Z., Walulu, R., Mao, C.: Tensegrity: Construction of rigid DNA triangles with flexible four-arm dna junctions. J. Am. Chem. Soc. 126, 2324–2325 (2004)
12. Mao, C., Sun, W., Seeman, N.: Designed two-dimensional DNA holliday junction arrays visualized by atomic force microscopy. J. Am. Chem. Soc. 121, 5437–5443 (1999)
13. Mao, C., Sun, W., Shen, Z., Seeman, N.: A DNA nanomechanical device based on the B-Z transition. Nature 397, 144–146 (1999)
14. Matsuda, D., Yamamura, M.: Cascading whiplash pcr with a nicking enzyme. In: Hagiya, M., Ohuchi, A. (eds.) DNA Computing. LNCS, vol. 2568, pp. 38–46. Springer, Heidelberg (2003)
15. Nishikawa, A., Hagiya, M.: Towards a system for simulating DNA computing with whiplash PCR. In: Angeline, P.J., Michalewicz, Z., Schoenauer, M., Yao, X., Zalzala, A. (eds.) Proceedings of the Congress on Evolutionary Computation, vol. 2, pp. 960–966, 6–9. IEEE Press, Washington (1999)
16. Park, S.H., Pistol, C., Ahn, S.J., Reif, J.H., Lebeck, A.R., Dwyer, C., LaBean, T.H.: Finite-size, fully addressable dna tile lattices formed by hierarchical assembly procedures. Angew. Chem. Int. Ed. 45, 735–739 (2006)
17. Reif, J.: The design of autonomous DNA nanomechanical devices: Walking and rolling DNA. In: Hagiya, M., Ohuchi, A. (eds.) DNA Computing. LNCS, vol. 2568, pp. 22–37. Springer, Heidelberg (2003), Published in Natural Computing, DNA8 special issue, vol. 2, pp. 439–461 (2003)
18. Reif, J.H., Sahu, S.: Autonomous programmable dna nanorobotic devices using dnazymes. Technical Report CS-2007-06, Duke University, Computer Science Department (2007)
19. Rose, J.A., Deaton, R.J., Hagiya, M., Suyama, A.: Pna-mediated whiplash pcr. In: Jonoska, N., Seeman, N.C. (eds.) DNA Computing. LNCS, vol. 2340, pp. 104–116. Springer, Heidelberg (2002)
20. Rothemund, P.: Generation of arbitrary nanoscale shapes and patterns by scaffolded DNA origami. Nature (2005)
21. Sherman, W., Seeman, N.: A precisely controlled DNA biped walking device. Nano Lett. 4, 1203–1207 (2004)
22. Shin, J., Pierce, N.: A synthetic DNA walker for molecular transport. J. Am. Chem. Soc. 126, 10834–10835 (2004)
23. Simmel, F., Yurke, B.: Using DNA to construct and power a nanoactuator. Phys. Rev. E 63, 41913 (2001)
24. Simmel, F., Yurke, B.: A DNA-based molecular device switchable between three distinct mechanical states. Appl. Phys. Lett. 80, 883–885 (2002)
25. Tian, Y., He, Y., Chen, Y., Yin, P., Mao, C.: Molecular devices - a DNAzyme that walks processively and autonomously along a one-dimensional track. Angew. Chem. Intl. Ed. 44, 4355–4358 (2005)
26. Tian, Y., Mao, C.: Molecular gears: A pair of DNA circles continously rolls against each other. J. Am. Chem. Soc. 126, 11410–11411 (2004)
27. Turberfield, A., Mitchell, J., Yurke, B., Mills, J.A.P., Blakey, M., Simmel, F.: DNA fuel for free-running nanomachines. Phys. Rev. Lett. 90, 118102 (2003)
28. Winfree, E.: Whiplash pcr for o(1) computing. Technical Report 1998.23, Caltech (1998)
29. Winfree, E., Liu, F., Wenzler, L., Seeman, N.: Design and self-assembly of two-dimensional DNA crystals. Nature 394(6693), 539–544 (1998)
30. Yan, H., LaBean, T., Feng, L., Reif, J.: Directed nucleation assembly of DNA tile complexes for barcode patterned DNA lattices. Proc. Natl. Acad. Sci. USA 100(14), 8103–8108 (2003)

31. Yan, H., Park, S., Finkelstein, G., Reif, J., LaBean, T.: DNA-templated self-assembly of protein arrays and highly conductive nanowires. Science 301(5641), 1882–1884 (2003)
32. Yan, H., Zhang, X., Shen, Z., Seeman, N.: A robust DNA mechanical device controlled by hybridization topology. Nature 415, 62–65 (2002)
33. Yin, P., Yan, H., Daniell, X., Turberfield, A., Reif, J.: A unidirectional DNA walker moving autonomously along a linear track. Angew. Chem. Int. Ed. 43, 4906–4911 (2004)
34. Yurke, B., Turberfield, A., Mills, J.A.P., Simmel, F., Neumann, J.: A DNA-fuelled molecular machine made of DNA. Nature 406, 605–608 (2000)

Multi-fueled Approach to DNA Nano-Robotics

Akio Nishikawa[1,*], Satsuki Yaegashi[4], Kazumasa Ohtake[3,4], and Masami Hagiya[2,4]

[1] Department of Economics, Fuji University, Hanamaki, Iwate, Japan
[2] Department of Computer Science, University of Tokyo, Tokyo, Japan
[3] Department of Biochemistry, University of Tokyo, Tokyo, Japan
[4] CREST JST, Japan
nisikawa@fuji-u.ac.jp, yaegashi@lyon.is.s.u-tokyo.ac.jp,
ohtake@biochem.s.u-tokyo.ac.jp, hagiya@is.s.u-tokyo.ac.jp

Abstract. An approach to multi-fueled DNA nano-robotics is described. We propose three types of driving force (i.e., fuel for DNA nano-robots): thermal fuel, pH fuel, and light fuel. The thermal fuel controls the hybridization of DNA molecules around the melting temperature. The pH fuel controls the hybridization of the so-called i-motif by changing the pH condition. The light fuel controls the hybridization of DNA oligomers that are intercalated with azobenzene by irradiation with UV or visible light. These three fuels are not mutually exclusive. However, experimental conditions for the fueling of DNA nano-robots show efficacy. Concrete ideas for using these three fuel types are proposed and discussed.

Keywords: DNA nano-robotics, multi-fueled approach, thermal fuel, pH fuel, light fuel, azobenzene, i-motif.

1 Introduction

Beginning with the pioneering work of Yurke et al. [7], DNA nano-robotic systems have made steady progress. In particular, the notion of a DNA fuel has been used in many applications. For example, Pierce et al. [5] have constructed a DNA walking device, in which the steps are fueled by single strands of DNA, each corresponding to one step of the device. In fact, DNA fuel has become a versatile tool in DNA nano-robotics. One advantage of DNA fuel is that different types of fuel with different base sequences can be used separately, so that they control the hybridization of different DNA molecules independently of one another.

However, DNA fuel emits double-stranded DNA molecules as a waste product, which accumulates in the solution and eventually inhibits the desired reaction. Therefore, it is reasonable to seek other sources of fuel, ideally those that do not produce waste products, which can be used to control the hybridization of DNA molecules.

Clearly, one can control the hybridization of DNA molecules by changing the temperature of the solution around the melting temperature of the molecules (thermal

---

* Corresponding author.

M.H. Garzon and H. Yan (Eds.): DNA 13, LNCS 4848, pp. 79–88, 2007.
© Springer-Verlag Berlin Heidelberg 2007

fuel). Another source of fuel is light. Takahashi et al. [6] have constructed DNA nano-machines that can be controlled by light radiation. They used DNA molecules intercalated with azobenzene, which changes conformation from trans to cis under UV-light irradiation and from cis to trans under visible light [1, 2]. A modified DNA molecule can hybridize with its complementary counterpart if the intercalated azobenzene molecules take the trans conformation (light fuel). Yet another source of fuel are protons. Liu and Balasubramanian [3] have proposed the use of the so-called i-motif for DNA nano-machines. Under the appropriate acidic condition, the i-motif adopts a folded form and does not hybridize with its complementary counterpart (pH fuel).

Using these DNA molecules as different components that are controllable by different sources of fuel, one can confer complex behaviors on a nano-machine by controlling the injection of each type of fuel. Note that the thermal fuel and the light fuel do not produce waste. Although the pH fuel increases the salt concentration each time the pH is changed, it is considered more tractable (and is cheaper) than DNA fuel.

The crucial issue in using these different sources of fuel is whether they can work independently of one another. If they do work independently, one can imagine various applications. For example, we can imagine a walking device on a DNA trail (Fig. 1), in which three types of DNA molecule protrude from the trail in a cyclic order. Each type of DNA is controlled by the corresponding type of fuel, i.e., thermal, pH, or light.

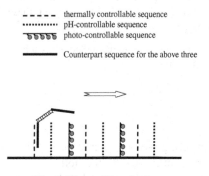

Fig. 1. The walking device

The motion of the device can be controlled if the three types of fuel work independently. For example, the walking device can be designed to move in a single direction as follows:

1. DNA oligomers that are thermally controllable, pH-controllable, and photo-controllable, and which can hybridize with their counterpart oligomers, are prepared. Controllable signifies that it is possible to control the hybridization and denaturation of the target oligomer and its counterpart. The sequences of the thermally controllable oligomers are designed using melting temperature predictions, the sequences of the pH-controllable oligomers are designed based on the i-motif [3], and the photo-controllable oligomers are prepared by azobenzene intercalation [1, 2].

2. The walking device has two counterpart oligomers as the 'feet'.

3. The three types of oligomers are immobilized repeatedly on the DNA trail in a cyclic order.
4. The initial condition is set up, e.g., the pH is set to 5.0 (acidic), the temperature is around 25°C (low), and the solution has been radiated with UV light (i.e., azobenzene is cis-formed). Then counterpart oligomers can only hybridize with the thermally controllable oligomers, i.e., one foot of the device hybridizes with a thermally controllable foot (Fig. 1).
5. To move the device in the right direction, we first change the pH to 7.0 (neutral). The other foot then hybridizes with the adjacent pH-controllable oligomer. By raising the temperature to 45°C (high), the first foot is denatured. At this point, if we irradiate the solution with visible light, the first foot will hybridize with the adjacent photo-controllable oligomer.

Figure 2 shows a similar DNA device based on the three types of fuel. In this system, the DNA trail forms a small triangle, and the device rotates the triangle like a motor.

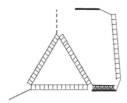

Fig. 2. The triangular trail

As mentioned above, in order to realize such DNA devices, it is crucial that the three types of fuel can be controlled independently. The goal of the present study is to investigate how the three types of fuel can be controlled independently. In the present paper, we report the results of our preliminary experiments in which we have prepared three oligomers that are thermally controllable, pH-controllable, and photo-controllable, as well as their counterparts, and observed their hybridization profiles under various conditions.

Unfortunately, in the current design, the three types of fuel were not always independent of each another, although we observed some independence. For example, the pH fuel and the light fuel can be controlled almost independently. Even if the three types of fuel are not completely independent, the information gathered regarding independence should be useful for the future development of DNA nano-robots.

2 Materials and Methods

2.1 Materials

We prepared four oligomers. The oligomer termed RG-motor is the so-called i-motif, as described by Liu and Balasubramanian [3]. This oligomer has four CCC motifs and folds into a specific form under acidic pH conditions. The oligomer named Y-A12-BHQ2, which is the complementary counterpart of RG-motor, is also taken from Liu and Balasubramanian [3], except that we replaced one T with A to slightly break the symmetry. Thus, it is complementary to a sub-sequence of RG-motor with two mismatches, which lower the melting temperatures of RG-motor and Y-A12-BHQ2. Given that RG-motor prefers the folded form, under acidic pH conditions, the hybrid of RG-motor and Y-A12-BHQ2 is denatured. Note that RG-motor has rhodamine green at its 5'-end, whereas Y-A12-BHQ2 has BHQ2 at its 3'-end. Therefore, while these oligomers hybridize to each together, the fluorescence associated with RG is quenched.

The thermally controllable oligomer, Cy5-YY8, is complementary to the 3'-end 10-mer segment of Y-A12-BHQ2. The length of Cy5-YY8 was adjusted to control the hybridization between it and Y-A12-BHQ2 at temperatures between 25°C and 45°C. Note also that Cy5-YY8 has Cy5 at its 5'-end, the fluorescence of which is quenched by BHQ2.

The photo-controllable oligomer, TAMRA-YY7-AZ, has five azobenzenes intercalated into its side-chain. Without azobenzene, the oligomer is complementary to the 3'-end 13-mer segment of Y-A12-BHQ2. TAMRA-YY7-AZ also has TAMRA at its 5'-end, the fluorescence of which is quenched by BHQ2.

The sequences of the oligomers are listed below. In the design of these sequences, we sometimes used the Hyther program, which is available through the web interface [4], for predicting the oligomer melting temperatures.

RG-motor (pH-controllable sequence):
5'-rhodamine green-CCCTAACCCTAACCCTAACCC-3'

Cy5-YY8 (thermally controllable sequence):
5'-Cy5-CTAACTCTAA-3'

TAMRA-YY7-AZ (photo-controllable sequence):
5'-TAMRA-CTAXACXTCXTAXACXAC-3'; X = azobenzene

Y-A12-BHQ2 (counterpart sequence):
5'-GTTAGTGTTAGAGTTAG-BHQ2-3'

2.2 Selection of Fluorescent Groups and Buffers

Before the preliminary experiments, we had to decide which fluorescent groups to attach to oligomers, as the multi-fueled approach needs efficient and stable fluorescence under acidic environments and irradiation of UV light.

In general, as the pH decreases, the fluorescence becomes weaker. Furthermore, if the UV irradiation light is strong, the fluorescence generally degrades. However, the efficiency of fluorescence under these conditions depends greatly upon the fluorescent groups attached to the oligomers. Under acidic pH conditions, many fluorescent groups lose fluorescence. For example, FAM and Cy3 are not suitable for use under acidic pH conditions, such as pH 5.0. Although rhodamine green retains fluorescence under these conditions, the buffer used is of crucial importance. After examining several types of buffer, we found that SSC buffer was optimal. The so-called Good buffers were not always adequate for our experiments. Even the best combination of rhodamine green and SSC buffer requires some data normalization. UV-light irradiation also affects the efficiency of fluorescence. TAMRA and Cy5 are tolerant to UV light, compared with FAM and Cy3. Therefore, we chose rhodamine green, TAMRA, and Cy5 for our experiments.

2.3 Methods

We conducted two experiments, the first with azobenzene-intercalated oligomers, and the second with thermally controllable and pH-controllable oligomers. All three oligomer types could be mixed in a single solution, but since UV-light radiation requires different protocols and devices, we conducted that experiment separately with the azobenzene-intercalated oligomer.

Experiment with Azobenzene-Intercalated Oligomers

The 13-mer azobenzene-intercalated oligomer with TAMRA attached to the 5'-end of the sequence, TAMRA-YY7-AZ (5'-TAMRA-CTAXACXTCXTAXACXAC-3'; where X = azobenzene), was prepared in a tube with 1× SSC buffer to a final concentration of 0.1 μM. UV light at 360 nm was applied to the tube through a UV-D36C glass filter (Asahi Techno Glass) with the UVP B-100AP 100-W lamp for 30 min. The irradiated sample (400 μl) was transferred to a quartz cell and placed in a Hitachi F-2500 Spectrophotometer, the temperature of which was maintained with the LAUDA RC-6 apparatus. While the fluorescence of TAMRA was measured for 300 seconds, the equivalent concentration of the quencher oligomer, Y-A12-BHQ2 (5'-GTTAGTGTTAGAGTTAG-BHQ2-3'), which is partially complementary (13-mer) to TAMRA-YY7-AZ, was added to the cell to measure the fluorescence change of TAMRA. The fluorescence of TAMRA in the trans-form was also measured. Since the trans-form azobenzene allowed Y-A12-BHQ2 to hybridize, we also investigated whether TAMRA-YY7-AZ and Y-A12-BHQ2 were separated by UV irradiation for 30 min. This experiment was carried out at 25°C and 45°C under neutral and acidic (pH 5.0) pH conditions.

Experiment to Control the pH and Temperature

The Cy5-YY8 and RG-motor oligomers were first mixed in a tube, at final concentrations of 0.1 μM, with 1× SSC. The quencher, Y-A12-BHQ2, was added at four-fold higher concentration for the duration of the fluorescence measurement, in order to measure the effect of quenching on each type of fluorescence. This measurement was performed in the cycle of neutral (pH 7.6), acidic, and neutral pH at 25°C. In the cycle, 6 μl of 1 M HCl were added to the cell to produce the acidic pH condition, and 55μl of 0.1 M NaOH were applied to neutralize the acidic solution. In the same manner, each type of fluorescence was measured for the temperature cycle of 25°C , 45°C , and 25°C at neutral pH (pH 7.6).

3 Results

The purpose of the preliminary experiments was to check the feasibility of the multi-fueled approach. For this purpose, we examined the basic behaviors of the DNA oligomers under various conditions generated by combinations of fueling operations.

1. For the temperature of the solution, we examined the alternatives of 25°C and 45°C.
2. For the pH of the solution, we examined the neutral pH condition (around pH 7.0) and the acidic pH condition (around pH 5.0).
3. For UV-light irradiation, we examined three alternatives:
   (a) The solution was irradiated with UV light before the hybridization reaction.
   (b) The solution was irradiated with UV light after the hybridization reaction.
   (c) The solution was not irradiated with UV light.

   DNA oligomers that are intercalated with azobenzene can be controlled in two ways. One way is to block hybridization beforehand using UV-light irradiation. The other way is to denature the double-stranded hybridized DNA by irradiation. These alternatives are based on the conformational change of azobenzene from trans to cis that occurs under UV-light irradiation. Since the cis-form of azobenzene hinders hydrogen bonding of base pairs, the conformational change from trans to cis is expected to cause denaturation or blockage of hybridization.

Therefore, we examined various experimental conditions, each of which was a combination of one of the two thermal conditions, one of the two pH conditions, and one of the three light conditions. In total, we examined 12 ($2\times2\times3$) conditions. In addition, each condition was examined with the different fluorescent wavelengths of Cy5, rhodamine green, and TAMRA. As described in Materials and Methods, Cy5 was attached to the oligomer for thermal control (Cy5-YY8), rhodamine green was attached to the oligomer for pH control (RG-motor), and TAMRA was attached to the oligomer for photo-control (TAMRA-YY7-AZ). All these oligomers are (partially) complementary to the counterpart oligomer (Y-A12-BHQ2), which is modified to have BHQ2 at its 3'-end so that the fluorescence is quenched when the oligomers hybridize.

Table 1. Summary of the experimental results. The + symbol denotes that the spectrophotometer detected strong fluorescence coming from the fluorescent group of the DNA oligomers, which was not quenched by BHQ2 in the counterpart DNA oligomer. The − symbol denotes that the spectrophotometer did not detect strong fluorescence. The observation results in the meshed area do not coincide with the expected results. The ± symbol denotes a fluorescence level between + and −. The question mark symbol (?) indicates that fluorescence could not be measured, due to the extreme conditions.

		25°C			45°C		
		No UV	UV (before)	UV (after)	No UV	UV (before)	UV (after)
Neutral (pH 7.0)	Thermal	−	−		+	+	+
	pH	−	−	−	+	+	+
	Photo	−	+	+		+	±
Acidic (pH 5.0)	Thermal	−	−	−	+	+	+
	pH	+	+	+	+	+	+
	Photo	−	+	±	?	?	?

In the experimental results shown in Table 1, the + symbol indicates that the observed fluorescence is almost as strong as that of the fluorescent group alone, while the − symbol means that the observed fluorescence is much weaker than that of the fluorescent group alone. Other symbols are mentioned in the next section.

As explained in more detail in the next section, the observation results shown in the meshed area of Table 1 do not correlate with the expected outcomes. This means that the independence of these conditions is compromised.

Owing to space limitations, we mention only a few examples of the observed fluorescent data. Figure 3 shows the result of changing the temperature and measuring the fluorescence from Cy5 attached to the thermally controllable oligomer (Cy5-YY8). The observation was made with a mixture of Cy5-YY8, RG-motor, and Y-A12-BHQ2, as described in Section 2.3, under the neutral pH condition. At 25°C, fluorescence was not observed (?), which indicates that Cy5-YY8 hybridizes with Y-A12-BHQ2. At 45°C, fluorescence was observed (+), which indicates that Cy5-YY8 and Y-A12-BHQ2 are denatured. In Figure 3, the fluorescence data are not normalized, as Cy5 is not influenced by temperature.

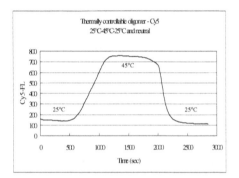

Fig. 3. Results for the thermally controllable oligomer

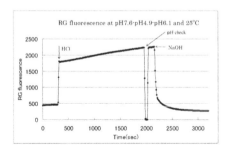

Fig. 4. The results for the pH-controllable oligomer

Figure 4 shows the results from changing the pH conditions and observing the fluorescence from RG attached to the pH-controllable oligomer. These observations were made for the mixture of Cy5-YY8, RG-motor, and Y-A12-BHQ2, as described in Section 2.3, at 25°C. The pH condition was first changed from neutral to acidic using HCl. After the pH was measured (pH 4.9), NaOH was added to give the final pH condition (pH 6.1) for this experiment.

Regarding the main focus of the present study, i.e., the independence of the three types of fuel, Figure 5 shows the results of one of the successful experiments. The fluorescence of TAMRA attached to the photo-controllable oligomer is shown under four different conditions. In the left panel, the fluorescence intensity in the absence of UV light is compared with that of the oligomer that was irradiated with UV light for 30 min. Both measurements were conducted under the neutral pH condition at 25°C. In the right panel, the same comparison is made under the acidic pH condition at 25°C. Although the fluorescence level under the acidic pH condition is lower, similar results were obtained under both conditions. These results indicate the independence of the light fuel from the pH fuel at 25°C.

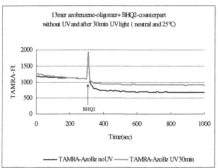
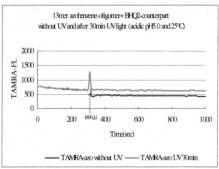

Fig. 5. Independence of the light fuel from the pH fuel at 25°C. The photo-controllable oligomer TAMRA-YY7-AZ was irradiated with UV light (black line) or not irradiated (gray line). Y-A12-BHQ2 was added 300 sec after the start of the measurement period. In both panels, the fluorescence levels are adjusted to the time at which Y-A12-BHQ2 was added.

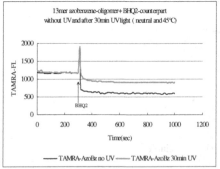
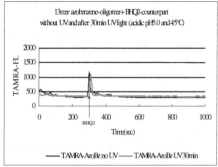

Fig. 6. An unsuccessful outcome regarding the independence of the light fuel from the pH fuel at 45°C

On the other hand, Figure 6 shows an experimental result under extreme condition. Under the acidic pH condition at 45°C (right panel), the fluorescence level was extremely low compared with the other conditions.

4 Discussion

As shown in the previous section, the levels of independence among the three types of fuel are incomplete. The following observation results (Table 1, meshed area) did not coincide with the expected outcomes.

1. The pH-controllable oligomer was denatured at 45°C even under the neutral pH condition. Although the pH-controllable sequence is seven bases longer than the thermally controllable oligomer, this difference in length appears to be insufficient, as the pH-controllable sequence contains some mismatches.
2. Under the acidic pH condition at 25°C and the neutral pH condition at 45°C , when the solution was irradiated with UV light after the hybridization reaction with the photo-controllable oligomer, the fluorescence level was not sufficiently strong. This means that the photo-controllable oligomer was not denatured completely.
3. Under the acidic pH-condition at 45°C, the fluorescence level of TAMRA was very low and unstable. This appears to be due to the severity of the conditions (pH 5.0 and 45°C) for the TAMRA fluorescence group. Although the hybridization reaction may have occurred as expected, methods other than fluorescence detection are required for reliable observations.

5 Concluding Remarks

We have proposed a multi-fueled approach to DNA nano-robotics and examined the feasibility of the approach in some preliminary experiments. Although the three types of fuel are not always independent of each other, combinations of the three fuels used under the appropriate experimental conditions have proven to be useful for DNA nano-robotics.

To ensure that the three types of fuel function more independently of one another, it is necessary to examine the possibility of controlling the hybridization under milder conditions. Such mild conditions would also solve the problem encountered with fluorescence detection, as described in the previous section.

Although some combinations of the three fuels have proven to be useful, their effectiveness was only shown qualitatively. In order to optimize experimental protocols and eventually construct motors and walkers, we need to make quantitatively estimation of the hybridization ratio in each combination. Calibration between the fluorescence level and the hybridization ratio is the first thing to be done.

The sequences used for the experiments are not considered optimal. In order to re-design them, it seems worthwhile to try in-vitro search of sequences in addition to ordinary free energy prediction based on the nearest neighbor model.

The search for an alternative source of fuel is critically important.

References

1. Asanuma, H., et al.: Photoregulation of the formation and dissociation of a DNA duplex by using the cis-trans isomerization of azobenzene. Angew. Chem. 38, 2293–2395 (1999)
2. Asanuma, H., et al.: Photo-regulation of DNA function by azobenzene-tethered oligonucleotides. Nucleic Acids Res. Suppl. 3, 117–118 (2003)
3. Liu, D., Balasubramanian, S.: A proton-fuelled DNA nanomachine. Angew. Chem. 115, 5912–5914 (2003)
4. SantaLucia Jr, J., et al.: HyTher, http://ozone3.chem.wayne.edu/
5. Shin, J.-S., Pierce, N.A.: A synthetic DNA walker for molecular transport. J. Am. Chem. Soc. 126, 10834–10835 (2004)
6. Takahashi, K., Yaegashi, S., Asanuma, H., Hagiya, M.: Photo- and thermoregulation of DNA nanomachines. In: Carbone, A., Pierce, N.A. (eds.) DNA Computing. LNCS, vol. 3892, pp. 336–346. Springer, Heidelberg (2006)
7. Yurke, B., Turberfield, A.J., Mills Jr., A.P., Simmel, F.C., Neumann, J.L.: A DNA-fuelled molecular machine made of DNA. Nature 406, 605 (2000)

Experimental Validation of the Transcription-Based Diagnostic Automata with Quantitative Control by Programmed Molecules

Miki Hirabayashi[1], Hirotada Ohashi[1], and Tai Kubo[2]

[1] Department of Quantum Engineering and Systems Science, The University of Tokyo,
7-3-1 Hongo, Bunkyo-ku, Tokyo 113-8656, Japan
[2] Neuroscience Research Institute, National Institute of Advanced Industrial Science and
Technology (AIST), AIST Tsukuba Central 6, 1-1-1 Higashi, Tsukuba, Ibaraki 305-8566, Japan
miki@crimson.q.t.u-tokyo.ac.jp, ohashi@q.t.u-tokyo.ac.jp,
tai.kubo@aist.go.jp

Abstract. Biomolecular computing using the artificial nucleic acid technology is expected to bring new solutions to various health problems. We focus on the noninvasive transcriptome diagnosis by salivary mRNAs and present the novel concept of transcription-based diagnostic automata that are constructed by programmed DNA modules. The main computational element has a stem shaped promoter region and a pseudo-loop shaped read-only memory region for transcription regulation through the conformation change caused by targets. Our system quantifies targets by transcription of malachite green aptamer sequence triggered by the target recognition. This algorithm makes it possible to realize the cost-effective and sequence-specific real-time target detection. Moreover, in the *in-vivo* therapeutic use, this transcription-based system can release RNA-aptamer drugs multiply at the transcription stage, different from the digestion-based systems by the restriction enzyme which was proposed previously. We verified the sensitivity, the selectivity and the quantitative stability of the diagnostic automata in basic conditions. Our approach will provide promising applications of autonomous intelligent systems using programmed molecules.

Keywords: Biomolecular computing, nucleic acid detection systems, DNA computing, molecular programming, autonomous diagnostic devices, molecular circuits.

1 Introduction

Salivary transcriptome diagnostics is expected as a novel clinical approach for early disease detection [1]. We describe herein a potential general approach to the rational construction of an intelligent sensor to detect the fluctuation of salivary mRNA biomarkers by transcriptional regulatory systems using molecular computation [2-9].

Recently, it was found that the combination of several disease-related salivary mRNA biomarkers yielded sensitivity (91%) and specificity (91%) to oral squamous cell carcinoma (OSCC) distinguishing from the controls [1]. The existing methods to

M.H. Garzon and H. Yan (Eds.): DNA 13, LNCS 4848, pp. 89–98, 2008.

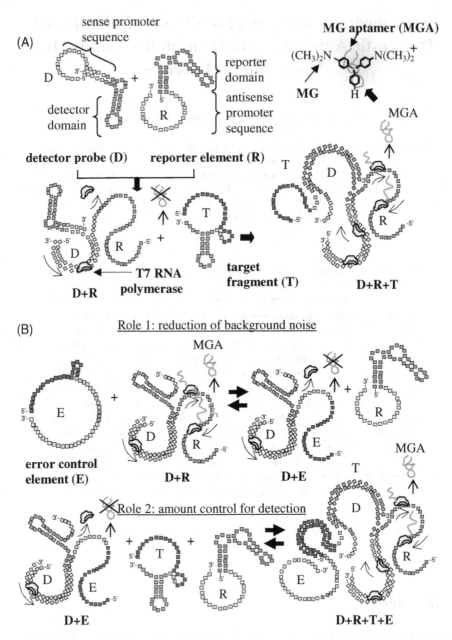

Fig. 1. (A) Operation principle of the detection system. (B) Expected main roles of error control elements.

quantify these transcripts have several problems. For example, three methods of quantitative polymerase chain reaction (qPCR), which is commonly employed to validate a subset of differently expressed transcripts identified by microarray analysis, have some drawback and advantage [10]. The method through agarose gel electrophoresis is less expensive but less accurate than others. The method using sequence-specific fluorescent reporter probes is accurate and reliable, but expensive. The method using SYBR Green dye labels all double-stranded (ds)DNA including any unexpected PCR products, however, it has the advantage in the cost performance. Here we present the cost-effective and sequence-specific intelligent systems using molecular computation for the transcriptome diagnosis and demonstrate the concept of new diagnostic automata.

Our system consists of several computational elements (Fig. 1A). The detector probe has one stem for transcriptional regulation using the sense T7 promoter sequence and one read-only memory loop for target recognition. The stem is designed to significantly enhance their specificity in transcription regulation. The reporter element has an anti-sense promoter domain and a malachite green (MG) aptamer sequence domain. It can transcribe the aptameric sensor if the stem opens through the binding of a target oligonucleotide complementary to the memory sequence.

As for the computational elements, Stojanovic's group proposed molecular automata combined the standard oligonucleotide recognition element, a stem-loop molecular beacon with the MG aptamer itself as a part of computational elements and used on deoxyribozyme-based logic gates [11]. Here we utilize the transcription process of MG aptamer to detect the targets. As for the autonomous diagnostic and therapeutic systems using molecular computation, Shapiro's group demonstrated digestion-based therapeutic systems by the restriction enzyme (*Fok*I), which can release a single DNA molecule for therapy per therapeutic element according to diagnosis of biomarkers [9]. Our transcription-based system can transcribe RNA aptamer drugs [12] instead of the MG aptamer as much as needed under the control of programmed molecules, when it will be applied to the *in-vivo* treatment in the future. As for the additional computational elements, we introduced an error control element in order to increase the quantitative stability (Fig. 1B). This element has a target recognition domain and an inhibitor domain for detection probes, which was designed to reduce the background noise without inhibition of target recognition. Because our system adopts the conformation change to detect targets, we can control the conditions flexibly by using the additional programmed molecules such as this error control element.

We confirmed the significant sensitivity, the selectivity and the quantitative stability of in basic conditions. These proof-of-concept results will contribute to bring us the realization of autonomous intelligent diagnostic and therapeutic devices using molecular-scale computation.

2 Materials and Methods

2.1 Preparation of Oligonucleotides

DNA sequences of the oligonucleotide used for the construction of the three molecular computer components and two inputs are shown in Tables 1-3. The three components consist of the diagnostic system: detector probes, reporter elements, and error control elements. Oligonucleotides were custom-made and DNase/RNase free

HPLC purified by (Operon Biotechnologies, Tokyo, JAPAN) or (Hokkaido System Science, Sapporo, JAPAN) and used as received. Each sequence was folded using mFold server v 3.2 (URL: http://www.bioinfo.rpi.edu/applications/mfold) and visually examined to find sequences of low secondary structure.

Computational Elements. (1) detector probe: The detector probe is a detector of a target sequence. It has a stem shaped sense T7 promoter region for the transcription regulation of reporter molecules and a pseudo-loop shaped read-only memory region for the target detection (Table 1). The detector probe could receive information from target inputs at the memory domain and transfer signals to the promoter domain through the conformation change by opening the stem. **(2) reporter element:** The reporter element is a output-producing element. It has an anti-sense T7 promoter

Table 1. Single-stranded DNA models for computational elements

Name	DNA sequences (5'→3')	Length
Actin detector	5'-AGCTTAATACGACTCACTATAGGAC CTGAGGCTCTTTTCCAGCCTTTCCTAT AG-3'	54
MG aptamer reporter	5'-GGATCCATTCGTTACCTGGCTCTCGC CAGTCGGGATCCTATAGTGAGTCGTAT TAAGCT-3'	59
Actin error control	5'-TCTTGGGTATGGAATCCTGTGGAAA AAAAAAAAAATCCTATAGTGAGTCGTA TTAAGCT-3'	56

Table 2. Profiling of computational elements

Name	T_m (calculated)	GC %
Actin detector domain	63.6 ℃	54.0
Actin error control domain	60.6 ℃	50.0
T7 promoter domain	57.0 ℃	37.5

Table 3. Single-stranded DNA models for input molecules

Name	DNA sequences (5'→3')	Length
β Actin	5'-CCACAGGATTCCATACCCAAGAAGG AAGGCTGGAAAAGAGCCTCAGG-3'	47
IL8	5'-CACCGGAAGGAACCATCTCCATCCC ATCTCACTGTGTGTAAACATGACTTCC AAGCTG-3'	47

domain and an MG RNA aptamer sequence domain (Table 1). An MG aptamer increases the fluorescence of MG when bound [13-18], allowing us to know that the transcription occurs. The hybridization of a target at the memory region in the detector probe triggers the stem open and then the promoter region form a double strand with the reporter element. Consequently, the transcription of the MG aptamer sequence is active and fluorescence is observed by the addition of MG. These successive reactions will enable us to recognize the existence of targets. **(3) error control element:** The error control element is a supporting element to control the computing cascade. It consists of the sense promoter domain and the target recognition domain for the reduction of the background noise and the introduction of the threshold in the transcription process (Table 1). When the target recognition domain of this element does not bind to inputs, the sense promoter region has more accessibility to the promoter module and inhibits the transcription of the MG aptamer sequence. Consequently, it is expected that the element can reduce the background noise and increase the quantitative stability.

Input Molecules. We used single-stranded (ss)DNAs to represent disease-related mRNA based on precedents in Ref. [9]. Two concentrations to represent mRNA levels: 0 µM for low level and 2-4 µM for high level at the detection stage. As disease-related biomarker models, β-actin and *IL8* mRNAs were selected based on reported cancer association [1]. The β-actin gene is a representative house-keeping gene and the transcript of *IL8* is one of salivary mRNA biomarkers for OSCC. DNA sequences used for the construction of the input models are shown in Table 3. These input ssDNA models include two recognition modules: one for detector probes and the other for error control elements.

2.2 Instrumental

Fluorescent spectra were taken on a microplate spectrofluorometer (Japan Molecular Devices, Tokyo, JAPAN, SpectraMax Gemini). Experiments were performed at the excitation wavelength (λ_{ex}) of 620 nm and emission wavelength (λ_{em}) scan of 650-700 nm. The spectra were exported to Microsoft Excel files.

2.3 Diagnostic Computations

Diagnostic computations consist of three steps: 1) mixing the detector probes for each input disease-related biomarker models and other computational elements and equilibrating them. 2) processing of the diagnostic string by T7 RNA polymerase supplementation. 3) quantifying of the fluorescence by MG supplementation.

Step 1. Control of DNA Hybridization. Detector probes, input molecules and reporter elements were mixed in that order and diluted in annealing buffer (50 mM NaCl, 100 mM HEPES pH 7.4) to 3 µM concentration each. The reaction mixtures were incubated for 22 h at 45 °C following denaturation at 94 °C for 2 min in a PCR machine block.

Step 2. Detection of Memory Recall. Hybridization mixture was subjected to transcription reaction using Ambion MEGAscript T7 Kit. The mixtures were incubated at 37 °C for up to 6 hours.

Step 3. Observation of MG Binding. Two µL of the reaction mixtures and MG were mixed in binding buffer (50 mM Tris-HCl, pH = 7.4, 50 mM $MgCl_2$, 1mM NaCl) with the final concentration of 10 µM of MG and the fluorescent spectra were taken.

3 Results

We investigated fundamental properties of our transcription-based diagnostic systems: sensitivity, selectivity, quantitative stability and scalability. The sensitivity is served by the stem shaped promoter region and the selectivity is served by the pseudo-loop shaped recognition domain for the target sequence in the detector probe. The stable quantitative scalability is realized by introduction of the error control element.

3.1 Sensitivity and Selectivity

Figure 2 shows fluorescence time scans at the transcription stage using the detector probe (D), the reporter element (R), the target ssDNA (T), and the non-target ssDNA (NT). Each spectrum and data point represents the average of ten consecutive scans at λ_{em} = 675 nm. It is confirmed that the system can generate about two-fold fluorescence when it recognizes targets (Fig. 2A). Nonzero fluorescence without targets is attributed to the fact that the reporter probe itself has a function as an opener of the stem-shaped promoter region and induces the transcription of reporter molecules. By the reducing of this background noise, the sensitivity would be further improved.

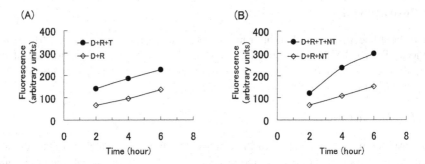

Fig. 2. Fluorescence time scans at the transcription stage. (A) Sensitivity of reporter probes. (B) Selectivity of detector probes in mixed conditions. Each data point represents the average of ten consecutive scans at λ_{em} = 675 nm.

Figure 2B shows that the detector probes can recognize target fragments in the mixture conditions basically. However, it seems that the increase of molecular species or the total amount of oligonucleotides introduces decrease of temporal quantitative stability in the amount of fluorescence increase.

3.2 Quantitative Stability and Scalability

To find the solution of the quantitative stability problems, we introduced an error control element to the system.

Figure 3 shows the increase of quantitative stability and controllability by the error control element (E). Figures 3A and 3B show that the error control element exhibits the fluorescent reduction effects without losing the ability of fluorescence recovery and the selectivity although the sensitivity is not enough compared with Fig. 2. On the other hand, Fig. 3C shows that half standard detector probes can not decrease the background noise without losing the ability of fluorescence recovery. In Fig. 3D, the error control element exhibits the fluorescent reduction effects, which are dependent on the amount of elements. The case of half amount of standard error control elements reduces basic fluorescence as in the case of half amount of standard detector probes without losing the ability of fluorescence recovery. The case of full standard error control elements can decrease the larger amount of basic fluorescence and shows no fluorescent recovery by targets. These features make it possible to set threshold for the detectable amount of targets by adjustment of error control elements. In addition, the case of half standard error control elements with targets in Fig. 3B shows the temporal quantitative stability in the amount of fluorescence increase compared with Fig. 2B and shows the improvement of sensitivity compared with the Fig. 3A. This sensitivity improvement may be brought by the promotion of the programmed reaction due to entropy increase. Moreover it is expected that the increase of reaction efficiency by using higher level of reaction mixture also improves the system sensitivity.

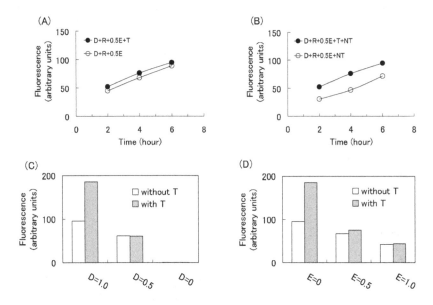

Fig. 3. Quantitative stability and controllability tests using error control elements. (A) Noise reduction effects. (B) Examination of selectivity and quantitative stability. Each data point represents the average of ten consecutive scans at λ_{em} = 675 nm. (C) Background noise control by adjustment of the detector probe. (D) Background noise control by the error control element. Fluorescence at λ_{em} = 675 nm measured at 6 hours after the transcription starts. Each datum represents the average of ten consecutive measurements.

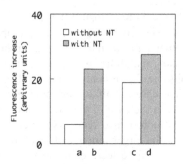

Fig. 4. Scalability of sensitivity with error control elements. Fluorescence increase by target detection at $\lambda_{em} = 675$ nm measured at 6 hours after the transcription starts. Low concentration: D+R+0.5E (a) or D+R+0.5E+NT (b) for T, and high concentration: 2D+2R+E (c) or 2D+2R+E+2NT (d) for 2T. Each datum represents the average of ten consecutive measurements.

Figure 4 shows that higher concentration improves the system sensitivity. This may be because the computational elements are short oligonucleotides and therefore the initiation becomes rare events when the concentration is low. Scalability remains intact in the relative complex cases with NT. This shows that the system can detect higher concentration of targets by using higher concentration of probes. The system sensitivity is improved with an increase in entropy as in the case of Fig. 3B. This phenomenon was observed through the preliminary experiments. We expect that this feature provides the promising consideration toward the practical use.

We showed that the error control element can adjust the detection amount of targets and improve the system stability for quantitative detection. These results indicate that it is possible that we set the threshold of detection amount of targets and perform the quantitative stable detection by choosing optimized combination and concentration of appropriate programmed molecules such as detector probes and error control elements.

4 Discussion

We tested the new molecular computation algorithm for the transcriptome diagnosis and confirmed that it can supply the system which shows the significant sensitivity, selectivity and quantitative stability in mixed conditions, in which none, one, or two input disease-related biomarker models are present.

Our transcription-based diagnostic automata have the following three remarkable features over existing quantitative methods for targets.

(1) flexibility in programming: For the target detection, our system utilizes the transcriptional regulation based on the conformational change of the detector probe that is triggered by the target recognition. This enables the flexible control by additional programmed molecules such as proposed error control elements. Thus the construction of complicated and intelligent automata becomes possible.

(2) cost-effective and sequence-specific real-time detection: The introduction of the MG-RNA-aptamer transcription for the fluorescence detection realizes the cost-effective real-time observation. Moreover the transcription regulated by the specific target sequence in the pseudo-loop shaped read-only memory region enables the sequence-specific detection. This is a substantial feature for the parallel quantitative diagnostic operation.

(3) potential ability for therapeutic automata: Because we adopts the transcription-based diagnostic systems instead of the digestion-based systems by the restriction enzyme, the controlled multiple release of RNA aptamer drugs is possible when this system is applied to the *in-vivo* therapeutic use. Our system has the prominent potential ability to organize the *in-vivo* therapeutic automata.

In this paper, we demonstrated the new concept of transcription-based diagnostic automata. By additional programming, we will be able to detect the target combinations of up-regulated disease-related biomarkers. For example, it was reported that by using the combination of up-regulated salivary mRNAs: *IL8* (24.3-fold), *SAT* (2.98-fold) and *H3F3A* (5.61-fold) for OSCC prediction, the overall sensitivity is 90.6% [1]. The introduction of molecular gates to our systems will enable this kind of autonomous one-step diagnosis for OSCC. In the near future, the accurate and reliable control by programmed molecules may offer the easy self-diagnostic tool using saliva.

These results would bring us one step closer to the realization of new intelligent diagnostic automata based on biomolecular computation.

References

1. Li, Y., John, M.A.R.St., Zhou, X., Kim, Y., Sinha, U., Jordan, R.C.K., Eisele, D., Abemayor, E., Elashoff, D., Park, N.-H., Wong, D.T.: Salivary transcriptome diagnostics for oral cancer detection. Clin. Cancer Res. 10, 8442–8450 (2004)
2. De Silva, A.P., McClenaghan, N.D.: Molecular-scale logic gates. Chemistry 10, 574–586 (2004)
3. Braich, R.S., Chelyapov, N., Johnson, C., Rothemund, P.W.K., Adleman, L.: Solution of a 20-variable 3-SAT problem on a DNA computer. Science 296, 499–502 (2002)
4. Winfree, E., Liu, F.R., Wenzler, L.A., Seeman, N.C.: Design and self-assembly of two-dimensional DNA crystals. Nature 394, 539–544 (1998)
5. Benenson, Y., Paz-Elizur, T., Adar, R., Keinan, E., Livneh, Z., Shapiro, E.: Programmable and autonomous computing machine made of biomolecules. Nature 414, 430–434 (2001)
6. Wang, H., Hall, J.G., Liu, Q., Smith, L.M.: A DNA computing readout operation based on structure-specific cleavage. Nat. Biotechnol. 19, 1053–1059 (2001)
7. Saghatelian, A., Voelcker, N.H., Guckian, K.M., Lin, V.S.-Y., Ghadiri, M.R.: DNA-based photonic logic gates: AND, NAND, and INHIBIT. J. Am. Chem. Soc. 125, 346–347 (2003)
8. Okamoto, A., Tanaka, K., Saito, I.: DNA logic gates. J. Am. Chem. Soc. 126, 9458–9463 (2004)
9. Benenson, Y., Gil, B.-D.U., Adar, R., Shapiro, E.: An autonomous molecular computer for logical control of gene expression. Nature 429, 423–429 (2004)
10. Ginzinger, D.G.: Gene quantification using real-time quantitative PCR: An emerging technology hits the mainstream. Exp. Hematol. 30, 503–512 (2002)

11. Kolpashchikov, D.M., Stojanovic, M.N.: Boolean control of aptamer binding states. J. Am. Chem. Soc. 127, 11348–11351 (2005)

12. Bunka, D.H.J., Stockley, P.G.: Aptamers come of age – at last. Nat. Rev. Microbiol. 4, 588–596 (2006)

13. Babendure, J.R., Adams, S.R., Tsien, R.Y.: Aptamers switch on fluorescence of triphenylmethane dyes. J. Am. Chem. Soc. 125(48), 14716–14717 (2003)

14. Baugh, C., Grate, D., Wilson, C.: 2.8 A° crystal structure of the malachite green aptamer. J. Mol. Biol. 301(1), 117–128 (2000)

15. Famulok, M.: Chemical biology: Green fluorescent RNA. Nature 430(7003), 976–977 (2004)

16. Grate, D., Wilson, C.: Laser-mediated, site-specific inactivation of RNA transcripts. Proc. Natl. Acad. Sci. USA 96(11), 6131–6136 (1999)

17. Stojanovic, M.N., Kolpashchikov, D.M.: Modular aptameric sensors. J. Am. Chem. Soc. 126(30), 9266–9270 (2004)

18. Hirabayashi, M., Taira, S., Kobayashi, S., Konishi, K., Katoh, K., Hiratsuka, Y., Kodaka, M., Uyeda, T.Q.P., Yumoto, N., Kubo, T.: Malachite green-conjugated microtubules as mobile bioprobes selective for malachite green aptamers with capturing/releasing ability. Biotechnol. Bioeng. 94(3), 473–480 (2006)

DNA Memory with 16.8M Addresses

Masahito Yamamoto[1,2], Satoshi Kashiwamura[3], and Azuma Ohuchi[1,2]

[1] Graduate School of Information Science and Technology, Hokkaido University,
Sapporo, Japan
[2] CREST, Japan Science and Technology Agency (JST), Japan
[3] HITACHI, Co. Ltd., Japan

Abstract. A DNA Memory with over 10 million (16.8M) addresses was achieved. The data embedded into a unique address was correctly extracted through addressing processes based on the nested PCR. The limitation of the scaling-up of the proposed DNA memory is discussed by using a theoretical model based on combinatorial optimization with some experimental restrictions. The results reveal that the size of the address space of the DNA memory presented here may be close to the theoretical limit. The high-capacity DNA memory can be also used in cryptography (steganography) or DNA ink.

Keywords: DNA Computing, DNA memory, NPMM, Theoretical capacity.

1 Introduction

Deoxyribonucleic acid (DNA) is well known as the blueprint of life, while it is also an attractive material because of its excellent properties such as minute size, extraordinary information density, and self-assembly. Focusing on these facts in recent years, various works, especially studies in the research field of DNA computing, have tried to develop a method for solving the combinatorial problems or for developing DNA machines such as a DNA automata [1][2][3][4].

One of the most promising applications of DNA computing might be a DNA memory. DNA molecules can store a huge amount of information in their sequences in extremely small spaces. The stored information on DNA can be kept without deteriorating for a long period of time because DNA is very hard to collapse [5]. Baum was the first to propose DNA memory, which can have a capacity greater than the human brain in minute scale [6]. The model enables a massively parallel associative search in a vast memory by utilizing the parallelism of DNA's hybridization. Rife et al. and Neel et al. have described DNA memory similar to that of Baum's model [7][8]. Recently, Chen et al. have proposed a DNA memory model that is capable of learning new data and recalling data associatively [9]. Although various research has been conducted on the construction of DNA memory, almost all of the works involve only proposals of models or only the preliminary experiments on a very small scale. Even if they could operate correctly on a small scale, it is doubtful that the operation would be successful in larger DNA memory because the efficiency and specificity of DNA's chemical

M.H. Garzon and H. Yan (Eds.): DNA 13, LNCS 4848, pp. 99–108, 2008.

reaction become much more severe. Therefore, it is very important to prove the technology through actual demonstration of the construction and addressing of DNA memory.

In this work, a DNA memory with 16.8M addresses is achieved. Our proposed DNA memory is addressable by using nested PCR and is named Nested Primer Molecular Memory (NPMM) [10][11][12]. The size of NPMM may be the largest pool of DNA molecules, which means that there are a large number of kinds of DNA sequences and any kind of DNA sequence can be extracted from the pool. The advantage of our memory is addressing based on amplification, which can amplify the target sequences and not amplify the non-target sequences. By using this amplification in several addressing steps, the probability of extraction of non-target DNA sequences can be very low. In fact, it is shown that any data can be retrieved with very high fidelity. The limitation of scaling-up of the proposed DNA memory is also discussed by using a theoretical model based on combinatorial optimization with some experimental restrictions. The results reveal that the size of the address space of the DNA memory presented here may be close to the theoretical limit. The high-capacity DNA memory can be also used in cryptography (steganography) or DNA ink [13][14][15][16][17].

2   Nested Primer Molecular Memory

NPMM is the pool of DNA strands such that each strand codes both data information and its address information. The data information (ex. binary data, strings, etc.) is expressed by encoded base sequence. The address information consists of several layers and each layer contains several components (specific DNA sequences) and is expressed by the combination of components on each layer. These layers are divided into two portions and are located on both sides of the data. In this work, we deal with the following one called 16.8M-NPMM: three layers on each side (named XY, $X \in \{A, B, C\}$, $Y \in \{L, R\}$) and sixteen sequences (20 mer) on each layer (named XYi, $i \in \{0, 1, 2 \ldots 15\}$). Each DNA molecule is structured such as $CL*$-$CLlink$-$BL*$-$BLlink$-$AL*$-$ALlink$-$Data$-$ARlink$-$AR*$-$BRlink$-$BR*$-$CRlink$-$CR*$ as shown in Fig 1. The notation '-' means the concatenation of DNA sequences. The $XYlink$ ($X \in \{A, B, C\}$, $Y \in \{L, R\}$) indicates a linker section (20 mer) and these are used to construct 16.8M-NPMM. The address information is expressed by the combination of XYi denoted by such as $[CLi, BLj, ALk, ARl, BRm, CRn]$ ($i, j, k, l, m, n \in \{0, 1, \ldots 15\}$). The address space of 16.8M-NPMM is about 16.8 million ($= 16^6 = 16,777,216$).

Operations for retrieving the stored data are executed by specifying each address layer based on PCR. For easy understanding, we will now explain how to retrieve the target information stored at $[CL0, BL2, AL4, AR5, BR3, CR1]$ from 16.8M-NPMM (Fig. 2). For the first operation, PCR is performed for 16.8M-NPMM using $CL0$ and $\overline{CR1}$ as primer pairs ($\overline{x}$ is the complementary DNA

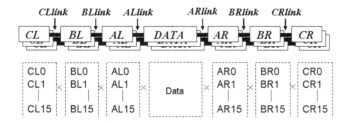

Fig. 1. Sequence structure of each DNA strand in 16.8M-NPMM. The area expressing the address information consists of six layers (XY, $X \in \{A, B, C\}$, $Y \in \{L, R\}$) and ten sequences are defined in each layer (named XYi, $i \in \{0, 1, 2 \ldots 15\}$). The address space of 16.8M-NPMM is over 10 million.

sequence of x). As a result, we can extract the collection of DNA molecules containing $CL0$ and $CR1$ from 16.8M-NPMM and exclude all DNA molecules without $CL0$ or $CR1$. This is because PCR yields a significant difference in the concentration between the amplified and non-amplified DNA molecules; therefore, we can disregard the non-amplified DNA. Next, we perform the second PCR using $BL2$ and $\overline{BR3}$ for the diluted solution after the first PCR. At this point, we can extract the DNA molecules containing $CR0$, $CR1$, $BL2$ and $BR3$. Next, we perform the third PCR for each diluted solution using $AL4$ and $\overline{AR5}$. After all PCRs are completed, we can extract only the DNA molecule expressing $[CL0, BL2, AL4, AR5, BR3, CR1]$ that codes the target data. Sequencing and decoding the extracted DNA molecules allows us to retrieve the target data.

3   Construction and Addressing of 16.8M-NPMM

We carried out laboratory experiments to verify the behavior of NPMM with over ten million address spaces. For simplicity and easy detection, the stored information in 16.8M-NPMM is either Data20 (20 mer), Data40 (40 mer) or Data60 (60 mer). Data40 is embedded into a unique address $[CL0, BL0, AL0, AR0, BR0, CR0]$, and Data60 is embedded into another unique address $[CL8, BL8, AL8, AR8, BR8, CR8]$, while Data20 is in all addresses. By observing the results of data extraction, the success of the addressing is evaluated. Note that Data40 and Data60 are stored in only one address among 16.8M addresses. We can detect the behavior very easily using only gel electrophoresis due to the difference in length. All DNA sequences in this work were designed by using a Two-Step Search Algorithm to avoid any mis-hybridization based on Hamming Distance [18]. Of course, another algorithm for designing DNA sequences can be used[19][20][21]. Moreover, to avoid a secondary structure, they are designed so that the free energy of each DNA molecule can reach a high score by using an m-$fold$ algorithm customized for DNA molecules [22][23][24][25].

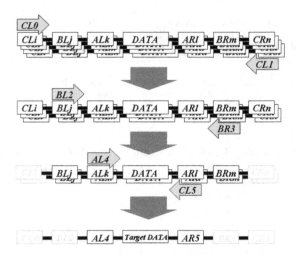

Fig. 2. Operations for retrieving the target DNA from NPMM are implemented in the nested PCR. In this case, the target address is [*CL0*, *BL2*, *AL4*, *AR5*, *BR3*, *CR1*]. Only if specifying the target address is completed will the target DNA be extracted from the large mixture of DNA molecules. If any other address is specified, non-target DNA will be extracted. The faint portions are the areas eliminated by the previous PCR.

3.1   Construction of 16.8M-NPMM

We prepare a DNA molecule such as Ci-$CLlink$, $CLlink$-Bj-$BLlink$, $BLlink$-Ak-$ALlink$, $ALlink$-$Data20$-$\overline{ARlink}$, $ALlink$-$Data40$-$\overline{ARlink}$, $BRlink$-ARl - $ARlink$, $CRlink$-BRm-$BRlink$ and CRn-$CRlink$ ($i, j, k, l, m, n \in \{0, 1 \ldots 15\}$). The first step is to perform PCR for DNA containing $Data20$ using $BLlink$-Ak-$ALlink$ and $BRlink$-ARl-$ARlink$ as primer pairs. This leads the AL and AR layers to Data20 based on the priming reaction of the linker sections. Similarly, other layers are also integrated using linker sections as knots. These steps produced about 16.8M whole addresses. Specific data (Data40 and Data60) are also created using the same method. There were no unwanted products through each experimental process (data not shown). Next, we measure the total amount of the whole pool of memory molecules and the Data40 and Data60 are mixed into the pool so that the number of each longer DNA and other DNA are equivalent.

3.2   Addressing from 16.8M-NPMM

We used sixteen kinds of addresses shown in Table 1 as test samples. The image of PolyAcrylamide Gel (Fig. 3) shows the results of laboratory experiments for the addressing operations. The three lanes in each surrounded area correspond to each addressing. The most right lane indicates the result after the final PCR. From the results of Fig. 3, we can see there was no minor band throughout the whole experiment. DNA product 120 bp long was obtained only in address All 0

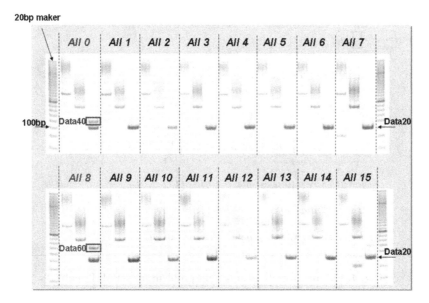

Fig. 3. Result of PolyAcrylamide Gel Electrophoresis for addressing: he three lanes in each surrounded area correspond to each addressing. Three lanes in each surrounded area correspond to the result of the first PCR, the second PCR and the final PCR, respectively. All$i$ means the address $[CLi, BLi, ALi, ARi, BRi, CRi]$.

Table 1. Addressing Samples

Label	Address	Data
All 0	$[CL0, BR0, AL0, AR0, BL0, CR0]$	Data20, Data40
All 8	$[CL8, BL8, AL8, AR8, BR8, CR8]$	Data20, Data60
All i $(i = 0, \cdots, 15, i \neq 0, 8)$	$[CLi, BLi, ALi, ARi, BRi, CRi]$	Data20

and DNA product 140 bp long was obtained only in address All 8, while all other addresses output product 100 bp long. Therefore, we can successfully extract the corresponding data of each address among over 10 million addresses. Based on these facts, although we did not confirm all addresses, we assume that NPMM could reliably screen and extract a target DNA molecule from over 10 million DNA mixtures. In DNA memory, it is most important to extract the target data with high fidelity. Therefore, this experiment strongly supports the effectiveness of NPMM.

4   Theoretical Analysis of Capacity Limitation

It is obvious that the scaling up of NPMM is limited due to some physical or chemical restrictions such as the limitation of the amount of DNA in solution

$$\textit{maximize} \quad \prod_{i=1}^{L} n_i^2$$

$$\textit{sub.to} \qquad n_i \geq 2 \; (i \in \{1, 2, \ldots L\}) \tag{1}$$

$$N \geq \sum_{i=1}^{L} 2n_i \tag{2}$$

$$v(2^{c_i} - 1) \prod_{j=i+1}^{L} n_j^2 + 2vc_i(n_i - 1) \prod_{j=i+1}^{L} n_j^2 \leq P \quad \forall i \in \{1, 2, \ldots L\} \tag{3}$$

$$amp_{c_i} > \alpha \times non\_amp_{c_i} \qquad \forall i \in \{1, 2, \ldots L\} \tag{4}$$

$$0 \leq c_i \leq max\_cycle \qquad \forall i \in \{1, 2, \ldots L\} \tag{5}$$

$$P, \, L, \, N, \, v, \, \alpha, \, max\_cycle : \text{integer (given)}$$

Fig. 4. Combinatorial optimization problem for NPMM analysis

and the number of DNA sequences available for the layer area. Therefore, it is important to discern how far the capacity of NPMM can be extended. We propose an addressing model for analyzing NPMM capacity limitations based on the combinatorial optimization problem. By solving this optimization problem, the maximum capacity of NPMM and the optimal assignment of address sequences to each address area can be obtained.

The capacity of NPMM depends on the address space and size of each datum. However, since enlarging address space is more difficult experimentally, we mainly discuss the enlargement of address space here. We deal with such models as L layers (L is integer grater than one) and n_i DNA sequences located on the ith layer ($n_i \geq 2, i \in \{1, 2, \ldots L\}$, i indicates a layer location from left side). For simplicity, the data area is abbreviated.

First, we consider the requirements for operating NPMM correctly and then write out a mathematical formula as constraints. Then, based on the constraints, we establish an expression to calculate the maximum capacity of NPMM (Fig. 4). The expression outputs the arrangements of n_i ($i \in \{1, \ldots L\}$) maximizing the address space while satisfying the constraints when input parameters are given. The input parameters should be the ones used in laboratory experiments to reflect the actual environment.

4.1   Model of the Addressing of NPMM

Input Parameters: P is the total number of primer molecules and should range from 1.2×10^{13} to 5.0×10^{13}. These values are practically appropriate for $100\mu l$

of PCR mixture. L is the number of layers. N is the total number of DNA sequences for address layers, which are designed so carefully that they can be available as PCR primer. v expresses the initial amount of each DNA strand in NPMM and v must be greater than one. Otherwise, it means that several address units are missing in memory. α expresses the rate between the total number of amplified DNA strands and that of non-amplified DNA strands after PCR. To succeed with each addressing process, α should be large enough so that the non-amplified DNA strands can be eliminated. max_cycle is the upper limit of PCR cycles.

Objective function: The target function is calculated as follows:

$$Capacity = \prod_{i=1}^{L} n_i^2.$$

Capacity, which is the width of address-space, depends on n_i. That is, this problem is an optimization problem to explore an arrangement n_i to maximize *Capacity* while satisfying the following constraints.

Constraints: These constraints are provided to ensure NPMM's behavior. Constraint (1) is obviously to express the address structure of NPMM. Constraint (2) ensures that all sequences located on each layer must be well-designed DNA sequences that can avoid any mis-hybridization or mis-priming. Constraints (3), (4) and (5) are established to make each addressing operation possible physically, and these three constraints must be satisfied in each ith addressing operation ($i \in \{1, 2, \ldots L\}$). c_i is the number of PCR cycles for addressing the ith layer. Constraint (3) ensures that each addressing operation is able to work. The terms amp_{c_i} and $non\_amp_{c_i}$ are calculated as follows:

$$amp_{c_i} = v \prod_{j=i+1}^{L} n_j^2 + v(2^{c_i} - 1) \prod_{j=i+1}^{L} n_j^2$$

$$non\_amp_{c_i} = v(n_i^2 - 1) \prod_{j=i+1}^{L} n_j^2 + 2vc_i(n_i - 1) \prod_{j=i+1}^{L} n_j^2.$$

amp_{c_i} and $non\_amp_{c_i}$ indicate the total amount of amplified and non-amplified DNA strands after the ith PCR, respectively. Therefore, amp_{c_i} must greatly exceed $non\_amp_{c_i}$[26][27]. (Here, we use α.) Constraint (4) expresses the total amount of amplified DNA molecules in the ith PCR. According to the principle of PCR, the total amount of amplified DNA in the PCR is never greater than that of the PCR primer. The former term is the increment of target DNA strands, which are amplified exponentially. The latter term is the increment of non-target DNA strands, which are linearly amplified from only one side priming. Constraint (5) is defined to avoid excess PCR cycles because excess PCR cycles cause unwanted reaction. That is, constraints (3), (4) and (5) ensure that a large difference in concentration is acquired between amplified and non-amplified DNA after PCR for addressing the ith layer.

4.2 Computational Result

According to this expression, we explore the theoretical maximum capacity of NPMM by exhaustive search. The input parameters are shown in Table 2. The computational results are shown in Table 3.

Table 2. Parameter settings

P	5.0×10^{13}
L	2,3,4
N	200
α	1000
v	100
max_cycle	30

P is set to the standard number of primers used in 100 μl of PCR mixture. L is the very important parameter for defining capacity. However, a large L makes the laboratory experiments cumbersome and complicated. In this paper, we selected $L = 2, 3, 4$ and analyzed in these cases. As for α and v, what values are appropriate for a successful addressing operation are as yet unclear. Therefore, we negatively set these values to 1,000 and 100, respectively. N is determined based on works in the research field of DNA word design problems.

Table 3. Theoretical capacity of case of $L = 2, 3, 4$. The numbers in brackets means the assignment of the number of address sequences, In the case of $L = 2$, it is shown that the optimal assignment of the address sequences to (BL, AL, AR, BR) is (50, 50, 50, 50).

Layer	L=2	L=3	L=4
Capacity	6,250,000	274,233,600	297,666,009
	(50,50,50,50)	(69,16,15,15,16,69)	(71,9,9,3,3,9,9,71)

A chemical reaction inevitably includes fluctuations (for example, the deviation of the number of DNA strands in NPMM and that of the amplification efficiency), so v should be greater than 100 to accomplish operations of NPMM with a high degree of fidelity. Therefore, the limitation of NPMM is expected up to MEGA order. However, this is an ideal one when a PCR reaction is carried at the maximum efficiency (amplification efficiency is always twice, and the unwanted reaction does not occur). Probably, practicable marginal capacity will be slightly smaller, and so the 16.8M-NPMM of $L = 3$ we constructed here has a large capacity considerably close to the practical limit.

5   Concluding Remarks

In this work, we dealt with NPMM, which is our proposed DNA memory. We constructed NPMM with over 10 million address spaces (16.8M-NPMM), and then several addresses were operated. From the experimental results, they showed completely correct behavior. Furthermore, since we showed that 16.8M-NPMM works with very high fidelity, we can conclude that DNA memory with 16.8M address spaces is achieved. We established a technology that selects only specific DNA from many kinds of DNA in mixture. The achievement of 16.8M DNA memory may be the largest pool of DNA mixture so far.

The latter part mainly discusses the theoretical limitations of NPMM's capacity. The behavior of NPMM was expressed by mathematical formula. By solving a combinatorial optimization problem, we could estimate that the theoretical limitation is MEGA order. Taking the efficiency of a chemical reaction into consideration, the process will become less practical.

Acknowledgements

We thank M. Hagiya, A. Suyama, A. Kameda and S. Yaegashi for helpful advice and discussions. The work presented in this paper was partially supported by a Grant-in-Aid for Scientific Research on Priority Area No.14085201 and a Grant-in-Aid for Young Scientists (A) No. 17680025, Ministry of Education, Culture, Sports, Science and Technology, Japan.

References

1. Adleman, L.M.: Molecular Computation of Solutions to Combinatorial Problems. Science 266, 1021–1024 (1994)
2. Lipton, R.: DNA solution of hard combinatorial problems. Science 268, 542–545 (1995)
3. Braich, R.S., Chelyapov, N., Johnson, C., Rothemund, P.W.K., Adleman, L.: Solution of a 20-Variable 3-SAT Problem on a DNA Computer. Science 296, 499–502 (2002)
4. Benenson, Y., Paz-Elizur, T., Adar, R., Keinan, E., Livneh, Z., Shapiro, E.: Programmable and autonomous computing machine made of biomolecules. Nature 414, 430–434 (2001)
5. Wong, P.C., Wong, K.K., Foote, H.: Organic Data Memory Using the DNA Approach. Communications of the ACM 46(1), 95–98 (2003)
6. Baum, E.B.: Building an Associative Memory Vastly Larger Than the Brain. Science 268, 583–585 (1995)
7. Reif, J.H., LaBean, T.H., Pirrung, M., Rana, V.S., Guo, B., Kingsford, C., Wickham, G.S.: Experimental Construction of Very Large Scale DNA Databases with Associative Search Capability. In: Jonoska, N., Seeman, N.C. (eds.) DNA Computing. LNCS, vol. 2340, pp. 231–247. Springer, Heidelberg (2002)
8. Neel, A., Garzon, M.H., Penumatsa, P.: Improving the Quality of Semantic Retrieval in DNA-Based Memories with Learning. In: Negoita, M.G., Howlett, R.J., Jain, L.C. (eds.) KES 2004. LNCS (LNAI), vol. 3213, pp. 18–24. Springer, Heidelberg (2004)

9. Chen, J., Deaton, R., Wang, Y.Z.: A DNA-based memory with in vitro learning and associative recall. Natural Computing 4(2), 83–101 (2005)
10. Kashiwamura, S., Yamamoto, M., Kameda, A., Shiba, T., Ohuchi, A., Hierarchical, D.N.A.: Memory based on Nested PCR. In: Hagiya, M., Ohuchi, A. (eds.) DNA Computing. LNCS, vol. 2568, pp. 112–123. Springer, Heidelberg (2003)
11. Kashiwamura, S., Yamamoto, M., Kameda, A., Shiba, T., Ohuchi, A.: Potential for enlarging DNA memory: The validity of experimental operations of scaled-up nested primer molecular memory. BioSystems 80, 99–112 (2005)
12. Kashiwamura, S., Yamamoto, M., Kameda, A., Ohuchi, A.: Experimental Challenge of Scaled-up Hierarchical DNA Memory Expressing a 10,000-Address Space. In: Preliminary Proceeding of 11th International Meeting on DNA Based Computers. vol. 396 (2005)
13. Clelland, C.T., Risca, V., Bancroft, C.: Hiding message in DNA microdots. Nature 399, 533–544 (1999)
14. Hashiyada, M.: Development of Biometric DNA Ink for Authentication Security. Tohoku J. Exp. Med. 204, 109–117 (2004)
15. Hashiyada, M., Itakura, Y., Nagashima, T., Nata, M., Funayama, M.: Polymorphism of 17 STRs by multiplex analysis in Japanese population. Forensic Sci. Int. 133, 250–253 (2003)
16. Itakura, Y., Hashiyada, M., Nagashima, T., Tsuji, S.: Proposal on Personal Identifiers Generated from the STR Information of DNA. Int. J. Information Security 1, 149–160 (2002)
17. Kameda, A., Kashiwamura, S., Yamamoto, M., Ohuchi, A., Hagiya, M.: Combining randomness and a high-capacity DNA memory. DNA13 (submitted 2007)
18. Kashiwamura, S., Kameda, A., Yamamoto, M., Ohuchi, A.: Two-Step Search for DNA Sequence Design. IEICE TRANSACTIONS on Fundamentals of Electronics, Communications and Computer Sciences E87-A (6), 1446–1453 (2004)
19. Deaton, R., Murphy, R.C., Garzon, M., Franceschetti, D.R., Stevens Jr, S.E.: Good Encoding for DNA-Based Solutions to Combinatorial Problems. In: Landweber, L.F., Baum, E.B. (eds.) DNA Based Computers II. DIMACS Series in Discrete Mathematics and Theoretical Computer Science, vol. 44, pp. 247–258 (1999)
20. Tanaka, F., Kameda, A., Yamamoto, M., Ohuchi, A.: Design of nucleic acid sequences for DNA computing based on a thermodynamic approach. Nucleic Acids Research 33, 903–911 (2005)
21. Tulpan, D.C., Hoos, H.H., Condon, A.: Stochastic Local Search Algorithms for DNA word Design. In: Hagiya, M., Ohuchi, A. (eds.) DNA Computing. LNCS, vol. 2568, pp. 229–241. Springer, Heidelberg (2003)
22. Lyngso, L.B., Zuker, M., Pedersen, C.N.: Fast evaluation of internal loops in RNA secondary structure prediction. Bioinformatics 15, 440–445 (1999)
23. SantaLucia, J., Allawi, H.T., Seneviratne, P.A.: Improved nearest-neighbor parameters for predicting DNA duplex stability. Biochemistry 35, 3555–3562 (1996)
24. Sugimoto, N., Nakano, S., Yoneyama, M., Honda, K.: Improved thermodynamic parameters and helix initiation factor to predict stability of DNA duplexes. Nucleic Acids Research 24, 4501–4505 (1996)
25. Zuker, M., Stiegler, P.: Optimal computer folding of large RNA sequences using thermodynamics and auxiliary information. Nucleic Acids Research 9, 133–148 (1981)
26. McPherson, M.J., Hames, B.D., Taylor, G.R.: PCR A Practical Approach. IRL Press (1995)
27. McPherson, M.J., Hames, B.D., Taylor, G.R.: PCR2 A Practical Approach. IRL Press (1993)

Combining Randomness and a High-Capacity DNA Memory

Atsushi Kameda[1], Satoshi Kashiwamura[2], Masahito Yamamoto[1,2],
Azuma Ohuchi[1,2], and Masami Hagiya[1,3]

[1] Japan Science and Technology Corporation (JST-CREST)
[2] Graduate School of Information Science and Technology, Hokkaido University
[3] Graduate School of Information Science and Technology, University of Tokyo
hagiya@is.s.u-tokyo.ac.jp

Abstract. In molecular computing, it has long been a central focus to
realize robust computational processes by suppressing the randomness of
molecular reactions. To this end, several methods have been developed
to control hybridization reactions of DNA molecules by optimizing DNA
sequences and reaction parameters. However, another direction in the
field is to take advantage of molecular randomness rather than avoid
it. In this paper, we show that randomness can be useful in combination
with a huge-capacity molecular memory, and demonstrate its application
to an existing technology — DNA ink.

Keywords: DNA memory, molecular memory, DNA ink, randomness.

1 Introduction

A central focus in the field of molecular computing is to suppress randomness of molecular reactions to realize robust computational processes. Methods that employ this strategy include designing precise DNA sequences based
on energy predictions of DNA secondary structures, and carefully controlling
DNA hybridization by tuning reaction conditions such as temperature and salt
concentration.

For example, errors in the algorithmic self-assembly of DNA tiles occur because sticky ends that are not completely complementary may hybridize with
nonzero probability. Suppressing such errors is a central challenge in DNA nanotechnology. In addition to designing DNA sequences that reduce error probability, new machineries such as proof-reading and self-healing tiles have been
proposed [3,12].

Some molecular machines, such as Yurke's tweezers [16] or our photo-regulated
hairpin machine [14], change their conformation according to inputs from the external environment. As a result, they can make state transitions, move toward
a specified direction, or produce outputs to the environment. However, conformational change cannot occur with 100% probability, and thus to design robust
molecular machines, it is crucial to design DNA sequences that promote intended conformational change and prohibit unintended changes. In the case of

M.H. Garzon and H. Yan (Eds.): DNA 13, LNCS 4848, pp. 109–118, 2008.
© Springer-Verlag Berlin Heidelberg 2008

DNA, because transformation of secondary structures roughly determines conformational change, predicting the energy landscape of secondary structures is extremely important [10].

However, applications can also take advantage of randomness. In the seminal work by Adleman [1], which initiated the field of DNA computing, the random generation of paths in a directed graph was achieved via random hybridization reactions. However, random generation was not essential, but only substituted *complete enumeration,* which was required for solving the Hamiltonian path problem and should be implemented by molecules. Therefore, our goal was to create an application in which molecular randomness is essential.

In previous work, we developed a huge-capacity DNA memory, in which each molecular address consists of 6 hexadecimal digits, and the entire address space is about 16.8 million (words) [7,9]. The whole memory is managed as a solution of about $1\mu l$, using standard experimental techniques such as PCR. In the present work, we employ molecular randomness caused by statistical fluctuations in a sample with a low copy number of molecules. By simply diluting the solution, one can easily obtain a unique memory state that cannot be replicated. Moreover, such a memory state can be amplified by PCR. Therefore, by combining randomness and a huge-capacity memory, it is possible to construct DNA ink, for example, which can never be reconstructed.

Here, we briefly describe our huge-capacity DNA memory, called Nested Primer Molecular Memory (NPMM). Then we summarize the concept of DNA ink and explain how to produce it using our proposed strategy. Finally, we report the results of a preliminary experiment we conducted to demonstrate the feasibility of our method.

2 NPMM

We developed the NPMM under the JST CREST Molecular Memory Project [9]. Our goal was to overwhelm the random pool of size 2^{20} (for solving a 20-variable SAT problem) realized by Adleman's group in 2002 [2], and construct a DNA memory that can easily be managed by well established and standard experimental techniques, such as PCR [7].

The result was a DNA memory composed of memory molecules with the structure shown in Fig. 1 [15]. A memory molecule is a double strand of DNA consisting of data at its center surrounded by 3-digit addresses. Because each digit is chosen from among 16 sequences, and the address of each memory molecule is 6 digits long, the whole address space is about 16.8 million (words), i.e., the whole set of 6-digit hexadecimal numbers.

The DNA memory is managed as a solution of about $1\mu l$, as shown in Fig. 2.

To access each memory molecule, PCR (the established method for copying DNA) is repeated, using address digits as *nested primers*. First, PCR is performed using the outermost two digits denoted by C (CL and CR) as primers, so

6-hierarchy 16-sequence NPMM (16.8M ≈ 16,777,216)

| CL | ▪ | BL | ▪ | AL | ▪ | DATA | ▪ | AR | ▪ | BR | ▪ | CR |

CL0		BL0		AL0				AR0		BR0		CR0
CL1	×	BL1	×	AL1	×	Data	×	AR1	×	BR1	×	CR1
|		|		|				|		|		|
CL15		BL15		AL15				AR15		BR15		CR15

Fig. 1. The structure of a memory molecule in NPMM

Fig. 2. NPMM in a 1-μl solution

that the molecules with CLi and CRj specified in their address are extracted and amplified. Then the process is repeated for BL and BR (BLi and BRj), and AL and AR (ALi and ARj).

It took a few years to construct the 16.8-million-address DNA memory [15] because we had to develop new technologies or refine existing ones. A summary follows.

Designing 16 sequences for each of 6 digits plus bridge sequences: We designed 16 different 20-mer sequences for each of the 6 address digits, i.e., 96 different sequences in total. In addition, we designed bridge sequences, which are placed between adjacent address digits and also used to bridge data and the innermost digits. These sequences should not hybridize with one another, but only with their complementary counterparts. We developed a new design method for these sequences, called a *two-step search* [8].

Protocol for constructing memory molecules for each address: Molecules of 16.8 million addresses should be synthesized as uniformly as possible. NPMM is constructed in three steps: concatenating address digits with data from the innermost to the outermost, performing PCR on DNA molecules whose bridge sequences are complementary, and hybridizing these together. Regarding uniformity, the ratio between the most and least concentrated types of molecules was estimated between 2 and 3 during each step.

Protocol for accessing memory molecules with the specified address:
To selectively amplify the molecules with a specified address, we should optimize
various reaction parameters for PCR, such as the number of cycles. As a result,
we successfully accessed a single address out of 16.8 million addresses. To date,
we have accessed 16 different addresses.

Consequently, each address in the DNA memory consists of about 200 to 250
molecules. Using a mathematical model of our PCR reaction, we determined that
the current capacity is almost maximal for correctly accessing memory molecules
by their addresses.

3   DNA Ink

To the best of our knowledge, the concept of DNA ink was first proposed and
tested by Tsujii *et al.* in 2001 [5,6]. DNA molecules of a given sequence are diluted
in ink; after the ink is applied to paper, then extracted, the DNA molecules in the
ink can be analyzed to determine the sequence. DNA ink can be used for various
purposes related to encryption, steganography, and authentication. For example,
a secret key for encrypted communication can be sent in DNA ink. Contracts
can be signed using DNA ink, and paper money can be printed with DNA ink
to avoid counterfeiting. Brand-name products, such as Chanel or Prada, can be
made of strings dyed with DNA ink to ensure authenticity. It is also possible
to spray DNA ink over brand-name foods, such as Kobe beef, because DNA is
completely harmless in foods.

Recently, Suyama *et al.* developed a practical DNA ink under the JST CREST
Molecular Memory Project. The ink is composed of about 300 pairs of orthonormal sequences, although it could be expanded to more than 10,000 sequences,
and thus can store 300 bits of information by including or excluding specific sequences. The orthonormal sequences employed in their DNA ink were originally
developed for DNA computing [13,11] and gene expression profiling [4].

The original DNA ink developed by Tsujii *et al.* [5,6] was limited in that it
required DNA sequencing to analyze the extracted ink. More importantly, the
contents of the ink could be exposed by attaching primers to DNA molecules
and amplifying them by PCR. Suyama *et al.* solved these problems by mixing
DNA ink with dummy DNA (which has primers in common with true DNA
and serves to mask the latter) and noise DNA (which consists of unused orthonormal sequences that hinder whole DNA amplification). Suyama *et al.* also
prepared a detection kit for their DNA ink, which contains dummy primers so
that even if the kit is analyzed, the true primers will not be revealed. To detect the primers, one must use a special DNA chip, which itself also contains
dummy probes so that if the chip alone is analyzed, the true probes will not be
revealed.

Suyama *et al.* succeeded in recovering the correct information in DNA ink
extracted from paper using their detection kit and special DNA chip.

4   DNA Ink Constructed by Inducing Randomness at the Molecular Level

With our method, molecular randomness is introduced by a very simple reaction: diluting the solution. For example, if we dilute a given solution twice and dispense the results into two tubes, A and B, then a single molecule in the original solution is either in tube A or in tube B, with a probability of $1/2$. If the original solution contained 100 molecules and was diluted 100 times, then the probability that the resulting tube contains one or more molecules is about $1 - e^{-1} = 0.63$, because the number of chosen molecules roughly follows the Poisson distribution with the average value 1 and the probability that no molecule is chosen is approximately equal to e^{-1}. Similarly, if the original solution was diluted 200 times, then the probability that the resulting tube contains one or more molecules is about $1 - e^{-0.5} = 0.39$.

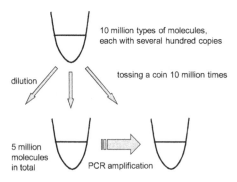

10 million types of molecules, each with several hundred copies

tossing a coin 10 million times

dilution

5 million molecules in total

PCR amplification

Fig. 3. DNA ink made by inducing randomness at the molecular level

Assume that a given solution contains 10 million types of molecules, each with several hundred copies (Fig. 3; note that NPMM actually has 16.8 million types of molecules, each with about 200 copies), and it is diluted to 5 million molecules. The probability that at least one molecule of any given type remains in the diluted solution is about $1 - e^{-0.5} = 0.39$, as described above. This process of dilution is akin to tossing a coin 10 million times, when one side of the coin turns up with a probability of 0.39.

After dilution, the solution is uniformly amplified by PCR. In NPMM, amplification requires common outermost primers in each memory molecule, or all outermost digits as primers.

After amplification, each molecule that survived should have a reasonable number of copies. Therefore, even if the amplified solution is further diluted, those copies will remain. Let us call the solution obtained by PCR *the master ink*. Each solution obtained by diluting the master ink should have the same ingredients as those of the master ink. Therefore, each diluted solution can be used as DNA ink corresponding to the master ink. The molecules extracted from inked paper are then compared with those in the master ink. To this end, DNA ink should contain an enough number of copies of each molecule.

The contents of the master ink can never be revealed completely; checking for each type of molecule among millions of molecules is prohibitively time-consuming. However, correctly verifying whether a given DNA ink was copied from the master ink is much simpler, by checking for a reasonable number of addresses in the given DNA ink and the master ink. In this way, DNA ink can be created by combining molecular randomness with a huge-capacity DNA memory. Note that the contents of the master ink are not known even to the creator of the ink.

One note of caution: if the DNA ink extracted from paper is amplified as a whole, then the forged ink can be used as if it were obtained from the master ink. To avoid this problem, one can apply various methods as in Suyama's DNA ink. For example, one can add short junk DNA molecules as noise, which are more easily amplified by PCR.

5   Preliminary Experiment

To test the feasibility of our method, we conducted a preliminary experiment. The overall setting of the experiment is shown in Fig. 4. We employed a partial instance of NPMM obtained by completing the first two steps in the construction process. This partial instance had four address digits, each with 10 sequences. Thus, we had a total of 10,000 addresses, each with about 3×10^5 copies in a $1\text{-}\mu l$ solution. Because molecules in this memory have common bridge sequences at both ends, they can be uniformly amplified using the bridge sequences as primers. This greatly simplified the experiment.

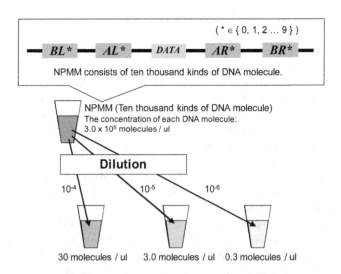

Fig. 4. Setting of the preliminary experiment

A. First PCR amplification

1. In a PCR reaction tube, prepare the following.
- distilled water $4.4\mu l$
- 10 x PCR buffer $1.0\mu l$
- 2 mM dNTP mix $1.0\mu l$
- 25 mM $MgSO_4$ $0.4\mu l$
- diluted NPMM solution $1.0\mu l$ (dilution ratios: 10^{-4}, 10^{-5} and 10^{-6})
- 5 μM Primer [BL0] and c[BR0] $1.0\mu l$ each ('c' denotes complementary sequence)
- KOD plus DNA polymerase (TOYOBO) $0.2\mu l$
- total volume $10\mu l$
2. Perform 20 cycles of PCR
- preheat (94°C for 2 min.)
↓
- denature (94°C for 20 sec.)
- anneal and extension (65°C for 10 sec.)
(Repeated 20 cycles)
PCR was performed by PTC-200 peltier thermal cycler (MJ Research).

B. Secondary PCR

1. Dilute the first PCR products by the ratio 10^{-3}.
2. In a 96-well PCR reaction plate, prepare the following.
- distilled water $3.4\mu l$
- 10 x PCR buffer $1.0\mu l$
- 2 mM dNTP mix $1.0\mu l$
- 25 mM $MgSO_4$ $0.4\mu l$
- Diluted first PCR product $1.0\mu l$
- 10 pM Primer set $1.0\mu l$ each
- KOD plus DNA polymerase (TOYOBO) $0.2\mu l$
- 1/1000 diluted SYBR green I $1.0\mu l$
- total volume $10\mu l$
The primer pairs used in the experiment were ([AL0] and c[AR0]), ([AL0] and c[AR1]), ([AL1] and c[AR0]), and ([AL1] and c[AR1]).
2. Perform 40 cycles of realtime PCR.
- preheat 95°C for 2 min.
↓
- denature 94°C for 20 sec.
- anneal and extension 65°C for 10 sec.
(Repeated 40 cycles)
3. After PCR, perform melting curve analysis.

The confirmation of the existence of each address was made by the results of amplification check comprised of measuring the SYBR green I fluorescence and analyzing the melting curve. Realtime PCR was performed by DNA Engine OPTICON 2 (MJ Research).

Fig. 5. Details of the PCR protocol

The details of the PCR protocol employed in the experiment are described in Fig. 5.

Results: We diluted the solution by 10^{-4}, 10^{-5}, and 10^{-6}, and then checked for four addresses. First, we performed an initial PCR using the two digits BLOBRO as primers, then performed a second PCR using ALOARO, ALOAR1, AL1ARO, and AL1AR1 as primers. This trial was repeated eight times for each set of primers (Fig. 6).

At dilutions of 10^{-4} and 10^{-5}, we detected molecules for each of the four addresses. Note that the probability of having one or more molecules in the diluted solution is about $1 - e^{-30.0} = 0.999999 = 1$ for 10^{-4} and about $1 - e^{-3.0} = 0.95$ for 10^{-5}.

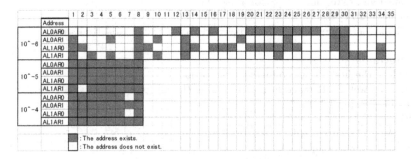

	Address	Exists	Does not exist	Total	The probability of existence of the address
	ALOARO	14	21	35	0.400
10^{-6}	ALOAR1	9	26	35	0.257
	AL1ARO	18	17	35	0.514
	AL1AR1	10	25	35	0.286
	ALOARO	8	0	8	1.000
10^{-5}	ALOAR1	8	0	8	1.000
	AL1ARO	8	0	8	1.000
	AL1AR1	7	1	8	0.875
	ALOARO	7	1	8	0.875
10^{-4}	ALOAR1	8	0	8	1.000
	AL1ARO	7	1	8	0.875
	AL1AR1	8	0	8	1.000

Fig. 6. Results of the preliminary experiment

At a dilution of 10^{-6}, only a random detection of addresses was achieved. Therefore, we conducted more trials at this dilution, 35 in total. As a result, we obtained the detection ratios for the four addresses: 14/35=0.400, 9/35=0.257, 18/35=0.514, and 10/35=0.286. The probability of having one or more molecules in the diluted solution was estimated at about $1 - e^{-0.3} = 0.26$ in this case. Although some correlation existed among addresses (for ARj in particular), the resulting pattern was nearly random.

It may be questionable whether PCR succeeded in amplifying a single molecule because the actual number of molecules that remained in the diluted solutions was not known. We carefully excluded wrong PCR products in the experiment. However, we plan to perform DNA sequencing to assess the correctness of the amplified solution.

One of the diluted solutions was amplified by the outermost primers, and the result was further diluted. We checked each address in each diluted solution, and verified that the detection pattern was uniform, which indicates that the detection pattern using the outermost primers was correctly amplified by PCR.

6 Concluding Remark

Molecular authentication is considered to have a potential of becoming a killer application of DNA and molecular computation. On the other hand, constructing a random pool of DNA molecules has long been an active research topic in the field. This paper proposed to apply achievements of the latter research to the former application by inducing randomness at the molecular level.

We plan to eventually apply the 16.8-million-address molecular memory to DNA ink. However, there is much to be done in analyzing the results of the preliminary experiments. As noted above, the correctness of the PCR products should be confirmed by DNA sequencing.

Acknowledgments

The work presented in this paper was partially supported by Grand-in-Aid for Scientific Research on Priority Area No.14085202, Ministry of Education, Culture, Sports, Science and Technology, Japan.

References

1. Leonard, M.: Adleman: Molecular Computation of Solutions to Combinatorial Problems. Science 266, 1021–1024 (1994)
2. Braich, R.S., Chelyapov, N., Johnson, C., Rothemund, P.W.K., Adleman, L.: Solution to a 20-Variable 3-SAT Problem on a DNA Computer. Science 296, 499–502 (2002)
3. Fujibayashi, K., Murata, S.: A Method of Error Suppression for Self-assembling DNA Tiles. In: Ferretti, C., Mauri, G., Zandron, C. (eds.) DNA Computing. LNCS, vol. 3384, pp. 113–127. Springer, Heidelberg (2005)
4. Gotoh, O., Sakai, Y., Mawatari, Y., Gunji, W., Murakami, Y., Suyama, A.: Normalized molecular encoding method for quantitative gene expression profiling. In: Carbone, A., Pierce, N.A. (eds.) DNA Computing. LNCS, vol. 3892, p. 395. Springer, Heidelberg (2006)
5. Itakura, Y., Hashiyada, M., Nagashima, T., Fukuyama, M.: Validation Experiment Report on DNA Information for Personal Identification (Part I). Technical Report of IEICE ISEC2001-12 (2001-05), The Institute of Electronics, Information and Communication Engineers, pp. 1–8 (2001) (in Japanese)

6. Itakura, Y., Hashiyada, M., Nagashima, T., Tsuji, S.: Validation Experiment Report on DNA Information for Personal Identification (Part II). Technical Report of IEICE ISEC2001-13 (2001-05), The Institute of Electronics, Information and Communication Engineers , 9–16 (2001) (in Japanese)
7. Kashiwamura, S., Yamamoto, M., Kameda, A., Shiba, T., Ohuchi, A.: Hierarchical DNA Memory Based on Nested PCR. In: Hagiya, M., Ohuchi, A. (eds.) DNA Computing. LNCS, vol. 2568, pp. 112–123. Springer, Heidelberg (2003)
8. Kashiwamura, S., Kameda, A., Yamamoto, M., Ohuchi, A.: Two-Step Search for DNA Sequence Design. IEICE E87-A(6), 1446–1453 (2004)
9. Kashiwamura, S., Yamamoto, M., Kameda, A., Shiba, T., Ohuchi, A.: Potential for Enlarging DNA Memory: The Validity of Experimental Operations of Scaled-up Nested Primer Molecular Memory. BioSystems 80, 99–112 (2005)
10. Kubota, M., Hagiya, M.: Minimum Basin Algorithm: An Effective Analysis Technique for DNA Energy Landscapes. In: Ferretti, C., Mauri, G., Zandron, C. (eds.) DNA Computing. LNCS, vol. 3384, pp. 202–214. Springer, Heidelberg (2005)
11. Nitta, N., Suyama, A.: Autonomous biomolecular computer modeled after retroviral replication. In: Chen, J., Reif, J.H. (eds.) DNA Computing. LNCS, vol. 2943, pp. 203–212. Springer, Heidelberg (2004)
12. Soloveichik, D., Winfree, E.: Complexity of Compact Proofreading for Self-assembled Patterns. In: Carbone, A., Pierce, N.A. (eds.) DNA Computing. LNCS, vol. 3892, pp. 305–324. Springer, Heidelberg (2006)
13. Suyama, A.: Programmable DNA computer with application to mathematical and biological problems. Preliminary Proceedings of the Eighth International Meeting on DNA Based Computers, 91 (2002)
14. Takahashi, K., Yaegashi, S., Asanuma, H., Hagiya, M.: Photo- and Thermoregulation of DNA Nanomachines. In: Carbone, A., Pierce, N.A. (eds.) DNA Computing. LNCS, vol. 3892, pp. 336–346. Springer, Heidelberg (2006)
15. Yamamoto, M., Kashiwamura, S., Ohuchi, A.: DNA Memory with 16.8M addresses. DNA13 (submitted 2007)
16. Yurke, B., Turberfield, A.J., Mills Jr., A.P., Simmel, F.C., Neumann, J.L.: A DNA-fuelled molecular machine made of DNA. Nature 406, 605–608 (2000)

Design of Code Words for DNA Computers and Nanostructures with Consideration of Hybridization Kinetics

Tetsuro Kitajima, Masahiro Takinoue, Ko-ichiroh Shohda,
and Akira Suyama

Department of Life Science and Institute of Physics,
Graduate school of Arts and Sciences
The University of Tokyo
3-8-1 Komaba, Meguro-ku, Tokyo 153-8902, Japan
{kitajima, takinoue}@genta.c.u-tokyo.ac.jp, suyama@dna.c.u-tokyo.ac.jp

Abstract. We have developed a method for designing rapidly-hybridizing orthonormal DNA sequences. Two conditions were used in the prediction method. One condition concerned the stability of the self-folded secondary structures of forward and reverse strands. The other condition concerned the nucleation capability of complementary strands at the tails of their self-folded secondary structures. These conditions were derived from the complementary strands' experimentally-determined hybridization rates' dependence on their stability and nucleation capability. These dependences were examined for 37 orthonormal DNA sequences randomly selected from our set of 300 orthonormal DNA sequences. By applying this new method to the set of 300 orthonormal DNA sequences, more than 100 rapidly-hybridizing sequences were obtained.

1 Introduction

DNA computing and DNA nanotechnology employ remarkable features unique to DNA and RNA molecules. The interactions and structures of DNA and RNA molecules can be most successfully designed in terms of their base sequences among various molecules. In DNA computing, programs and data are encoded into DNA/RNA sequences, while in DNA nanotechnology, structures and functions are encoded into DNA/RNA sequences. Thus, the design of DNA and RNA sequences is a crucial step for DNA computing and DNA nanotechnology.

Various methods have been developed to design DNA and RNA sequences for DNA computing and DNA nanotechnology[1,2,3,4,5,6,7,8,9]. Sets of DNA sequences of a uniform length and stability without mis-hybridizations and stable self-folded structures have been designed and applied to DNA computing, DNA probe sensors for genome analysis, and constructions of DNA nanostructures and nanodevices. The design methods so far developed are based on thermodynamic models. The stability of desirable and undesirable hybrids formed through inter-molecular base-pairing, and the stability of self-folded structures formed through

M.H. Garzon and H. Yan (Eds.): DNA 13, LNCS 4848, pp. 119–129, 2008.

intramolecular base-pairing are calculated from base sequences by using thermo-dynamic models and parameters.

In DNA computing and DNA nanotechnology, however, not only thermody-namic properties but also the kinetic properties of DNA/RNA sequences sub-stantially affect the results of computations and nanostructure constructions. Non-uniform hybridization rates will make the speed at which instruction codes are executed dependent on the content of the data, especially for computation on autonomous DNA computers running under isothermal conditions. This data-dependent execution speed makes the computation less reliable, while the non-uniformity of the rates will also make the construction of DNA nanostructures more complicated. Therefore, design methods in which kinetic properties are also considered are essential for the further development of DNA computing and DNA nanotechnology.

In this study, we have explored the DNA sequence dependence of hybridization rates in order to develop a DNA sequence design method that takes the kinetic properties of DNA/RNA hybridization into consideration. The rate of hybridiza-tion of two complementary strands was measured for 37 DNA sequences randomly chosen from our set of orthonormal DNA sequences 23 nucleotides long. The or-thonormal sequences were designed to be orthogonal and normalized. This orthog-onality means that every sequence in the set significantly hybridizes neither with any sequences in the set other than its compliment nor with any concatenated se-quences made of sequences in the set without its complement. Normalization in this context means that every sequence has a uniform length and a uniform du-plex stability, and has no very stable self-folded secondary structures that may significantly hinder rapid hybridization with its complement. The orthonormal DNA sequences were assured to have a uniform duplex stability not only by an equal melting temperature but also by the equal free energy changes accompany-ing duplex formation. However, the hybridization rate significantly depended on the DNA sequences. The stability and the shape of the sequences' self-folded sec-ondary structures were examined to elucidate their relationships to the hybridiza-tion rates. Based on these relationships, a method to design a DNA sequence for rapidly-hybridizing orthonormal DNA sequences has been proposed.

2   Materials and Methods

2.1   DNA Sequences

The DNA sequences used in hybridization experiments and secondary structure predictions were 37 orthonormal DNA sequences randomly selected from our 23-mer orthonormal sequence set containing more than 300 DNA sequences. Their sequences were as follows: 5'-GCATCTACACTCAATACCCAGCC-3' ,
5'-CGTCTATTGCTTGTCACTTCCCC-3' , 5'-GGCTCTATACGATTAAACTCCCC-3' ,
5'-GAAGGAATGTTAAAATCGTCGCG-3' , 5'-GCACCTCCAAATAAAAACTCCGC-3' ,
5'-GAGAAGTGCTTGATAACGTGTCT-3' , 5'-GCATGTGTAGTTATCAGCTTCCA-3' ,
5'-CTAGTCCATTGTAACGAAGGCCA-3' , 5'-GTCCCGGAAAATACTATGAGACC-3' ,

5'-GAGTCCGCAAAAATATAGGAGGC-3' , 5'-CATCTGAACGAGTAAGGACCCCA-3' ,
5'-CGCGATTCCTATTGATTGATCCC-3' , 5'-GGTGGCTTATTTACAGGCGTTAG-3' ,
5'-TTCGGTTCTCTCCAAAAAAAGCA-3' , 5'-GGCGCTTAAATCATCTTTCATCG-3' ,
5'-CCGTCGTGTTATTAAAGACCCCT-3' , 5'-CGAGAGTCTGTAATAGCCGATGC-3' ,
5'-TGGCACTTATAGCTGTCGGAAGA-3' , 5'-GGCTGTTTACAAAATCGAGCTAG-3' ,
5'-TGCGAAATTTGAAAAATGGCTGC-3' , 5'-GCATTGAGGTATTGTTGCTCCCA-3' ,
5'-GGCTGTCAATTTATCAGGGAGGC-3' , 5'-GCCTCAAGTACGACTGATGATCG-3' ,
5'-GAAGCCCTATTTTGCAATTCCCC-3' , 5'-CGCGGGTACGTTGATGTAACAAA-3' ,
5'-ATGGGAACCTAAAAGTGTGGCTA-3' , 5'-GAGTCAATCGAGTTTACGTGGCG-3' ,
5'-TTCGCTGATTGTAGTGTTGCACA-3' , 5'-GCCTCACATAACTGGAGAAACCT-3' ,
5'-CCATCAGGAATGACACACACAAA-3' , 5'-GGGATAGAACTCACGTACTCCCC-3' ,
5'-CCATATCCGATTATTAGCGACGG-3' , 5'-GGGATCAGTTGTACACTCCCTAG-3' ,
5'-CTGTGATGATACCGTTCTTCACC-3' , 5'-CGCGGTTGAAATAACTAATCGCG-3' ,
5'-GGTCGAAACGTTATATTAACGCG-3' , 5'-TAGCACCCGTTAAAACGGAAATG-3' .

The DNA strands of 38 orthonormal sequences and their complements were synthesized and purified by HPLC commercially (SYGMA genosys, Japan).

2.2 DNA Hybridization

The time course of the hybridization of two complementary DNA strands was measured on a fluorescence spectrophotometer LS 55 (Perkin Elmer, USA) equipped with a stopped-flow apparatus RX-2000 (Applied Photophysics, UK). Two complementary DNA strands at 50 nM each in 1xSSC (0.015 M Na_3 − citrate, 0.15 M NaCl) containing PicoGreen® (Invitrogen, USA) were mixed rapidly (a dead-time of 8 ms) at 25°C through the use of the stopped-flow apparatus. The DNA duplex formation was followed by the fluorescence emission from PicoGreen® at 523 nm exited at 502 nm. PicoGreen® is a dye for quantitating double-stranded DNA (dsDNA) in the presence of single-stranded DNA/RNA. The linear detection range of the dye extends over more than four orders of magnitude in dsDNA concentration (from 25 pg/ml to 1,000 ng/ml), allowing the precise observation of the time course of DNA hybridization.

2.3 Determination of Hybridization Rates

The time courses of the hybridization of two complementary DNA strands were fitted to a single-exponential model:

$$I(t) = I_\infty + A \exp(-t/\tau),$$

where $I(t)$ is the observed fluorescence intensity at time t , I_∞ is the final fluorescence intensity, A is the amplitude of hybridization, and $1/\tau$ is the rate of hybridization. The best-fit-values of the hybridization rates were obtained by nonlinear regression using Origin 7.0 and Microsoft Excel solver. The goodness of fit was measured by the value of $R2$, and was also confirmed visually through every plot overlapping the observed data and the best-fit-curve.

2.4 Secondary Structure Prediction

The self-folded secondary structures of the single DNA strands of 37 orthonormal sequences were predicted by using the mfold system, which is a tool for predicting RNA/DNA secondary structures by free energy minimization. For the prediction, a condition of 0.195 M Na^+ and 0 M Mg^{2+} was employed. This condition is equivalent to the 1xSSC at which the hybridization experiments were performed.

3   Results and Discussion

3.1 Hybridization Rates of Orthonormal DNA Sequences

The hybridization rate of two complementary DNA strands was determined for 37 orthonormal DNA sequences randomly selected from our set of 300 orthonormal sequences 23 nucleotides long. The time course of the hybridization was followed by measurement of the fluorescence intensity, which is proportional to the concentration of hybrid duplexes formed (i.e., to that of the base-pair stacks formed). The observed fluorescence intensity data were fitted to a single-exponential model to determine the hybridization time τ. Figure 1 shows a typical time course of the hybridization and the best-fitting single exponential curve. For all of the 37 orthonormal sequences, the hybridization rate was determined from three independent hybridization experiments, which were highly reproducible.

Figure 2 shows the distribution of the hybridization time determined for the 37 orthonormal sequences. The rate of hybridization significantly depended on DNA sequences although all DNA duplexes have a uniform thermodynamic stability. Most of the sequences finished hybridizing in 200 s, but some of the sequences indicated a very slow hybridization (i.e., a hybridization time of more than 15 min).

3.2 Hybridization Rates and the Stability of Self-folded Secondary Structures

The hybridization rates of complementary strands of nucleic acids are affected by the stability of the strands' self-folded structures. The strands' stable secondary structures are shown to significantly reduce the rate of hybridization [10,11]. Therefore, those sequences that form very stable secondary structures were discarded in the design of the set of 300 orthonormal DNA sequences. However, the threshold value of the free energy change of secondary structure formation below which sequences should be discarded was not evident, so that the set contains secondary structure sequences with a wide range of stabilities: for some sequences the value of the free energy change in secondary structure formation ΔG is positive, and for some other sequences ΔG is as low as -5 kcal/mol. Consequently, the slow hybridization observed for some sequences may be due to the formation of slightly stable secondary structures. Thus we examined the dependence of the observed hybridization time τ on the stability of predicted secondary structures.

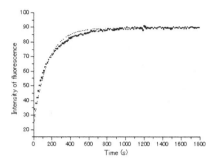

Fig. 1. Typical time course of the hybridization of orthonormal DNA sequence. Two complementary DNA strands at 50 nM each were rapidly mixed at 25 °C in 1xSSC in the presence of PicoGreen®. The sequence of DNA was 5f-CCATATCCgATTATTAgCgACgg-3f. The closed squares are observed fluorescence intensities and the broken line is the best-fitting single-exponential curve obtained by non-linear regression. The hybridization time, which is the reciprocal of the hybridization rate, determined by the best-fitting curve was 1.5×10^2 s.

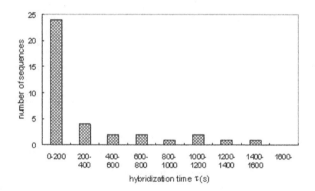

Fig. 2. Distribution of hybridization times of 37 orthonormal DNA sequences

Figure 3 shows the relationship between the stability of self-folded secondary structure predicted using the mfold method and the observed hybridization time τ. The stability of the secondary structure was measured in terms of the free energy change ΔG in secondary structure formation. When the value of ΔG is positive, the secondary structure is less stable than the unstructured coil. When the value of ΔG is negative, the secondary structure is more stable than the unstructured coil. Only the lowest energy structures were considered in this study, though some sequences were predicted to have more than one possible structure. Figure 3 indicates that two complementary strands, i.e., a forward and a reverse strand, hybridized rapidly ($\tau < 240$ s) when either of the two strands had a positive value of ΔG. Especially when both of them had a positive ΔG value, the hybridization was more rapid ($\tau < 120$ s). When either of the two strands had a negative ΔG value, some of the DNA sequences showed slow hybridization rates, while the other sequences still hybridized rapidly. Therefore, an increase in the stability of the self-folded secondary structures actually decreased the hybridization rates of two complementary strands. However, there must exist factors that affect the rate of hybridization other than the stability of secondary structures, because some DNA sequences with largely negative ΔG values still hybridized rapidly.

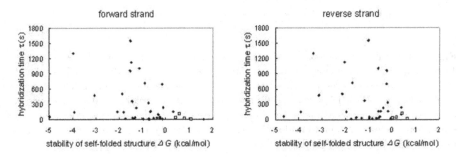

Fig. 3. Dependence of the hybridization time τ on the stability of self-folded structures of forward and reverse strands. The stability of the structure was measured in terms of the predicted free energy change ΔG in secondary structure formation. Open squares designate data of the orthonormal DNA sequences that have positive ΔG values for both forward and reverse strands.

3.3   Hybridization Rates and the Nucleation Capability of Self-folded Structures

The hybridization of nucleic acid strands requires the formation of a nucleus composed of at least three contiguous base-pairs. As soon as nucleation occurs, each duplex zips up to completion instantly. In the hybridization of short DNA strands at low concentrations such as those studied here, it has long been accepted that the nucleation step is rate-limiting [12]. We thus examined how the nucleation step affects the hybridization rate of complementary strands of the 37 orthonormal DNA sequences.

In the process of nucleation, each strand tries to find unpaired-base stretches of complementary sequences. Those stretches are found only in the loop and tail (end-coil) regions of self-folded strands. The length of the orthonormal sequences studied here was as short as 23 nucleotides, so that the size of the loops found in their secondary structures may not be large and flexible enough to perform a rapid nucleation. Tails are, in contrast, more flexible, so that unpaired bases found in the tails would more easily be involved in nucleation. We thus focused on the length of tails and examined whether the self-folded secondary structures of the orthonormal sequences have tails that are long enough for nucleation.

The orthonormal DNA sequences were classified into two groups, 'nucleation-inhibited' and 'nucleation-allowed' sequences, according to the length of contiguous unpaired-bases at the ends. A DNA sequence was defined as 'nucleation-

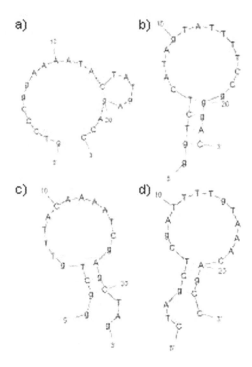

Fig. 4. Self-folded structures of forward and reverse strands of two orthonormal DNA sequences A and B. The most stable secondary structures of the forward (a) and the reverse (b) strand of sequence A. Those of the forward (c) and the reverse (d) strand of sequence B.

inhibited' if neither tail at the 5f- nor the 3f-end of its forward strand can be involved in the formation of a nucleus of 3 or 4 base-pairs. A DNA sequence was defined as 'nucleation-allowed' if either tail at the 5f- or the 3f-end of its forward strand can be involved in nucleation. The orthonormal DNA sequence shown in Figures 4a and 4b, for example, is nucleation-inhibited. The 5f- end of the forward strand has 13 contiguous unpaired-bases, which is long enough to form a nucleus of 3 or 4 base-pairs (Fig. 4a). However, the 5f-end cannot be involved in nucleation because the 3f-end of the reverse strand has no unpaired -base (Fig. 4b). The 3f-end of the forward strand cannot be involved in nucleation either because the 3f-end of the forward strand has 3 contiguous unpaired-bases and the 5f-end of the reverse strand has only one unpaired-base. Therefore, for this orthonormal DNA sequence nucleation is inhibited. On the other hand, the orthonormal DNA sequence shown in Figures 4c and 4d is nucleation-allowed. The 5f- end of the forward strand has one unpaired base and the 3f-end of the reverse strand has also one unpaired base (Figs. 4c and 4d). The 5f-end of the forward strand, therefore, cannot be involved in nucleation. However, the 3f-end can be involved in the formation of a nucleus of 3 base-pairs because the 3f-end of the forward strand and the 5f-end of the reverse strand, both ends have 3 contiguous unpaired-bases. Therefore, for this orthonormal sequence nucleation is allowed.

Figure 5 indicates how the number of nucleation-inhibited and nucleation-allowed sequences varied with the hybridization time τ. The number of nucleation-inhibited sequences increased with the increase in hybridization time. In contrast, the number of the nucleation-allowed sequences increased as the hybridization time decreased. Therefore, nucleation at the tails of self-folded secondary structures should be one of the critical factors affecting the hybridization rate.

3.4 Prediction of Orthonormal DNA Sequences Rapidly Hybridizing with Complementary Strands

The effect of the self-folded secondary structures' thermodynamic stability on the hybridization rate (Fig. 3) and that of the nucleation capability at the tails of self-folded secondary structures on the hybridization rate (Fig. 5) have provided the basic concept of a method for the design of DNA code word sequences rapidly hybridizing with complementary strands. Each of these effects by itself was not sufficiently significant to determine the hybridization rate. Therefore, in the design method both of these factors were taken into consideration.

The table summarizes how the number of rapidly-hybridizing orthonormal DNA sequences and the number of slowly-hybridizing ones depended on the self-folded secondary structures' thermodynamic stability and the nucleation capability at the tails of the self-folded secondary structures. In the table, DNA sequences with a hybridization time of less than 300 s are categorized as rapidly-hybridizing sequence and those with 300 s or more are categorized as slowly-hybridizing sequences. The hybridization-time threshold of 300 s was determined according to the experimental conditions of our autonomous DNA computing system RTRACS, and thus it is not exclusive to this design method.

From the results shown in the table, two conditions were derived to predict rapidly-hybridizing orthonormal DNA sequences. In Condition 1, both the forward and reverse strands of a sequence have positive ΔG values. Condition 1 self-evidently assures that the sequence is nucleation-allowed. In Condition 2, the sum of the ΔG values of a sequences' forward and reverse strands is larger than -1 kcal/mol and the sequence is also nucleation-allowed. Condition 2 is always applied after Condition 1; that is, Condition 2 is applied only to those DNA sequences whose forward or reverse strands have a negative ΔG value.

By applying Condition 1 to the set of 37 orthonormal DNA sequences studied here, 4 sequences were selected. Then by applying Condition 2, 12 sequences were further selected. A total of 16 sequences out of the 37 orthonormal DNA sequences were predicted as rapidly-hybridizing sequences. According to the table, only one sequence out of the 16 sequences predicted is not a rapidly-hybridizing sequence, while its hybridization time (330 s) was close to the threshold time (300 s). The false positive rate of prediction, therefore, is as small as 6%. Our orthonormal DNA sequence set contains 300 sequences. We then applied

(A)   (B)

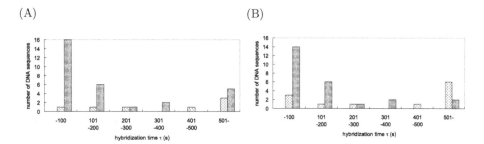

Fig. 5. Change of the number of the nucleation-inhibited and the nucleation-allowed orthonormal DNA sequences with the hybridization time τ. Hatched bars indicate the number of the nucleation-allowed sequences, and the dashed bars that of nucleation-inhibited sequences. The size of the nucleus allowed to form is 3 base-pairs (a) and 4 base-pairs (b).

Table 1. Summary of the dependence of the number of orthonormal DNA sequences on the stability of self-folded secondary structures and the nucleation of complementary strands at their tails

$\Delta G_{fwd} + \Delta G_{rev}$	nucleation-inhibited		nucleation-allowed	
	hybridization rate		hybridization rate	
(kcal / mol) (#)	slow	rapid	slow	rapid
Case 1: Both of ΔG_{fwd} and ΔG_{rev} are positive				
0		0	0	4
Case 2: Either of ΔG_{fwd} or ΔG_{rev} is negative				
$+2 \sim +1$	0	0	0	1
$+1 \sim 0$	0	0	0	2
$0 \sim -1$	1	0	1	8
$-1 \sim -2$	0	1	2	1
$-2 \sim -3$	0	1	3	4
$-3 \sim -4$	1	0	0	2
$-4 \sim -5$	0	0	1	0
$-5 \sim -6$	0	0	0	0
$-6 \sim -7$	1	0	0	0
$-7 \sim -8$	1	0	0	1
$-8 \sim -9$	0	0	0	0
$-9 \sim -10$	0	0	0	1

#) ΔG_{fwd} and ΔG_{rev} stand for ΔG of a forward strand and that of a reverse strand, respectively.

both Conditions 1 and 2 to the set of 300 orthonormal DNA sequences. A set of 108 rapidly-hybridizing sequences, which may contain 7-8 slowly-hybridizing sequences, was obtained.

The present prediction method using Conditions 1 and 2 may be satisfactory because 108 sequences are sufficient for most studies using rapidly-hybridizing orthonormal DNA sequences. In a set of 300 orthonormal DNA sequences, however, many sequences may still be predicted as slowly-hybridizing while they could actually be hybridizing rapidly, because the table contains many nucleation-allowed sequences hybridizing rapidly with largely negative ΔG values of less than -1 kcal/mol. If those sequences can be distinguished from other nucleation-allowed sequences hybridizing slowly by using additional conditions, the predictability of the method will be much increased. One promising condition would concern the stability of short duplexes adjacent to the tails involved in nucleation. It is conceivable that even when a sequence has a largely negative ΔG value indicating a globally-stable self-folded secondary structure, its complementary strands should hybridize rapidly if the adjacent short duplexes are unstable. Such an additional condition would increase the number of rapidly-hybridizing sequences predicted by the method as keeping the false positive rate substantially low.

4   Conclusion

We have developed a method for designing rapidly-hybridizing orthonormal DNA sequences. Two conditions are used in the prediction method. One condition concerns the stability of the self-folded secondary structures of forward and reverse strands, while the other concerns the nucleation at the tails of their self-folded secondary structures. More than 100 rapidly-hybridizing orthonormal DNA sequences were obtained by the present prediction method.

Acknowledgements

This work was supported by a grant for SENTAN (Development of System and Technology for Advanced Measurement and Analysis) from the Japan Science and Technology Agency (JST), and by a grant-in-aid for the 21st Century COE program "Research Center for Integrated Science" and for Scientific Research on Priority Areas "Molecular Programming" from the Ministry of Education, Culture, Sports, Science, and Technology of Japan.

References

1. Tulpan, D., Andronescu, M., Chang, S.B., Shortreed, M.R., Condon, A., Hoos, H.H., Smith, L.M.: Thermodynamically based DNA strand design. Nucleic Acids Res. 33, 4951–4964 (2005)
2. Shortreed, M.R., Chang, S.B., Hong, D., Phillips, M., Campion, B., Tulpan, D.C., Andronescu, M., Condon, A., Hoos, H.H., Smith, L.M.: A thermodynamic approach to designing structure-free combinatorial DNA word sets. Nucleic Acids Res. 33, 4965–4977 (2005)
3. Dirks, R.M., Lin, M., Winfree, E., Pierce, N.A.: Paradigms for computational nucleic acid design. Nucleic Acids Res. 32, 1392–1403 (2004)

4. Arita, M., Kobayashi, S.: Sequence design using template. New Generation Computing 20, 263–277 (2002)
5. Jonoska, N., Kephart, D., Mahalingam, K.: Generating DNA code words. Congressus Numerantium 156, 99–110 (2002)
6. Penchovsky, R., Ackermann, J.: DNA library design for molecular computation. J. Comp. Biol. 10, 215–229 (2003)
7. Feldkamp, U., Rauhe, H., Banzhaf, W.: Software Tools for DNA Sequence Design. Genetic Programming and Evolvable Machines 4, 153–171 (2003)
8. Garzon, M., Deatorthonormal, J.: Codeword design and information encoding in DNA ensembles. Natural Computing 3, 253–292 (2004)
9. Kari, L., Konstantinidis, S., Sosík, P.: Preventing undesirable bonds between DNA codewords. Lect. Notes Comput. Sc. 3384, 182–191 (2005)
10. Kushon, S.A., Jordan, J.P., Seifert, J.L., Nielsen, H., Nielsen, P.E., Armitage, B.A.: Effect of secondary structure on the thermodynamics and kinetics of PNA hybridization to DNA hairpins. J. Am. Chem. Soc. 123, 10805–10813 (2001)
11. Gao, Y., Wolf, L.K., Georgiadis, R.M.: Secondary structure effects on DNA hybridization kinetics: a solution versus surface comparison. Nucleic Acids Res. 34, 3370–3377 (2006)
12. Cantor, C.R., Schimmel, P.R.: Biophysical Chemistry: Part III: The Behavior of Biological Macromolecules. W. H. Freeman, San Francisco (1980)

Dynamic Neighborhood Searches for Thermodynamically Designing DNA Sequence*

Suguru Kawashimo, Hirotaka Ono, Kunihiko Sadakane,
and Masafumi Yamashita

Dept. of Computer Science and Communication Engineering, Kyushu University,
744 Motooka, Nishi-ku, Fukuoka, Fukuoka 819-0395, Japan
kawa@tcslab.csce.kyushu-u.ac.jp, {ono,sada,mak}@csce.kyushu-u.ac.jp

Abstract. We present a local search based algorithm designing DNA short-sequence sets satisfying thermodynamical constraints about minimum free energy (MFE) criteria. In DNA12, Kawashimo et al. propose a dynamic neighborhood search algorithm for the sequence design under hamming distance based constraints, where an efficient search is achieved by dynamically controlling the neighborhood structures. Different from the hamming distance based constraints, the thermodynamical constraints are generally difficult to handle in local-search type algorithms. This is because they require a large number of evaluations of MFE to find an improved solution, but the definition of MFE itself contains time-consuming computation. In this paper, we introduce techniques to reduce such time-consuming evaluations of MFE, by which the proposed dynamic neighborhood search strategy become applicable to the thermodynamical constraints in practice. In computational experiments, our algorithm succeeded in generating better sequence sets for many constraints than exiting methods.

Keywords: DNA Sequence Design Algorithm, Local Search, Statistical Thermodynamical Constraints.

1   Introduction

Designing DNA sequence sets is a fundamental issue in the fields of nanotechnology and nanocomputing, e.g., Adleman's DNA solution for the Hamiltonian path [1], DNA tiling with its self-assemble [22], hairpin-based state machine [10] and so on. One aspect of DNA computing / technology is to control the DNA molecules reactions. For a robust "computation", it is desirable that DNA molecules react only in expected ways, because unexpected secondary structures of DNA sequences may cause error, for example. Sequence design is an approach for a robust computation by designing DNA sequences that satisfy some constraints to avoid unexpected molecular reactions. Since expected or unexpected reactions depend on the applications or the purposes, several representative constraints

---

* This research partly received financial support from Scientific research fund of Ministry of Education, Culture, Sports, Science and Technology.

M.H. Garzon and H. Yan (Eds.): DNA 13, LNCS 4848, pp. 130–139, 2008.

are usually considered as below mentioned. Another requirement for DNA sequence sets is to be large. This is because designed DNA sequences are used as elemental components of computation; the amount of resources on DNA computation is proportional to the size of a sequence set. In summary, systematic methodologies are required to design large set of sequences, which satisfy certain types of constraints.

In the sequence design, constraints are introduced to prohibit unexpected secondary structures of DNA sequences, and several types of prohibition are proposed. Roughly speaking, the types of prohibitions are classified into combinatorial types and thermodynamical types. Combinatorial constraints are based on the idea that base conjugations of DNA sequences are regarded as a kind of combinatorial pattern matching, while the thermodynamical constraints are based on the thermodynamical property of the molecular reaction mechanism, in which conformations of small (resp., large) Gibbs standard free energies tend to be stable (resp., unstable). Although the thermodynamical ones seem to be more sophisticated, many algorithmic studies of the sequence design have treated combinatorial constraints due to their simplicity. Also the combinatorial properties help to bring efficient algorithms from combinatorics or combinatorial optimization fields [3,4,19,20]. On the other hand, there are few studies under thermodynamical constraints from the combinatorial algorithmic point of view; one example that the authors know is a Stochastic Local Search method by Tulpan et al. [21].

In this paper, we consider DNA sequence design algorithm under thermodynamical constraints from the viewpoint of the combinatorial optimization. More precisely, we propose a local-search type algorithm for the DNA sequence design under thermodynamical ones. A local search is a method to find a good solution by replacing a current solution with a better (improving) solution in its neighborhood until no better solution is found. In DNA12, the authors proposed a dynamic neighborhood search algorithm for DNA sequence design problem [11], which targets on short-sequence sets under combinatorial constraints. The algorithm is equipped with high search performance by changing the neighborhood structures dynamically. The computational experiments show a good design power of the algorithm; it succeeded in generating better sequence sets than exiting methods [3,4,19]. Also by the nature of local search methods, it has a good flexibility; we can finely adjust the constraints. Therefore, we attempt to implement the idea of our previous algorithm for the thermodynamical constraints, especially *Minimum Free Energy* (MFE, for short) constraints, in this paper.

However, such an implementation is nontrivial in general. The Gibbs standard free energy is an energy value associated with the conformation of a sequence or sequences given, and the MFE is the minimum value among free energies of all the possible structures. Namely, the definition of MFE itself contains a time-consuming calculation, and in fact its time complexity is $O(n^3)$ time where n is the length of a given sequence. That is, a large amount of evaluations of MFE values are not practical, which implies that a local-search type algorithms are not suitable since they need to repeatedly evaluate many solution values. A main contribution of this paper is to overcome this difficulty; we present two techniques to circumvent the heavy calculations. One is to realize an effective neighborhood search. For this purpose, we store extra data among bases of DNA sequences, by which we can find a base involved with the violation for MFE constraints. The other is to realize an efficient evaluation of MFEs. In neighborhood searches, most of neighbor solutions are apparently worse, and only a few of them are candidates of improving solutions. For screening such apparently worse solutions, we introduce a preprocessing phase in the search; instead of applying $O(n^3)$ time MFE calculation, we utilize an approximate calculation of the MFE.

By these techniques, our search framework introduced in [11] becomes applicable to the MFE-based constraints in practice. In order to see the performance of our approach, we conduct computational experiments for various settings of MFE constraints. The results show that we succeeded in designing a large set of sequences for many case. One virtue of our algorithm is that it is a practical local search: It is quite flexible and is easy to introduce a new constraint. Moreover, if a non-local-search type algorithm finds a (good) sequence set, then we may obtain an even better solution by applying our algorithm to the solution.

1.1 Related Work

Many studies consider the thermodynamical natures of DNA computing from various points of view (e.g., [14,16]), and the thermodynamical qualities of sequence sets are also discussed in several papers. Especially, Dirks et al. [7] discuss various thermodynamical criteria of designing secondary structures, and Rose et al. [15] propose a statistical thermodynamic error model in DNA computing.

Tulpan et al. succeeded in designing sequence sets under very complicated thermodynamical constraints by Stochastic Local Search method [21], though the running time is not clear because they evaluated the search time except the calculation of energy values in their experiments. They also proposed new thermodynamical constraints. One advantage of their method is that they can treat complicated constraints as well as ours, since it is a local-search type algorithm. Garzon et al. also designed sequence sets [9]. They designed sequence set under combinatorial constraint as preprocessing, and remove thermodynamically violated sequences from the set obtained in preprocessing by the reduction to the minimum vertex cover problem (actually, they consider the maximum independent set). However the minimum vertex cover problem itself is known to be NP-hard. Tanaka et al. used random-generation based method [18]. To reduce the calculating-time of evaluation, they proposed approximate method of calculating MFE by the greedy manner.

The remainder of the paper is organized as follows: Section 2 gives preliminaries of the paper, thermodynamical constraints, and basic definitions for local search. Section 3 discusses how the heavy MFE calculations can be embedded into our search framework. Section 4 shows the results of computational experiments, and then Section 5 concludes the paper.

2   Preliminaries

2.1   Definitions and MFE Constraints

A DNA sequences s is a string over $\{\mathtt{A},\mathtt{T},\mathtt{C},\mathtt{G}\}$. A DNA sequence or sequences form *secondary structures* by the Watson-Crick property, which are also called *conformations*. Each conformation of a sequence (or sequences) has a Gibbs standard *free energy*. The *Minimum Free Energy* (MFE, for short) of a sequence (resp., sequences) is the minimum value among free energies of all possible conformations of a sequence (resp., sequences). It is known that a conformation with a small Gibbs standard free energy is more stable than ones with larger Gibbs standard free energies. The Gibbs standard free energy values are measured through actual experiments and we can compute the value for one conformation in linear time of the length of the sequence.

Let s, s' be DNA sequences of length n, then $s, s' \in \{\mathtt{A},\mathtt{T},\mathtt{G},\mathtt{C}\}^n$. Sequences are represented by $s = s_1 s_2 \cdots s_n$, and $s' = s'_1 s'_2 \cdots s'_n$. In these representations, the left end of a sequence corresponds to $5'$ end of a DNA sequence. In addition, $wcc(s)$ denotes the Watson-Crick complement sequence of DNA sequence s, here, $wcc(s)$ is the sequence which reverse s and replaced each $\mathtt{A}$ in s by $\mathtt{T}$ and vice versa, replaced each $\mathtt{G}$ in s by $\mathtt{C}$ and vice versa.

Let S be the sequence set. In the context of the sequence design problems, we let "hybridization" refer to "the phenomenon that a sequence in S forms completely hydrogen bonds with its complement sequence", and "miss-hybridization" refer to "conformations which are not hybridization". The constraints described below are introduced in order to avoid miss-hybridization.

The MFE between s and s' is represented by $\Delta G(s, s')$ which can be calculated $O(n^3)$-time by the dynamic programming [2,12,23].

Let $wcc(S) = \{wcc(s) | s \in S\}$. Given threshold parameters t_{ww}, t_{wc}, and t_{cc}, we define the following constraints based on the MFE measure:

Word-Word Constraint: for all pairs of s, s' in S, $\Delta G(s, s') \geq t_{ww}$. That is, $\Delta G_{ww}(S) \overset{\text{def}}{=} \min_{s,s' \in S}\{\Delta G(s, s')\} \geq t_{ww}$.

Word-Complement Constraint: for all pairs of s in S, s' in $wcc(S)$, and $s \neq wcc(s')$, $\Delta G(s, s') \geq t_{wc}$. That is, $\Delta G_{wc}(S) \overset{\text{def}}{=} \min_{s \in S, s' \in wcc(S), s \neq wcc(s')}\{\Delta G(s, s')\} \geq t_{wc}$.

Complement-Complement Constraint: for all pairs of s, s' in $wcc(S)$, $\Delta G(s, s') \geq t_{cc}$. That is, $\Delta G_{cc}(S) \overset{\text{def}}{=} \min_{s,s' \in wcc(S)}\{\Delta G(s, s')\} \geq t_{cc}$.

Note that, in these constraints, self reactions of one sequence are under consideration. On the other hand, we do not concern with pseudo-knots.

In this paper, we adopt only three constraints for the sequence design, following the work by Garzon et al. [9]. This does not mean that our algorithm is specified to these constraints, and it is applicable to many other criteria based on MFE (e.g. energy gap [21]). For other criteria, such as melting temperature and DNA error rate [15], though we may need careful adjustments, it is also applicable.

By using these, our problem is described as "*find S such that $\Delta G_{ww}(S) \geq t_{ww}$, $\Delta G_{wc}(S) \geq t_{wc}$ and $\Delta G_{cc}(S) \geq t_{cc}$ for large t_{ww}, t_{wc}, and t_{cc}*".

2.2   Local Search, Neighborhood and Objective Functions

A local search is a method to find a solution by replacing a current solution with a solution which has better *objective function* value in its *neighborhood* until no better solution is found. In DNA12, we proposed a local search based algorithm for DNA sequence design problem under combinatorial constraints. In this paper, we apply this algorithm for thermodynamical constraints. We hope interested readers refer to [11], in which more details about our algorithm can be found[1].

We define the neighborhood of S (we represent it as $N(S)$) for the local search as follows: sequence sets obtained by flipping 1 base of a sequence belonging to S. Due to the simplicity of the definition, we can flexibly apply it to various constraints.

In this problem, we need to design the set such as $\Delta G_{ww}(S) \geq t_{ww}$, $\Delta G_{wc}(S) \geq t_{wc}$, and $\Delta G_{cc}(S) \geq t_{cc}$. Therefore, when we take together these constraints, the objective function is described as follows:

$$\Delta G_{min}(S) \overset{\text{def}}{=} \min\{\Delta G_{ww}(S) - t_{ww}, 0\} + \\ \min\{\Delta G_{wc}(S) - t_{wc}, 0\} + \\ \min\{\Delta G_{cc}(S) - t_{cc}, 0\}. \tag{1}$$

By definition, $\Delta G_{min}(S) = 0$ means that it satisfies the constraints, and it takes $O(m^2 n^3)$ time to evaluate $\Delta G_{min}(S)$.

3   Techniques to Reduce MFE Evaluations

In the local search, to determine if the neighbor solution is an improving solution or not, its solution value should be calculated. This operation is executed many times, since the size of neighborhood is usually very large. As mentioned above it takes $O(m^2 n^3)$ time to evaluate one solution, but that running time can be reduced in our neighborhood search, because all pairs of sequences for S and all pairs of sequences for $S' \in N(S)$ are overlapping. By reusing the calculation of $\Delta G_{min}(S)$, the calculation of $\Delta G_{min}(S')$ for $S' \in N(S)$ can be done in $O(mn^3)$ time.

However, it is still too time-consuming. That is, naive local search type algorithms may not work well. In this section, we explain two techniques by which we skip such a large amount calculations. One is a device to effectively check neighbor solutions, and the other is to screen bad solutions without calculating the exact $\Delta G(s, s')$.

---

[1] In this paper, we use the new framework which is simplified and improved from the previous one.

3.1   Effective Neighborhood Search

In the neighborhood search, we need to evaluate ΔG_{min} of neighborhood solutions to determine if we move to the solution or not. This means that evaluating ΔG_{min} for worse solutions is wasting time; by effectively finding an improving solution we can reduce the calculation of ΔG_{min} values. In this subsection, we explain how to realize a fast discovery of improving solutions. More concretely, we define a good order of checking the neighbor solutions, in which solutions to be likely improvements have high priorities.

To define the ordering, we use an array $min\_related$ as counters for bases in S; $min\_related$ is on all the bases in S, and $min\_related(s_i)$ for a base s_i of $s \in S$ stores the number of occurrences of base s_i for $\Delta G_x(S)$ where $x \in \{ww,wc,cc\}$. The idea itself was introduced in the previous work [11], but it is extended from the previous one. Here, an "occurrence of base s_i for $\Delta G_x(S)$" means the following two conditions are satisfied: (i) s containing the base s_i and another sequence s' have the MFE value equal to $\Delta G_x(S)$ and (ii) in the MFE structure of the $\Delta G_x(S)$, s_i forms hydrogen bonds. If a base has a large value of $min\_related$, the base may be critical for $\Delta G_{min}(S)$, therefore flipping such a base probably improves the solution value. On the other hand, flipping bases with $min\_related = 0$ does not change the solution value by definition. Therefore, we define the search order of neighbor solutions in $N(S)$ according to $min\_related$ values of the descending order. In case of ties, i.e., some bases have a same value of $min\_related$, we use another array $bond\_related$ to determine the order. The $bond\_related$ on all the bases similarly stores the number of occurrences of a base for hydrogen bonds about not MFE-structures of $\Delta G_{min}(S)$ but all MFE-structures. By a similar argument, we define the search order for ties in $min\_related$ according to $bond\_related$ of the descending order.

Table 1 shows results of preliminary computational experiments concerning the effectively of $min\_related$ and $bond\_related$. This result shows that the ordering based on $min\_related$ and $bond\_related$ apparently realizes an effective search.

Table 1. Result of the preliminary experiments for $min\_related$ and $bond\_related$

n	m	τ	time; with $min\_related$ average / standard deviation	time; random order average / standard deviation
8	30	-6.0(kcal/mol)	1.87(sec)/ 0.81(sec)	19.34(sec)/ 11.03(sec)
12	50	-10.0(kcal/mol)	5.79(sec)/ 4.54(sec)	120.61(sec)/116.17(sec)
15	20	-6.0(kcal/mol)	29.65(sec)/19.91(sec)	137.47(sec)/ 35.52(sec)
16	30	-8.0(kcal/mol)	34.31(sec)/ 9.40(sec)	255.44(sec)/ 87.65(sec)
20	40	-6.0(kcal/mol)	29.00(sec)/16.77(sec)	480.69(sec)/270.04(sec)

Give a length n, a size m, and τ. Randomly generate initial set which has m sequences, and apply our algorithm to improve the set until it satisfies $\Delta G_{ww}(S) \geq \tau$, $\Delta G_{wc}(S) \geq \tau$, and $\Delta G_{cc}(S) \geq \tau$. We measure the running-time to satisfy the constraint with $min\_related$ and $bond\_related$ or random order. We perform 50 trials for each condition.

3.2   Efficient Evaluation of MFEs

In neighborhood structures of the search, a good solution has a few good neighbor solutions and many worse neighbor solutions in general. This means that we need to check many neighbor solutions with worse solution values to find a neighbor solution with a better solution value. This means that the total evaluation-time is mainly occupied by evaluations of worse solutions; if we can quickly reject such bad solutions, we may greatly reduce the total evaluation-time. In this subsection, we explain how to screen bad solutions efficiently. We introduce a preprocessing phase that computes an approximate MFE values. A similar approach is also used in [18], but in a little different context[2].

We define the approximate MFE as the minimum Gibbs standard free energy under the restriction in which self reaction in one sequence is forbidden and the size of loop is bounded[3]. The approximate MFE is denoted by $\Delta G_{app}(s, s')$. This value itself can be computed in $O(ln^2)$ time by dynamic programming in theory, where l is the maximum-loop-size. Clearly $\Delta G_{app}(s, s') \geq \Delta G(s, s')$ holds.

We define $\Delta G_{app}(S)$ as an approximate of $\Delta G_{min}(S)$ (equation (1)) in which $\Delta G_{app}(s, s')$ is used instead of $\Delta G(s, s')$. To check if a neighbor solution S_{new} is an improvement from S_{old}, we perform the following operation using $\Delta G_{app}(S)$ in the preprocessing phase:

(1) Calculate $\Delta G_{app}(S_{new})$. If $\Delta G_{app}(S_{new}) \geq \Delta G_{min}(S_{old})$, we can determine that S_{new} is not an improvement of S_{old}. (End this routine.) Otherwise, go to (2).

(2) Calculate $\Delta G_{min}(S_{new})$. If $\Delta G_{min}(S_{new}) \geq \Delta G_{min}(S_{old})$, we determine that S_{new} is not an improvement of S_{old}. Otherwise, we determine that S_{new} is an improvement. End this routine.

The performance of this operation depends on the approximation quality of $\Delta G_{app}(s, s')$. If we set sufficiently large maximum-loop-size l, then the approximation quality is good enough but it takes large time, since the value can be computed in $O(ln^2)$ time. Therefore, to see the quality of $\Delta G_{app}(s, s')$ for a small l, we perform preliminary computational experiments. In the experiments, we randomly generate 10000 pairs of sequences, and calculate $\Delta G(s, s')$ and $\Delta G_{app}(s, s')$ for each pair. Table 2 shows the results.

As shown in this table, the calculating-times of $\Delta G_{app}(s, s')$ are much faster than $\Delta G(s, s')$, while the approximation ratios are good for short lengths. This might be because short sequences hardly take self reactions. In particular, the calculating-times of $\Delta G_{app}(s, s')$ are 1/4 to 1/5 compared with that of $\Delta G(s, s')$, while the ratio of $\Delta G_{app}(s, s') = \Delta G(s, s')$ are very high especially for small lengths. This result if preferable when we design sets of short sequences. Therefore, we adopt the screening phase by $\Delta G_{app}(s, s')$ in our search strategy.

---

[2] In [18], they introduce a notion of "degree" k, and an $O(kn^2)$ time greedy algorithm for the approximation is proposed.

[3] The function `pairfold_mfe_nointra` included `PairFold` package [2] can calculate this.

Table 2. Preliminary Experiments for $\Delta G_{app}(s, s')$

n	match ratio	$\Delta G_{app}(s, s') - \Delta G(s, s')$ average / standard deviation	time of $\Delta G(s, s')$	time of $\Delta G_{app}(s, s')$
8	99.91%	0.00043(kcal/mol) / 0.01646(kcal/mol)	1.69(sec)	0.45(sec)
12	97.33%	0.01472(kcal/mol) / 0.11472(kcal/mol)	6.50(sec)	1.25(sec)
15	91.70%	0.05370(kcal/mol) / 0.14155(kcal/mol)	13.61(sec)	2.81(sec)
16	88.52%	0.07915(kcal/mol) / 0.30228(kcal/mol)	14.06(sec)	3.24(sec)
20	68.50%	0.30268(kcal/mol) / 0.66313(kcal/mol)	30.7(sec)	6.30(sec)

We set the parameters $(n,l)=(8,2),(12,2),(15,4),(16,5),(20,6)$ in this experiment. Column "match ratio" shows the ratio of pairs satisfying $\Delta G_{app}(s, s') = \Delta G(s, s')$.

4 Computational Experiments

We implement the algorithm, and perform computational experiments. We use PairFold package [2] for calculation of MFEs. The setting temperature is 37°C. The cpu-times of experiments are between 2 hours and 24 hours.

We compare our results with Garzon et al. [9], also we compare with Deaton et al. [6][4]. Penchovsky et al. [13], Shortreed et al. [17], Braich et al. [5] and Faulhammer et al. [8].

Table 3. Results of Computational Experiments

No	n	$\Delta G_{ww}(S)$	$\Delta G_{wc}(S)$	$\Delta G_{cc}(S)$	size of ours	size of compared set
1	8	-3.3	-5.4	-3.9	**237**	(Garzon)132
2	8	-5.3	-6.5	-5.5	**233**	(Garzon)173
3	12	-3.5	-9.3	-4.5	**152**	(Shortreed)64
4	12	-5.9	-9.9	-5.9	321	(Garzon)**617**
5	12	-9.2	-11.2	-10.0	689	(Garzon)**1424**
6	15	-4.3	-8.3	-4.3	**80**	(Braich)40
7	15	-3.7	-10.4	-4.5	**85**	(Faulhammer)20
8	15	-6.0	-14.9	-7.3	**92**	(Garzon)42
9	15	-12.3	-15.3	-12.3	**224**	(Garzon)96
10	16	-7.5	-8.1	-8.5	**141**	(Shortreed)64
11	16	-1.5	-8.7	-3.9	**53**	(Penchovsky)24
12	20	-7.7	-7.2	-10.2	**88**	(Deaton)40

Table 3 shows the result. Column "size of ours" (resp., "size of compared set") shows size of sets obtained by our method (resp., the size of sets reported in [5,6,8,9,13,17]). Columns $\Delta G_{ww}(S)$, $\Delta G_{wc}(S)$, and $\Delta G_{cc}(S)$ represent parameters used in the experiment, which are obtained by the compared method[5]. For example, in No.1, we design S such as $\Delta G_{ww}(S) = -3.3$, $\Delta G_{wc}(S) = -5.4$,

---

[4] Deaton's set includes nucleotide "h", we treat "h" as "g".

[5] Except Garzon's set, values are calculated by us with published PairFold package.

$\Delta G_{cc}(S) = -3.9$, which are based on the result of Garzon's set with size 132. Sequence sets with greater sizes are better if the values of $\Delta G_x(S)$ are same.

For the cases of No.1–3, and 6–12, our sets have the same $\Delta G_x(S)$ values as the sets generated by the existing methods, however sizes of these are greater. That is to say, in spite that the sets generated by our method are larger, our sets cause "miss-hybridization" as well as compared sets. This implies that our method can design good sets and efficient for thermodynamical constraints.

Only for the cases of No.4 and 5, although these have same $\Delta G_x(S)$ values, the sizes of sets generated by our method are smaller than these generated by Garzon et al. Thus, we consider that our method is suitable for designing short sequence sets which are relatively small. However our method does not lose its worth even for the longer sequences, because we can treat the set generated by Garzon et al. as the initial set and may improve it.

5   Conclusion

In this paper, we present a local search type algorithm for short-sequence design under the thermodynamical constraints. Since local search type algorithms are not practical under the thermodynamical constraints due to time-consuming operations, we propose two thechniques for efficient computations. One of the techniques is the effective order in neighborhood search, and the other is a bounding technique to skip the search for apparently bad solution, by the preprocessing phase with approximate MFE. In the computational experiments, we succeeded in designing better sequence sets than the existing methods in the case where the sizes of sets are relatively small. Also for larger sets, our method is easy to be combined with non-local search methods such as [9].

As future work, further reduction of computational time is considered. For example, recalculation of MFE values for a new solution is time-consuming, but it can be reduced because the difference between the new and the previous solutions is very small; many internal calculations for MFE values can be reused. Applying our method to more complicated constraints, such as hairpin state machine [10] which have properties of "sequence set design" and "reverse folding problem", is another interesting issue.

References

1. Adleman, L.: Molecular Computation of Solutions to Combinatorial Problems. Science 266(5187), 1021–1024 (1994)
2. Andronescu, M., Zhang, Z., Condon, A.: Secondary Structure Prediction of Interacting RNA Molecules. J. of Molecular Biology 345(5), 987–1001 (2005), www.rnasoft.ca/download.html
3. Arita, M., Kobayashi, S.: DNA Sequence Design Using Templates. New Generation Computing 20(3), 263–273 (2002)
4. Asahiro, Y.: Simple Greedy Methods for DNA Word Design. Proc. of 9th World Multi-Conference on Systemics, Cybernetics and Informatics 3, 186–191 (2005)

5. Braich, R., Chelyapov, N., Johnson, C., Rothemund, P., Adleman, L.: Solution of a 20-Variable 3-SAT Problem on a DNA Computer. Science 296(5567), 499–502 (2002)
6. Deaton, R., Kim, J., Chen, J.: Design and test of noncrosshybridizing oligonu-cleotide building blocks for DNA computers and nanostructures. Applied Physics Letters 82(8), 1305–1307 (2003)
7. Dirks, R., Lin, M., Winfree, E., Pierce, N.: Paradigms for computational nucleic acid design. Nucleic Acids Research 32(4), 1392–1403 (2004)
8. Faulhammer, D., Cukras, A., Lipton, R., Landweber, L.: Molecular computation: RNA solutions to chess problems. Proc. of the National Academy of Sciences of the United States of America 97(4), 1385–1389 (2000)
9. Garzon, M., Phan, V., Roy, S., Neel, A.: In Search of Optimal Codes for DNA Computing. In: Mao, C., Yokomori, T. (eds.) DNA Computing. LNCS, vol. 4287, pp. 143–156. Springer, Heidelberg (2006)
10. Kameda, A., Yamamoto, M., Ohuchi, A., Yaegashi, S., Hagiya, M.: Unravel Four Hairpins! In: Mao, C., Yokomori, T. (eds.) DNA Computing. LNCS, vol. 4287, pp. 381–392. Springer, Heidelberg (2006)
11. Kawshimo, S., Ono, H., Sadakane, K., Yamashita, M.: DNA Sequence Design by Dynamic Neighborhood Searches. In: Mao, C., Yokomori, T. (eds.) DNA Comput-ing. LNCS, vol. 4287, pp. 157–171. Springer, Heidelberg (2006)
12. Lyngsø, R., Zuker, M., Pedersen, C.: Fast evaluation of internal loops in RNA secondary structure prediction. Bioinfomatics 15, 440–445 (1999)
13. Penchovsky, R., Ackermann, J.: DNA Library Design for Molecular Computation. J. of Computational Biology 10(2), 215–229 (2003)
14. Reif, J., Sahu, S., Yin, P.: Complexity of Graph Self-assembly in Accretive Systems and Self-destructible Systems. In: Carbone, A., Pierce, N.A. (eds.) DNA Comput-ing. LNCS, vol. 3892, pp. 257–274. Springer, Heidelberg (2006)
15. Rose, J., Deaton, R., Suyama, A.: Statistical thermodynamic analysis and design of DNA-based computers. Natural Computing 3, 443–459 (2004)
16. Shiozaki, M., Ono, H., Sadakane, K., Yamashita, M.: A Probabilistic Model of DNA Conformational Change. In: Mao, C., Yokomori, T. (eds.) DNA Computing. LNCS, vol. 4287, pp. 274–285. Springer, Heidelberg (2006)
17. Shorteed, M., Chang, S., Hong, D., Phillips, M., Campion, B., Tulpan, D., An-dronescu, M., Condon, A., Hoos, H., Smith, L.: A thermodynamic approach to designing struct-free combinatorial DNA word set. Nucleic Acids Research 33(15), 4965–4977 (2005)
18. Tanaka, F., Kameda, A., Yamamoto, M., Ohuchi, A.: Design of nucleic acid se-quences for DNA computing based on a thermodynamic approach. Nucleic Acids Research 33(3), 903–911 (2005)
19. Tulpan, D., Hoos, H., Condon, A.: Stochastic Local Search Algorithms for DNA Word Design. In: Hagiya, M., Ohuchi, A. (eds.) DNA Computing. LNCS, vol. 2568, pp. 229–241. Springer, Heidelberg (2003)
20. Tulpan, D., Hoos, H.: Hybrid Randomized Neighborhoods Improve Stochastic Local Search for DNA Code Design. In: Xiang, Y., Chaib-draa, B. (eds.) Advances in Arti-ficial Intelligence. LNCS (LNAI), vol. 2671, pp. 418–433. Springer, Heidelberg (2003)
21. Tulpan, D., Andronescu, M., Changf, S., Shortreed, M., Condon, A., Hoos, H., Smith, L.: Thermodynamically based DNA strand design. Nucleic Acids Re-search 33(15), 4951–4964 (2005)
22. Winfree, E., Liu, F., Wenzler, L., Seeman, N.: Design and self-assembly of DNA crystals. Nature 394, 539–544 (1998)
23. Zuker, M., Stiegler, P.: Optimal computer folding of large RNA sequences using ther-modynamics and auxiliary information. Nucleic Acids Research 9, 133–148 (1981)

Sequence Design Support System
for 4 × 4 DNA Tiles

Naoki Iimura[1], Masahito Yamamoto[2], Fumiaki Tanaka[3], and Azuma Ohuchi[2]

[1] NTT DoCoMo Hokkaido, Inc.
`iimura@complex.eng.hokudai.ac.jp`
[2] Graduate School of Information Science and Technology, Hokkaido University
`{masahito,ohuchi}@complex.eng.hokudai.ac.jp`
[3] Graduate School of Information Science and Technology, Univercity of Tokyo
`fumi95@is.s.u-tokyo.ac.jp`

Abstract. A DNA computation model by DNA tiles needs sequence design in order to correctly form tile structure and self-assembly. We design sequence, demonstrate biochemical experiments by a trial and error approach, and, repeatedly analyze tiles. Because no integrated sequence design system computes data that indicates properties of sequences, we must analyze designed sequences by hand and many types of software. In this paper, we develop a sequence design support system for 4× 4 DNA tiles that analyzes and optimizes tile sequences to support sequence design. The most remarkable feature of this system is optimization based on free energy. The optimization strategy is developed so that the energy of perfect tile is the stablest.

1 Introduction

New computation models and DNA nanotechnology by DNA tiles based on Watson-Crick complementarity pairing [1, 2] have been proposed. We have to design a stable tile sequence that minimizes mis-hybridization because these computation models by DNA tiles presupposes correct hybridization. However, sequence design is never easy due to the repetition of the trial and error approach. In DNA computing, sequence design that forms a wanted structure or does not form an unwanted structure has been researched in terms of various indexes, including melting temperature, GC content, free energy, and so on. Conventional work of DNA tile sequence design minimizes the reuse of the fragment of sequence without these indexes [3, 4]. This method allows sequence design that does not cause unnecessary hybridization regardless of the simple algorithm. However, it has problems. It does not consider the distinction of stability by base pair or loop structure and does not quantify the evaluation of tiles. Thus, if we design and analyze sequences by other evaluation indexes to solve these problems, we need to use many types of software[5–8].

In this paper, we develop a sequence design support system for a 4× 4 DNA tile. The 4×4DNA tile developed by Yan *et al*, consists of nine sequences: one CORE, four SHELL, and four ARM sequences (Fig. 1). This tile has four-way

M.H. Garzon and H. Yan (Eds.): DNA 13, LNCS 4848, pp. 140–145, 2008.

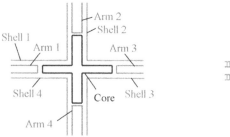

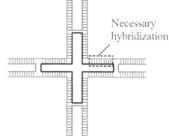

Fig. 1. Definition of sequence name

Fig. 2. Necessary hybridizations for 4×4 DNA tile

arms (north, south, east, and west) made of a single strand DNA molecule called "sticky-end" and forms a DNA self-assembly by connecting with other tiles. Applications of the 4×4 DNA tile have been proposed and demonstrated by several research groups[9, 10].Our system has two features for design support. First, it can analyze existing sequence data that are often used when designing sequences. Second, it can design them using optimization based on the stability evaluation of DNA tiles by free energy. Free energy is introduced because it can reduce mis-hybridization by making necessary hybridization stable and unnecessary hybridization unstable. Our proposed system are designed so that the evaluation function of tile structures can be easily replaced.

2   Design Strategy

The stability of tile structure allows sequence design that minimizes mis-hybridization. From the standpoint of tile stability, correct tile structure without mis-hybridization is considered the most stable; that is, the free energy of the correct tile must be the lowest. We apply free energy, which is an index that can evaluate the stability of loop structure and base pairs in DNA computing, to the stability of a 4×4 DNA tile. There are two kinds of free energy: of the secondary structure within a single strand DNA molecule, and of the hybridization between two single strand DNA molecules, however, there is no effective prediction method of free energy of a DNA tile. Here we are trying to quantify the stability of tile structure by these free energies.

We suppose that the whole tile structure becomes more stable as each necessary hybridization portion stabilizes, therefore, the summation of necessary hybridization calculated by the free energy between two sequences can indicate tile stability. Necessary hybridization is a base pair to form tile structure, as shown in Fig. 2. Because these necessary hybridization portions may incorrectly form loop structure, bulge loop, and so on, evaluation of the bond strength of the base pair is used to judge correct hybridization. Furthermore, it is desirable that the stablest structure is only the one structure of any and all structure with the potential to be formed by tile sequences. The reason for this is that sequences decrease the possibility of forming correct tiles if the structure that does not form tile is as stable as the tile structure.

Desirable sequences have lower energy if the tile is formed correctly and higher energy if the tile is not formed. In other words, desirable sequences have a big difference between the lowest and second lowest energy. A device is needed to design these sequences as well as to stabilize necessary hybridization. Our method incorporates inhibitory factors to avoid forming the secondary structure of a single strand DNA molecule and unnecessary hybridization. The inhibitory factor calculated by the free energy within the sequence and between two sequences reduces tile stability. Sequence forming correct tile are designed by optimization based on tile stability by free energy.

3   Support System

We developed a sequence design system with the previous strategy based on free energy. The system also has an analysis function of existing tiles besides sequence design by optimization because sequence design comprehensively uses not only free energy but also various indexes. These functions should reduce computational costs.

3.1   Analysis Module

We often comprehensively judge tile sequences by amount of data. Before analyzing a tile, the system requests sequence length and SHELL sequences from users, who input sequence data following the input forms on the screen. Figs. 3 and 4 are examples of the screen. Fig. 3 shows the input form of the sequence length of a CORE fragment that is hybridized to the SHELL sequence. This system deals with any tile size and any bulge loop in the corner of CORE. Fig. 4 shows the input form to enter the SHELL sequences. We adopt nucleic code, which has A, G, T, C, S(G or C), H(A,T or C), N(A,T,C or C), and so on by IUPAC, because input forms need to accept existing sequences and constraints for allocating bases. The input of sequence by nucleic code can directly, enter existing sequences, and randomly allocate bases if GC pairs or AT pairs are fixed. Additionally, a user can avoid including specific sequence fragments in a sequence when the system randomly allocates bases.

Fig. 5 shows a screen of the analysis result. The system analyzes the following data after a user inputs the essential data:

(1) GC content of each sequence
(2) Melting temperature of each sequence
    The value indicates the melting temperature between each sequence and its complementary sequence.
(3) Free energy of each sequence
    This value indicates the stability of the secondary structure within a single strand DNA molecule and that is calculated with MFOLD [5, 6].

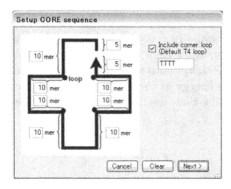

Fig. 3. Input form of CORE sequence data

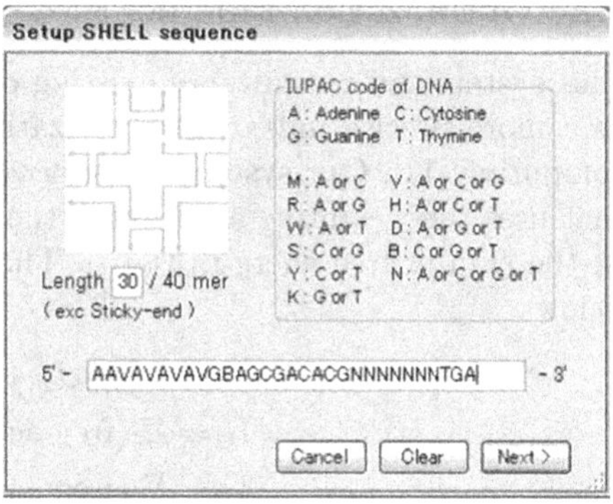

Fig. 4. Input form of SHELL sequences

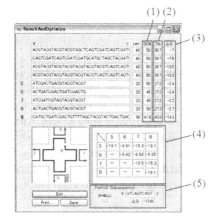

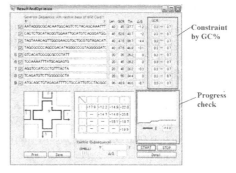

Fig. 5. Sytem interface

Fig. 6. Optimization interface

(4) Free energy between sequences that do not require hybridization
    This value indicates the stability of hybridization between sequences that should not be hybridized to form tiles. The system calculates the free energy between CORE–CORE sequences, between SHELL[i]–SHELL[j] sequences ($i \neq j$), and between ARM[i]–ARM[j] sequences ($i \neq j$) with PAIRFOLD [7] (Fig. 1). The combination of two sequences in all sequences is not calculated in terms of computation cost.

(5) Free energy between sequences that require hybridization to form tiles
    This value indicates the stability of hybridization between sequences that should be hybridized and stabilized to form tiles. A 4×4 DNA tile has 16 necessary hybridization parts (Fig. 2). The system uses a fragment of sequence along with a motif of the tile.

The values of (4) and (5) are displayed by click on the motif.

3.2   Optimization Module

This system can optimize an existing or a brand-new tile by free energy known as a more precise index of hybridization stability than other indexes in DNA computing [11]. Our system introduces free energy into the evaluation function that uses the weighting addition of the free energy in the previous section; that is, the sum of free energy (3)–(5). The evaluation function for any tile x is as follows.

$$E(x) = I_1 + \alpha I_2 + \beta I_3$$
$$I_1 = \Sigma \text{ (free energy values of (3))}$$
$$I_2 = \Sigma \text{ (free energy values of (4))}$$
$$I_3 = -\Sigma \text{ (free energy values of (5))}$$

Terms $I_1, I_2, and I_3$ indicate the bond strength of the secondary structure within a single strand, the bond strength of mis-hybridization between non-objective sequences, and the bond strength of hybridization between fragments of objective sequences, respectively. A sequence qualifies as a stable tile as each value increases.

The optimization algorithm adopts a hill-climbing algorithm. The system retains the nucleic code and the avoidance fragment in principle while optimizing sequences; furthermore, it can set constraints of GC content in each sequence. Optimization steps are initially 1,000, which a user can increase to 2,000. Running time is about nine minutes on a PC with 3.0 GHz Pentium4 processor and 512 MB RAM running Windows XP, if optimization steps are 2,000. Fig. 6 shows the interface of optimization. We can confirm the optimization progress and stop it if required. The system displays the sequence and analysis results at that time.

3.3   I/O Module

The system has the following convenient additional functions. I/O module inputs and outputs sequence data. The system can save and read sequences, their length, and the result of analysis or optimization by an XML document. This function not only save data but also alleviates input. The preparation of XML documents as templates of tile size facilitates various optimizations. Furthermore, the system can print these data.

4   Discussions and Concluding Remarks

We have designed sequences by minimizing the reuse of fragment of sequence and have analyzed their melting temperature, free energy, and so on by many types of software. Our system can analyze these data of existing sequences, optimize sequences by free energy, and analyze and design sequences from scratch. Optimization results have verified that optimized sequences can form tile correctly in terms of free energy. However, the actual verification of sequence needs biochemical experiments *in vitro*D Our system provides sequences that users want to design by inputting by nucleic code, setting specific fragments to avoid, setting the constraints of GC content, and changing the parameters of the evaluation function. We consider that this reflects knowledge gained by biochemical experiment into the system. Additionally, the system can design sequences that are not designed by the conventional algorithms.

We suggest that sequences by this system form a tile and self-assembly more precisely. However, we may need to consider the concentration of each sequences, curvature of tiles, other optimization methods and constraints in the future.

References

1. Winfree, E., Liu, F., Wenzler, L.A., Seeman, N.C.: Design and self-assembly of two-dimensional DNA crystals. Nature 394, 539–544 (1998)
2. Yan, H., Feng, L., LaBean, T.H., Reif, J.H.: Parallel Molecular Computations of Pairwise Exclusive-Or (XOR) Using DNA "String Tile" Self-Assembly. J. Am. Chem. Soc. 125(47), 14246–14247 (2003)
3. Seeman, N.C.: De Nove Design of Sequence for Nucleic Acid Structual Engineering. Jornal of Biomolecular Structure & Dynamics 8, 739–1102 (1990)
4. Wei, B., Wang, Z., Mi, Y.: Uniquimer: Software of De Novo DNA Sequence Generation for DNA Self-Assembly -An Introduction and the Related Applications in DNA Self-Assembly. Journal of Computational and Theoretical Nanoscience 4(1), 133–141 (2007)
5. Zuker, A.M., Mathews, B.D.H., Turner, C.D.H.: Algorithms and Thermodynamics for RNA Secondary Structure Prediction: A Practical Guide. In: Barciszewski, J., Clark, B.F.C. (eds.) RNA Biochemistryand Biotechnology. NATO ASI Series, Kluwer Academic Publisers, Dordrecht (1999)
6. Zuker, M.: Mfold web server for nulceic acid folding and hybridization prediction. Nucleic Acids Reserch 31(13), 3406–3415 (2003)
7. Andronescu, M., Aguirre-Hernandez, R., Condon, A., Hoos, H.H.: RNA soft: A suite of RNA secondary structure prediction and design software tools. Nucleic Acids Research 31(13), 3416–3422 (2003)
8. Hofacker, I.L.: Vienna RNA secondary structure server. Nucleic Acids Research 31(13), 3429–3431 (2003)
9. Yan, H., Park, S.H., Finkelstein, G., Reif, J.H., LaBean, T.H.: DNA-Templated Self-Assembly of Protein Arrays and Highly Conductive Nanowires. Science 301, 1882–1884 (2003)
10. Park, S.H., Yan, H., Reif, J.H., LaBean, T.H., Finkelstein, G.: Electronic nanostructures templated on self-assembled DNA scaffolds. Nanotechnology 15, S525–S527 (2004)
11. Tanaka, F., Kameda, A., Yamamoto, M., Ohuchi, A.: Design of nucleic acid sequences for DNA computing based on a thermodynamic approach. Nucleic Acids Research 33(3), 903–911 (2005)

DNA Codes Based on Stem Similarities Between DNA Sequences

Arkadii D'yachkov[1], Anthony Macula[2], Vyacheslav Rykov[3],
and Vladimir Ufimtsev[3]

[1] Moscow State University, Moscow 119992, Russia
agd-msu@yandex.ru
[2] Air Force Res. Lab., IFTC, Rome Research Site, Rome NY 13441, USA
macula@geneseo.edu
[3] University of Nebraska at Omaha, 6001 Dodge St., Omaha, NE 68182-0243 USA
vrykov@mail.unomaha.edu

Abstract. DNA codes consisting of DNA sequences are necessary for DNA computing. The minimum distance parameter of such codes is a measure of how dissimilar the codewords are, and thus is indirectly a measure of the likelihood of undetectedable or uncorrectable errors occurring during hybridization. To compute distance, an abstract metric, for example, longest common subsequence, must be used to model the actual bonding energies of DNA strands. In this paper we continue the development [1,2,3] of similarity functions for q-ary n-sequences The theoretical lower bound on the maximal possible size of codes, built on the space endowed with this metric, is obtained. that can be used (for $q = 4$) to model a thermodynamic similarity on DNA sequences. We introduce the concept of a stem similarity function and discuss DNA codes [2] based on the stem similarity. We suggest an optimal construction [2] and obtain random coding bounds on the maximum size and rate for such codes.

1 Introduction

In order to accomplish DNA computing, it is necessary to have DNA libraries, also known as DNA codes, of large size and small energies of hybridization between the DNA sequences. The ultimate criterion for the value of a metric for DNA codes is the degree to which it approximates actual bonding energies, which in turn determines the degree to which distance approximates the likelihood of one codeword mistakenly binding to the reverse complement of another codeword. We can use a branch of mathematics known as coding theory, that was initiated around the same time that the structure of DNA was discovered, to study the space of DNA sequences endowed with a measure of distance (metric). The introduced measure of distance between DNA sequences has an immediate application in determining the similarities between genes, expressed as DNA sequences, in any existing genome. Codes built on spaces of DNA sequences can be implemented in Biomolecular Computing and could have other important applications.

M.H. Garzon and H. Yan (Eds.): DNA 13, LNCS 4848, pp. 146–151, 2008.

2   Notations, Definitions

The symbol $\triangleq$ denotes definitional equalities and the symbol $[n] \triangleq \{1, 2, \ldots, n\}$ denotes the set of integers from 1 to n. Let $q = 2, 4, \ldots$ be an arbitrary even integer, $A \triangleq \{0, 1, \ldots, q-1\}$ be the standard q-nary alphabet. Consider two arbitrary q-nary n-sequences $\mathbf{x} = (x_1, x_2, \ldots, x_n) \in A^n$ and $\mathbf{y} = (y_1, y_2, \ldots, y_n) \in A^n$. By symbol $\mathbf{z} = (z_1, z_2, \ldots, z_\ell) \in A^\ell$, $\ell \in [n]$, we will denote a *common subsequence* [5] of length $|\mathbf{z}| \triangleq \ell$ between $\mathbf{x}$ and $\mathbf{y}$. The *empty* subsequence $\mathbf{z}$ of length $|\mathbf{z}| \triangleq 0$ is a common subsequence between any sequences $\mathbf{x}$ and $\mathbf{y}$.

Definition 1. *Let $1 \leq b \leq r \leq n$ be arbitrary integers. A fixed r-sequence $\mathbf{a} = (a_1, a_2, \ldots, a_r)$, $a_i \in A = \{0, 1, \ldots, q-1\}$, $i \in [r]$, is called a common block for sequences $\mathbf{x}$ and $\mathbf{y}$ (briefly, common $(\mathbf{x}, \mathbf{y})$-block) of length r if sequences $\mathbf{x}$ and $\mathbf{y}$ (simultaneously) contain $\mathbf{a}$ as a subsequence consisting of r consecutive elements of $\mathbf{x}$ and $\mathbf{y}$. We will say that a common $(\mathbf{x}, \mathbf{y})$-block $\mathbf{a}$ yields $r - (b-1)$ common b-stems $a_i, a_{i+1}, \ldots, a_{i+(b-1)}$, $i \in [r - (b-1)]$, containing b adjacent symbols of the given common $(\mathbf{x}, \mathbf{y})$-block.*

Definition 2. *Let $1 \leq t \leq \ell \leq n$ be integers. A sequence $\mathbf{z} = (z_1, z_2, \ldots, z_\ell)$, $z_i \in A$, $i \in [\ell]$, is called a common t-block subsequence of length $|\mathbf{z}| \triangleq \ell$ between $\mathbf{x}$ and $\mathbf{y}$ if $\mathbf{z}$ is an ordered collection of non-overlapping (separated) common $(\mathbf{x}, \mathbf{y})$-blocks and the length of each common $(\mathbf{x}, \mathbf{y})$-block in this collection is $\geq t$.*

Let $\mathcal{Z}_t(\mathbf{x}, \mathbf{y})$ be the set of all common t-block subsequences between $\mathbf{x}$ and $\mathbf{y}$. For any $\mathbf{z} \in \mathcal{Z}_t(\mathbf{x}, \mathbf{y})$, we denote by $k(\mathbf{z}, \mathbf{x}, \mathbf{y})$, $1 \leq k(\mathbf{z}, \mathbf{x}, \mathbf{y}) \leq |\mathbf{z}|/t$, the *minimal number* of common $(\mathbf{x}, \mathbf{y})$–blocks which *constitute* the given subsequence $\mathbf{z}$.

Note that for any integer b, $2 \leq b \leq t$, the difference $|\mathbf{z}| - (b-1)\, k(\mathbf{z}, \mathbf{x}, \mathbf{y})$, $\mathbf{z} \in \mathcal{Z}_t(\mathbf{x}, \mathbf{y})$, is a total number of common b-stems containing adjacent symbols in common $(\mathbf{x}, \mathbf{y})$-blocks constituting $\mathbf{z} \in \mathcal{Z}_t(\mathbf{x}, \mathbf{y})$.

Definition 3. *For any fixed integer b, $2 \leq b \leq n$, we define*

$$S_b(\mathbf{x}, \mathbf{y}) \triangleq \max_{b \leq t \leq n} \max_{\mathbf{z} \in \mathcal{Z}_t(\mathbf{x}, \mathbf{y})} \{|\mathbf{z}| - (b-1)\, k(\mathbf{z}, \mathbf{x}, \mathbf{y})\}, \qquad S_b(\mathbf{x}, \mathbf{y}) \geq 0.$$

If $\mathcal{Z}_b(\mathbf{x}, \mathbf{y}) = \varnothing$, then we will say that $S_b(\mathbf{x}, \mathbf{y}) \triangleq 0$. The number

$$S_b(\mathbf{x}, \mathbf{y}) = S_b(\mathbf{y}, \mathbf{x}) \leq S_b(\mathbf{x}, \mathbf{x}) = n - (b-1), \quad \mathbf{x} \in A^n, \quad \mathbf{y} \in A^n,$$

is called an b-stem similarity between $\mathbf{x}$ and $\mathbf{y}$. For $b = 2$, the concept of 2-stem similarity and its biological motivation were suggested in [1].

Definition 4. *[1,2]. If $q = 2, 4, \ldots$, then $\bar{x} \triangleq (q-1) - x$, $x \in A = \{0, 1, \ldots, q-1\}$, is called a complement of a letter x. For $\mathbf{x} = (x_1, x_2, \ldots, x_{n-1}, x_n) \in A^n$, we define its reverse complement $\widetilde{\overline{\mathbf{x}}} \triangleq (\bar{x}_n, \bar{x}_{n-1}, \ldots, \bar{x}_2, \bar{x}_1) \in A^n$. If $\mathbf{y} \triangleq \widetilde{\overline{\mathbf{x}}}$, then $\mathbf{x} = \widetilde{\overline{\mathbf{y}}}$ for any $\mathbf{x} \in A^n$. If $\mathbf{x} = \widetilde{\overline{\mathbf{x}}}$, then $\mathbf{x}$ is called a self reverse complementary sequence. If $\mathbf{x} \neq \widetilde{\overline{\mathbf{x}}}$, then a pair $(\mathbf{x}, \widetilde{\overline{\mathbf{x}}})$ is called a pair of mutually reverse complementary sequences.*

Let $\mathbf{x}(1), \mathbf{x}(2), \ldots, \mathbf{x}(N)$, where $\mathbf{x}(j) \triangleq (x_1(j), x_2(j), \ldots, x_n(j)) \in A^n$, $j \in [N]$, be *codewords* of a *q-ary code* $X = \{\mathbf{x}(1), \mathbf{x}(2), \ldots, \mathbf{x}(N)\}$ of *length* n and *size* N, where $N = 2, 4, \ldots$ be an *even* number. Let b, $2 \le b \le n$, and D, $b \le D \le n - 1$, be arbitrary integers.

Definition 5. *A code X is called a DNA (n, D)-code based on b-stem similarity $S_b(\mathbf{x}, \mathbf{y})$ (briefly, (n, D)-code) if the following two conditions are fulfilled.*

(i). For any number $j \in [N]$ there exists $j' \in [N]$, $j' \ne k$, such that $\mathbf{x}(j') = \overline{\mathbf{x}(j)} \ne \mathbf{x}(j)$. In other words, X is a collection of $N/2$ pairs of mutually reverse complementary sequences.

(ii). For any $j, j' \in [N]$, where $j \ne j'$, the similarity

$$S_b(\mathbf{x}(j), \mathbf{x}(j')) \le n - D - 1, \qquad b \le D \le n - 1. \tag{1}$$

Definition 6. *Let $N_b(n, D)$ be the maximum size for DNA (n, D)-codes based on b-stem similarity. If d, $0 < d < 1$, is a fixed number, then*

$$R_b(d) \triangleq \varlimsup_{n \to \infty} \frac{\log_q N_b(n, \lfloor nd \rfloor)}{n} \tag{2}$$

is called a rate of DNA $(n, \lfloor nd \rfloor)$-codes based on b-stem similarity.

3   Random Coding Bounds

Let b, $2 \le b \le n$, and s, $0 \le s \le n - (b - 1)$, be arbitrary integers and

$$\mathcal{P}_b(n, s) \triangleq \{(\mathbf{x}, \mathbf{y}) \in A^n \times A^n : S_b(\mathbf{x}, \mathbf{y}) = s\},$$

$$\overline{\mathcal{P}}_b(n, s) \triangleq \{\mathbf{x} \in A^n : S_b(\mathbf{x}, \tilde{\bar{\mathbf{x}}}) = s\},$$

be sets of pairs $(\mathbf{x}, \mathbf{y}) \in A^n \times A^n$ (sequences $\mathbf{x} \in A^n$) for which the given similarities be equal to s. Applying combinatorial arguments which are similar to the corresponding arguments of paper [2] for the block similarity function, one can check that the following statement is true.

Lemma 1. *The size*

$$|\mathcal{P}_b(n, s)| \le q^{2n-s} \cdot \sum_{k=1}^{\min\{s, (n-s)/(b-1)\}} q^{-(b-1)k} \binom{s-1}{k-1} \binom{n - s - (b-2)k}{k}^2. \tag{3}$$

The set $\overline{\mathcal{P}}_b(n, s)$ is empty if $s \ge 3$ is odd. If $s \ge 2$ is even, then the size

$$|\overline{\mathcal{P}}_b(n, s)| \le q^{n-s/2} \cdot \sum_{k=1}^{\min\{s, (n-s)/(b-1)\}} q^{-(b-1)k/2} \binom{s/2-1}{k/2-1} \binom{n - s - (b-2)k}{k}. \tag{4}$$

Lemma 1 and the standard random coding method [2] lead to Theorems 1 and 3 which give lower bounds on the size $N_b(n, D)$ and rate $R_b(d)$ of DNA codes based on b-stem similarity.

Theorem 1. *If $D \geq b \geq 2$ are fixed integers and $n \to \infty$, then*

$$N_b(n, D) \geq \frac{1}{4} \cdot \frac{(D_b - 1)! \cdot q^{(b-1)D_b}}{\left(\frac{D-(b-2)D_b}{D_b}\right)^2 \cdot q^D} \cdot \frac{q^n}{n^{D_b-1}} \cdot (1+o(1)), \quad D_b \triangleq \left\lfloor \frac{D}{b-1} \right\rfloor. \quad (5)$$

For the case $b = 2$, number $D_2 = D \geq 2$ and bound (5) has the form

$$N_2(n, D) \geq \frac{(D-1)!}{4} \cdot \frac{q^n}{n^{D-1}} \cdot (1 + o(1)), \qquad D \geq 2. \quad (6)$$

For the case $D = b \geq 3$, number $D_b = 1$ and bound (5) has the form

$$N_b(n, b) \geq \frac{q^{n-1}}{16} \cdot (1 + o(1)), \qquad b \geq 3. \quad (7)$$

An improvement of asymptotic lower bounds (6)-(7) follows from formula (8) for $N_b(n, b)$ presented in the theorem.

Theorem 2. *[2] If $n = qm$, $m = 1, 3, 5, \ldots$, then*

$$N_b(n, b) = \frac{q^{n-1} + q}{2}, \qquad 2 \leq b \leq n - 1. \quad (8)$$

Introduce the standard symbol

$$h_q(u) \triangleq -u \log_q u - (1 - u) \log_q(1 - u), \qquad 0 < u < 1, \quad (9)$$

for the binary entropy function.

Theorem 3. (i).   *The rate*

$$R_b(d) \geq \underline{R}_b(d) \triangleq \min_{0 \leq u \leq d} \{(1 - u) - E_b(u)\}, \quad (10)$$

where

$$E_b(u) \triangleq \max_{0 \leq v \leq \min\{\frac{u}{b-1}, 1-u\}} F_b(v, u), \quad (11)$$

$$F_b(v, u) \triangleq -(b - 1)v + (1 - u) h_q \left(\frac{v}{1-u}\right) +$$

$$+ 2\left[u - (b - 2)v\right] h_q \left(\frac{v}{u - (b - 2)v}\right). \quad (12)$$

(ii).   *Let d_b, $0 < d_b < 1$, be the unique root of equation $1 - d = E_b(d)$. If $0 < d < d_b$, then the rate $R_b(d) > 0$ and the following lower bound*

$$R_b(d) \geq \underline{R}_b(d) \triangleq (1 - d) - E_b(d), \qquad 0 < d < d_b, \quad (13)$$

holds.

We will say that the number d_b, $0 < d_b < 1$, is a *critical distance fraction* for the random coding bound $\underline{R}_b(d)$.

Maximization (11)-(12). The derivative of binary entropy function (9) is

$$h_q'(v) = \log_q \frac{1 - v}{v}, \qquad 0 < v < 1.$$

Thus, the partial derivative of function $F_b(v, u)$ is

$$\frac{\partial F_b(v, u)}{\partial v} = -(b - 1) + \log_q \frac{(1 - u) - v}{v} +$$

$$+2\left[-(b - 2) h_q \left(\frac{v}{u - (b - 2)v}\right) + \frac{u}{u - (b - 2)v} \log_q \frac{u - (b - 1)v}{v}\right]. \qquad (14)$$

Taking into account that $h_q \left(\frac{v}{u - (b-2)v}\right) =$

$$= \frac{v}{u - (b - 2)v} \log_q \frac{u - (b - 2)v}{v} + \frac{u - (b - 1)v}{u - (b - 2)v} \log_q \frac{u - (b - 2)v}{u - (b - 1)v},$$

one can easily check that (14) can be rewritten in the form

$$\frac{\partial F_b(v, u)}{\partial v} = -(b - 1) + 3 \log_q \frac{1}{v} + \log_q[(1 - u) - v] +$$

$$+2(b - 1) \log_q[u - (b - 1)v] - 2(b - 2) \log_q[u - (b - 2)v].$$

Therefore, for any fixed u, $0 < u < 1$, equation $\frac{\partial F_b(v,u)}{\partial v} = 0$ is equivalent to equation

$$\left(\frac{1 - u}{v} - 1\right) \left[\frac{u}{v} - (b - 1)\right]^{2(b-1)} \left[\frac{u}{v} - (b - 2)\right]^{-2(b-2)} = q^{b-1}, \quad \frac{u}{v} \geq b - 1. \tag{15}$$

Let $v = v(u)$ be the unique root of (15). This means that function

$$E_b(u) = F_b(v(u), u) = -(b - 1)v(u) + (1 - u) h_q \left(\frac{v(u)}{1 - u}\right) +$$

$$+ 2 [u - (b - 2)v(u)] h_q \left(\frac{v(u)}{u - (b - 2)v(u)}\right).$$

If we substitute parameter v for $w \triangleq u/v > b - 1$, then equation (15) has the form

$$\left(\frac{1 - u}{u} w - 1\right) [w - (b - 1)]^{2(b-1)} [w - (b - 2)]^{-2(b-2)} = q^{b-1}, \qquad w > b - 1.$$

Hence, the root $v = v(u)$ can be calculated using the following recurrent method:

$$w_1 \triangleq b, \quad w_{m+1} = (b-1) + \sqrt{q} \left\{ \frac{[w_m - (b-2)]^{2(b-2)}}{\frac{1-u}{u} w_m - 1} \right\}^{\frac{1}{2(b-1)}}, \quad m = 1, 2, \ldots,$$

$$v = v(u) = \frac{u}{\lim_{m \to \infty} w_m}. \tag{16}$$

If $q = 4$, then numerical values of critical distance fractions d_b, $b = 2, 3, \ldots 9$, along with the corresponding optimal parameters

$$v(d_b), \quad 0 \le v(d_b) \le \min \left\{ \frac{d_b}{b-1}, 1 - d_b \right\}, \quad b = 2, 3, \ldots 9,$$

for maximization (11)-(12) are given below:

b	2	3	4	5	6	7	8	9
d_b	0.4792	0.6676	0.7931	0.8768	0.9299	0.9618	0.9798	0.9896
$v(d_b)$	0.1903	0.1166	0.0744	0.0461	0.0272	0.0153	0.0082	0.0043

References

1. D'yachkov, A.G., Macula, A.J., Pogozelski, W.K., Renz, T.E., Rykov, V.V., Torney D.C.: A Weighted Insertion—Deletion Stacked Pair Thermodynamic Metric for DNA Codes. In: Proc. of 10th Int. Workshop on DNA Computing. Milan, Italy, pp. 90–103 (2004)
2. D'yachkov, A.G., Macula, A.J., Torney, D.C., Vilenkin, P.A., White, P.S., Ismagilov, I.K., Sarbayev, R.S.: On DNA Codes. Probl. Peredachi Informatsii (in Russian) 41(4), 57–77 (2005). English translation: Problems of Information Transmission 41(4), 349–367 (2005)
3. D'yachkov, A.G., Erdos, P.L., Macula, A.J., Rykov, V.V., Torney, D.C., Tung, C.S., Vilenkin, P.A., White, P.S.: Exordium for DNA Codes. J. Comb. Optimization 7(4), 369–379 (2003)
4. Levenshtein, V.I.: Binary Codes Capable of Correcting Deletions, Insertions, and Reversals. Dokl. Akad. Nauk USSR (in Russian) 163, 845–848 (1965), English translation: J. Soviet Phys.–Doklady 10, 707–710 (1966)
5. Levenshtein, V.I.: Efficient Reconstruction of Sequences from Their Subsequences and Supersequences. J. Comb. Th., Ser. A 93, 310–332 (2001)

Heuristic Solution to a 10-City Asymmetric Traveling Salesman Problem Using Probabilistic DNA Computing

David Spetzler, Fusheng Xiong, and Wayne D. Frasch

Molecular and Cellular Biology Graduate Program, and
Faculty of Biomedicine and Biotechnology, School of Life Sciences,
Arizona State University, PO Box 874501 Tempe, AZ 85287-4501, USA
frasch@asu.edu

Abstract. DNA hybridization was used to make a probabilistic computation to identify the optimal path for a fully connected asymmetric 10 city traveling salesman problem. Answer set formation was achieved using a unique DNA 20mer for each edge capable of hybridizing to half of each neighboring vertex. This allowed the vertex 20mers to be linked in all possible combinations to form paths through the network. Hybridization occurred in the presence of an excess of vertex 20mers, while edge 20mers were added in limiting amounts inversely proportional to the weight of each edge, resulting in the paths with the least cumulative weight being the most abundant. Correct answers, 230bp in length, contained a single copy of each vertex and were purified by PAGE and by successive magnetic bead affinity separations with probes for each vertex. Answer detection was accomplished using LCR of probes complementary to each vertex in a manner that identified the sequential order of vertices in each path by identifying vertex pairs. Optimal answer identification was accomplished using a conventional computer by normalizing the abundance of vertex pairings, and was found to be the same as that calculated by *in silico*.

Keywords: DNA computing, Traveling Salesman Problem, Ligation, Hybridization, Denaturing PAGE, Magnetic affinity.

1 Introduction

The use of DNA for making computations was first demonstrated by the successful computation of the solution to a 7 node Hamiltonian path problem (HPP) [1]. Methods to solve numerical optimization problems have been developed to expand the types of problems able to be solved using DNA computing [2-11]. Yamamoto et al. [12] accounted for the weight of each edge using a DNA concentration-dependent regime to design a computational method for a 6 node shortest path problem. Temperature has also been used to solve other numerical optimization problems with limited success. Although a subset of optimal solutions was purified, it was not possible to determine the optimal solution using this method [13].

One of the major limiting factors of DNA computing is that the number of molecules required to form every possible solution to an NP-complete problem is too large to generate [14]. Current methods consume a significant amount of the DNA

M.H. Garzon and H. Yan (Eds.): DNA 13, LNCS 4848, pp. 152–160, 2008.

forming incorrect solutions which must be thrown away, thus requiring more DNA to generate a complete solution set. Recent evidence also indicates that formation of secondary structures can occlude the correct solutions to make answer determination extremely difficult. Both of these factors have proved to be barriers preventing larger problems from being solved.

We now report an approach to DNA computing that generates a subset of the solution space composed of the solutions with the highest optimality rating, and does not require that the answer set includes every possible solution. Using this method, we have successfully solved a random instance of the fully connected 10 city asymmetric traveling salesman problem (Table 1) which has 3.3 million possible solutions. The NP completeness result for the Traveling Salesman problem implies that not all problem instances are hard . There are many different approaches to solve this type of problem, including exact and heuristic solutions. Conventional computers have completed hard instances of the problem, with thousands of cities, though at a cost of years of computing time. The TSP remains the standard optimization problem used to test new approaches as the struggle to solve NP problems continues.

Table 1. The distance matrix for the problem solved

	A[a]	B	C	D	E	F	G	H	I	J
A	***	55.2	34.05	31.75	53.85	39.95	36	39.9	36.55	52.6
B	63.95	***	54.25	54.95	72.6	45.05	71.65	50.55	52.75	52.15
C	51.35	47.6	***	41.45	39.8	57.8	55.2	32.75	34.85	37.05
D	46.65	46.25	54.6	***	49.4	45.55	55.9	52	57.35	54.6
E	49.9	39.1	42.65	52.4	***	25.9	39.85	38.85	37.95	33.1
F	59.7	49.15	47.8	56.9	58.05	***	48.05	46.6	48.4	47.7
G	51.25	36.7	43.95	43	42.45	40.25	***	64.2	47.8	46.95
H	58.1	35.85	53.7	45.05	47.3	43	84.25	***	42.8	41.9
I	52.9	38.2	40.35	33.45	36.5	65.2	35	29.7	***	30.95
J	60.05	39.1	40.65	55.75	41	41.1	45.1	58.65	43.95	***

[a] Each letter is a different vertex in the graph.
*** indicate paths that are not contained in the subset generated.

2 Methods

Computer design included three sequential steps to solve a problem that will find the most efficient path to visit all vertices once and only once, and then return to the starting vertex. In step 1, answer set formation was achieved using a combination of unique DNA 20mers for each vertex and edge. Hybridization of the edges occurs between the first half of one vertex and the second half of another. This allowed the unique vertex 20mers to be linked sequentially in all possible combinations to form paths through the network upon addition of ligase. The start and end vertex sequences contained an additional GC end cap which raised the melting point which eliminated nonspecific annealing during PCR. The computation was performed by adding all vertex 20mers in excess, while edge 20mers were added in limiting amounts that varied relative to the efficiency factor determined for that edge.

A set of 10 unique 20-mer sequences were designed using the software developed by Tanaka [15] to represent each vertex which were synthesized by Invitrogen. An additional 90 oligomers were synthesized containing all possible combinations of the complementary sequences to join any two city sequences together. The oligo sequences were designed to minimize cross hybridization, self-assembly and secondary structure formation and have similar thermal properties (melting temperature, between 61.3-61.8°C and GC content (25-30%). The yield of the sequences synthesized was used to define the distance matrix for the problem solved.

The initial answer DNA pool was generated by combining saturating amounts of vertex sequences with limiting amounts of edge sequences. The concentration of each edge was inversely proportional to the cost of that edge. The result was a population of heterogeneous sequences that formed upon hybridization and ligation of the sequences. The approximate concentration ratio for the hybridization and ligation reaction was set at 10:1 (vertex:edge). This ensured that vertex oligo concentrations were saturated while edge concentrations were limiting and varied in concentration so that a potential linking between any pair of vertices was dependent upon concentration of the corresponding edge sequence. Table 1 shows concentration differences for all linkers that were used to solve the 10-city problem.

The initial answer pool was generated using a two step process. First, initial hybridization/ligation was conducted in the absence of the ending vertex sequence and all linkers to the ending vertex. This greatly reduced formation of shorter answer sequences, thus improving the hybridization/ligation efficiency. Secondly, the hybridization/ligation was allowed to continue with the addition of fresh ligase, the ending vertex sequence and corresponding linkers. Hybridization-ligation products were purified through sequential PCR amplifications using the 5'-starting and the 3'-ending vertex primers with the target DNA templates that were extracted from the profiled PAGE gel containing previous PCR products.

Unless specified otherwise, PCR was performed in a 0.5 ml microcentrifuge tube with a total volume of 50 µl reaction mixture containing 10 mM Tris-HCl, pH 8.3, 50 mM KCl, 1.5 mM $MgCl_2$ and 0.001% gelatin, 200 µM each dNTP (ATP, GTP, CTP and TTP), 0.4 µM each of the starting and the ending primer, 2.5 U Taq DNA polymerase (New England BioLabs, MA, USA) and 50-100 ng DNA template. PCR reactions were carried out using a PJ2000 DNA thermal cycler that was programmed for a "Hot start" at 94°C for 2.5 min followed by 35 cycles. Each cycle consisted of a denaturing step at 94°C for 0.5 min, an annealing step at 68-70°C for 40 sec, and an extension step at 72°C for 30 sec. These cycles were concluded by a final extension for 3 min at 72°C.

The PCR products were profiled on a 6% denaturing polyacrylamide gel (the ratio of acrylamide to bisacrylamide was 29:1) in 1 x TBE buffer (90 mM Tris-borate, pH 8.3, 83 mM boric Acid, and 2 mM EDTA) at room temperature under 10 volts/cm. After electrophoresis, the gel was stained with ethidium bromide (1 mg/ml) for 10 min. The image profile was visualized and photographed under a UV transilluminator (UVP BioDoc-It[TM] system, UVP).

In Step 2, Answer sorting was achieved in two stages. First, answer sequences were separated by PAGE, the 230mer band was excised from the gel, and the DNA was amplified by PCR using the procedures described by Xiong et.al [16]. This PAGE separation, PCR amplification step was repeated five times to insure that only 230mer DNA was present. Second, the purified 230mer answer DNA was probed for the presence of each vertex sequence sequentially using magnetic affinity separation. Specifically, magnetic affinity separations were carried out by incubating 0.75 μl of vertex probe (400 μM), 149 μl 5x Binding/Washing (B/W) buffer with 150 μl of the M-280 beads (~1.5 mg, Dynal Biotech ASA, Osho, Norway), which were pre-washed 3 times with 500 μl B/W buffer for each time. After 45min at RT, the beads were separated using a magnetic separator. After washing 3 times using B/W buffer (500 μl each time), 90 μl of the ssDNA solutions was added along with 30 μl 20x SSC. After being gently vortexed and incubated at room temperature for 50-60 min, the DNA solutions hybridized to the immobilized probes were retained on the magnetic beads through biotin-streptavidin interaction. Those strands missing single or multiple vertex sequences were washed away (2 x SSC, 2 times with 500 μl each time, and then 0.5 x SSC, once, 500 μl). The captured duplexed answer sequences were dissociated from the biotinylated vertex probe in 80 μl of 0.1 N NaOH for 6-10 min. After separation using a magnetic separator, the collected supernatant containing the screened ssDNA solutions was neutralized by adding 8.2 μl of 1 N HCl, 10 μl of H_2O, 2 μl of 0.25 M Tris-Cl, pH 7.5.

In Step 3, ligation chain reaction (LCR) [17] combined with PAGE gel electrophoresis was implemented to characterize the answer pool. The answers were determined by performing a series of LCR reactions to determine the number of times one vertex preceded another. Complementary sequences to two vertex sequences were added at a time to determine the abundance of ordered pairs in the answer sequences. Since only one of the primers was phosphorylated, the probe that was phosphorylated on the 5' end dictated the order that the primers could link. Ligation chain reactions

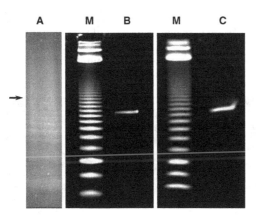

Fig. 1. A: The result of the two step hybridization and ligation reaction. The arrow shows where the 230 bp band was excised. B: Result of repeated rounds of PCR amplification on the 230bp band excised from the page gel for size separation. C: Result of PCR amplification of isolated 230-mer band after magnetic affinity purification containing the solutions to the problem. M: is the molecular marker, the brightest band is 100 bp.

were run for each possible vertex-vertex pairing, 90 reactions, and the products were profiled on PAGE gel. The relative abundance of each product measured as total density using UVP GDS-8000 BioImaging system.

Quantitative determination of the yield of LCR product was accomplished by the following procedure: (1) measure total density of the upper DNA band (the LCR product) and the lower DNA band (the PCR probes); (2) measure total density of the 100-mer band (the brightest one) from the DNA ladder; (3) normalize the total density of LCR product and probes against the 100-mer DNA ladder; (4) calculate the ratio of the normalized total density of PCR product over the normalized total density of PCR probe was calculated. This value represents a global measure of the abundance of each particular city pairing, and is used to determine the answer to the DNA calculation.

3 Results

Using a two-step hybridization/ligation protocol, an initial answer pool was generated with an amount of DNA distributed in the 230-mer region visible by PAGE, the required length for correct answers to the 10-city problem (Figure 1, Lane A). The hybridization/ligation products were actually distributed over a wide range of sizes and DNA sequences as large as ~500-mers were observed. This implies larger hybridization/ligation products were generated and that the techniques and protocols developed here for the 10-city problem are sufficient for larger problems.

Sequences formed that did not correspond to correct answers to the traveling salesman problem being solved were removed in two stages. First, correct answers must contain a single copy of each city sequence and thus should be 230bp in length. The 230mer answer band was excised from a gel to eliminate incorrect answers with too few or too many cities and was amplified by PCR. Four successive excision-amplifications yielded pure correct answers only when the PAGE step was done at 65°C (Figure 1, Lane B). This band corresponding to the correct length for solutions was then collected for subsequent purification by magnetic affinity separation. Second, avidin-coated magnetic beads were bound to biotinylated oligo probes complementary

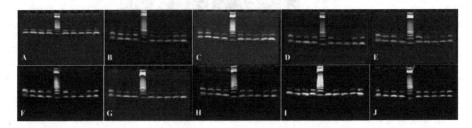

Fig. 2. PAGE profiles of ligation chain reaction product for all potential vertex pairings. For each LCR reaction, 230-mer DNA solutions plus two pairs of probes were included. The lower band is composed of probes that were not ligated. The upper bands are probes that were ligated and thus indicate the presence of that ordered pair in the answer pool.

the sequence for each vertex. These beads were used to probe the answer sequences sequentially to ensure every vertex was present. After all nine magnetic affinity separations, the remaining answer sequences that contained all vertices gave rise to a sharp 230mer band when separated by PAGE (Figure 1, Lane C).

The answer sequences were mixed with complementary sequences to two vertices that became covalently linked by ligase when the vertices were adjacent in a specific order. This was repeated for all combinations of vertex pairs, n^2-n required tests (Figure 2), and the relative abundance of all potential vertex pairings was determined. Thus, the concentration of each edge was determined from which the optimal pathway was deduced (Table 2). This was accomplished by normalizing the abundance of vertex pairings against the constant amount of city probes in each PAGE lane. Special attention was taken to perform the LCR with equal amounts of probe. The concentration of each probe was measured in triplicate using a NanoDrop ND-1000 spectrophotometer and a saturated concentration for each probe was used in the reaction. This ensured the yield of LCR product for a given link between two vertices was limited by the abundance of the corresponding answers.

Table 2. Matrix formed through the LCR gel read out

	A[a]	B	C	D	E	F	G	H	I	J
A	* * *	0.30	0.16	0.26	0.23	0.26	0.47	0.24	0.00	0.08
B	0.11	* * *	0.27	0.09	0.14	0.15	0.17	0.19	0.17	0.06
C	0.16	0.27	* * *	0.10	0.16	0.13	0.13	0.18	0.13	0.09
D	0.09	0.05	0.05	* * *	0.01	0.07	0.06	0.06	0.10	0.06
E	0.17	0.17	0.07	0.15	* * *	0.19	0.05	0.09	0.02	0.15
F	0.16	0.03	0.11	0.10	0.29	* * *	0.04	0.10	0.11	0.14
G	0.06	0.01	0.06	0.00	0.00	0.01	* * *	0.00	0.05	0.08
H	0.03	0.00	0.03	0.00	0.00	0.01	0.03	* * *	0.14	0.10
I	0.11	0.08	0.09	0.10	0.04	0.07	0.03	0.04	* * *	0.23
J	0.12	0.09	0.17	0.18	0.14	0.10	0.04	0.11	0.2	* * *

[a] Each letter is a different vertex in the graph.
*** indicate paths that are not contained in the subset generated.

The optimal answer obtained by the DNA computer was found to be the same as that calculated by a conventional computer. The DNA computer generated about 246,960 answer sets in total, which was 6.8% of the 3.3 million possible correct answers.

A conventional computer ranked the 1000 most optimal answers from best to worst and compared the answers generated by the DNA computer (Figure 3). The DNA computer successfully generated the 24 most optimal answers. The first answer not included was the 25th most optimal. The number of answers excluded by the DNA computer increased proportionately to the decrease in optimality. Of those sequences that did represent a correct answer, the majority of them corresponded to answers that had a high optimality rating. This occurred because the reaction mixture was

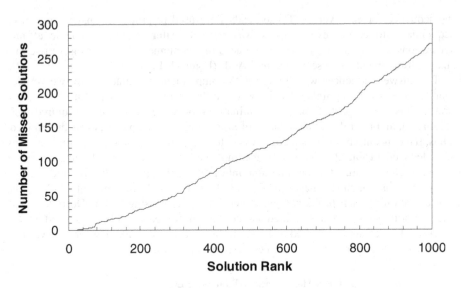

Fig. 3. The number of solutions that were missed as ranked by their optimality

composed of a variable amount of pathway sequences, such that the lower the cost of travel between two vertices, the more abundant the pathway. Thus, those sequences that represent good answers form a larger portion of the population.

4   Conclusion

These results demonstrate that our DNA computer presented here has successfully computed the 10 city problem. The first pathway through the network that was not contained by the subset of solutions generated by the DNA computer was the 25th most optimal solution. Thus, our method reduced the number of possible solutions while still retaining the most efficient pathway. Since this is a stochastic calculation, there is a chance that the optimal solution will not be created. However, since we do not readout particular solutions, but instead study the profile of the ordered pairs of vertices contained in the answer pool, it is likely that the paths that are involved in the optimal pathway will still be present. Thus, our readout method will allow the optimal path to be determined even though it might not be present in the DNA solution.

The distance matrix we chose to solve was not constrained, and solving it suffices to show that the technique can be used to solve any problem of lesser complexity. In the problem solved here, the number of possible solutions paths through the network of cities is limited by the rows and columns with the fewest possible transitions. Each row or column with fewer than 9 transitions limits the number of degrees of freedom that any path may travel. For example, vertex H may only be traveled to from vertices I or J, thus it has a degree of freedom of 6. To find the maximum number of potential solutions, we begin with the path that has the smallest degree of freedom and continue to the next smallest from there. In this case the number of possible solutions can be calculated by taking the minimum of the lowest degree of freedom and the number of remaining vertices to move to. In this way we can determine that the matrix generated by the DNA computer has at most $(7*7*7!) = 246,960$s or 6.8% of the possible solutions of the original problem. Thus the DNA computer has served to reduce a problem with 3.3 million possible solutions to one with 246,960. This is a statistical sampling of the total population that is weighted towards better solutions. Although there can be no guarantee that the optimal solution will be found, but it is highly likely that a near optimal solution will be. Since the number of solutions is small enough to be searched, we used a standard laptop computer to perform a brute search of all possible solutions to find the optimal solution, AFEJBCDIHGA. We compared this solution to the optimal solution of the original problem defined by the initial concentrations of all the pathways. The power of such a method to solve large optimization problems lies in the combination of biological and silicon computing and represents the most practical implementation of biological computing to date. However there are still significant obstacles to useful DNA computing. With all current methods as the number of vertices increase, the reaction volume increases as well. This problem is limiting due to the large percentage of sequences that may not form correct solutions. Until a method can be developed where each molecule forms a solution, acquiring a large enough sample of the solution population may be prohibitive for problems with more variables. Work is underway to establish this upper limit.

Acknowledgements

We would like to thank Justin York for helpful discussions. This work was supported by funding from DARPA/DSO and AFOSR to W.D.F.

References

1. Adleman, L.M.: Molecular computation of solutions to combinatorial problems. Science 266(5187), 1021–1024 (1994)
2. Reif, J.H., LaBean, T.H., Sahu, S., Yan, H., Yin, P.: Design, simulation, and experimental demonstration of self-assembled DNA nanostructures and motors. Unconventional Programming Paradigms 3566, 173–187 (2005)
3. Seeman, N.C., Wang, H., Yang, X.P., Liu, F.R., Mao, C.D., Sun, W.Q., Wenzler, L., Shen, Z.Y., Sha, R.J., Yan, H., Wong, M.H., Sa-Ardyen, P., Liu, B., Qiu, H.X., Li, X.J., Qi, J., Du, S.M., Zhang, Y.W., Mueller, J.E., Fu, T.J., Wang, Y.L., Chen, J.H.: New motifs in DNA nanotechnology. Nanotechnology 9(3), 257–273 (1998)
4. Schmidt, K.A., Henkel, C.V., Rozenberg, G., Spaink, H.P.: DNA computing using single-molecule hybridization detection. Nucleic Acids Research 32(17), 4962–4968 (2004)
5. Shin, S.Y., Lee, I.H., Kim, D., Zhang, B.T.: Multiobjective evolutionary optimization of DNA sequences for reliable DNA computing. Ieee Transactions on Evolutionary Computation 9(2), 143–158 (2005)
6. Qu, H.Q., Zhu, H., Peng, C.: New algorithms for some NP-optimization problems by DNA computing. Progress in Natural Science 12(6), 459–462 (2002)

7. Shao, X.G., Jiang, H.Y., Cai, W.S.: Advances in biomolecular computing. Progress in Chemistry 14(1), 37–46 (2002)
8. Tostesen, E., Liu, F., Jenssen, T.K., Hovig, E.: Speed-up of DNA melting algorithm with complete nearest neighbor properties. Biopolymers 70(3), 364–376 (2003)
9. Kim, D., Shin, S.Y., Lee, I.H., Zhang, B.T.: NACST/Seq: A sequence design system with multiobjective optimization. DNA Computing 2568, 242–251 (2003)
10. Lipton, R.J.: DNA Solution of Hard Computational Problems. Science 268(5210), 542–545 (1995)
11. Faulhammer, D., Cukras, A.R., Lipton, R.J., Landweber, L.F.: Molecular computation: RNA solutions to chess problems. Proceedings of the National Academy of Sciences of the United States of America 97(4), 1385–1389 (2000)
12. Yamamoto, M., Kameda, A., Matsuura, N., Shiba, T., Kawazoe, Y., Ohuchi, A.: A separation method for DNA computing based on concentration control. New Generation Computing 20(3), 251–261 (2002)
13. Lee, J.Y., Shin, S.Y., Park, T.H., Zhang, B.T.: Solving traveling salesman problems with DNA molecules encoding numerical values. Biosystems 78(1-3), 39–47 (2004)
14. Hartmanis, J.: Response to the Essays on Computational-Complexity and the Nature of Computer-Science. Acm Computing Surveys 27(1), 59–61 (1995)
15. Tanaka, F., Kameda, A., Yamamoto, M., Ohuchi, A.: Design of nucleic acid sequences for DNA computing based on a thermodynamic approach. Nucleic Acids Res. 33(3), 903–911 (2005)
16. Xiong, F., Spetzler, D., Frasch, W.: Elimination of Secondary Structure for DNA Computing. In: Proceedings of DNA13 (2007)
17. Yamanishi, K., Yasuno, H.: Ligase chain reaction (LCR). Hum. Cell. 6(2), 143–147 (1993)

An Approach for Using Modified Nucleotides in Aqueous DNA Computing

Angela M. Pagano[1] and Susannah Gal[2]

[1] SUNY Cortland, Department of Biological Sciences, P.O. Box 2000,
Cortland, NY 13045, USA
paganoA@cortland.edu
[2] Binghamton University, Department of Biological Sciences, P.O. Box 6000,
Binghamton, NY 13902, USA
sgal@binghamton.edu

Abstract. The concept of aqueous computing involves the use of large numbers of initially identical molecules to serve as memory registers in a fluid environment. Here, we consider a new approach to aqueous computing where modified nucleotides are used to 'write' on double-stranded DNA molecules to establish the logical values of true or false for a set of clauses. We introduce an implementation scenario where binding proteins specific to each modification can be used to selectively isolate DNA fragments with these modified nucleotides. In addition, we present initial results showing successful incorporation and detection of modifications as well as separation of modified molecules using binding proteins. As there are millions of molecules with corresponding binding proteins, this approach has the potential to yield unlimited computing power as compared with other aqueous computing methods.

1 Introduction

The successful use of molecular biology methods as computational tools by Adelman [1] has led to the development of a variety of DNA computing techniques. Aqueous computing as proposed by Head and Gal [2] (with an invitation to participate in Head *et al.* [3]) uses large numbers of initially identical molecules, such as DNA, which serve as memory registers in a fluid environment. Bit values (e.g., 0, 1) can be "written" on the molecule and subsequently "read" to determine solutions to computational problems. A major advantage of aqueous computing is that the fluid memory can be proportioned out and mixed back together such that problems requiring an exponential number of steps to solve conventionally, involve only a linear number of steps in this method.

Initial approaches to aqueous computing in our lab have involved the use of DNA and enzymes to perform the writing step [3]. Restriction enzymes cut DNA at specific sites along the molecule leaving overhangs at the cut site. Overhanging ends are filled in using a DNA polymerase, producing blunt ends. DNA fragments are then pieced back together using DNA ligase. The resultant "written" molecule is a longer DNA strand which can no longer be cut by the same restriction enzyme. This method

M.H. Garzon and H. Yan (Eds.): DNA 13, LNCS 4848, pp. 161–169, 2008.

has successfully been used to solve a 3-variable satisfiability (SAT) problem and a 3x3 Knight problem [3, 4]. However, this approach requires significant time – approximately two days for each "write" operation – as well as loss of DNA at each step, thereby limiting its utility in solving larger computations.

More recently, methylation of DNA [5] has been explored as a means to perform the writing step in an aqueous computing approach. In the "written" molecule, methylation of a restriction enzyme site prevents the enzyme from cutting the DNA. A site is assigned bit zero (false) if the site has been methylated and bit one (true) if unmethylated. The advantage here is the use of a single enzyme (vs. three) in the write step, although methylases are not as efficient in modifying DNA as restriction enzymes. Additionally, pairs of methylases are required to represent each state of a variable; for instance p and p'. Thus, for a 4-variable problem, eight methylases for eight different restriction enzyme sites would be required. Partial success in solving a 3-variable, 4-clause SAT problem has been demonstrated with this method [5]. Hence, there are still significant challenges in applying DNA methylation to computational problems.

Here we propose a new approach to aqueous computing where the "writing" step involves the labeling of DNA with different molecules incorporated into nucleotides. These modified nucleotides are used to selectively isolate DNA fragments using binding proteins specific to the label of interest. We demonstrate an implementation strategy for using these binding proteins in aqueous computing, the successful incorporation of 4 different modified nucleotides into DNA, and isolation of DNA labeled with two of these modified nucleotides using binding proteins. The commercial availability of modified nucleotides and corresponding binding proteins allows for potentially larger computations while the ease of incorporation, we felt, would likely speed up computation time.

2 Materials and Methods

The plasmid DNA, pBluescript SKII (Stratagene Incorporated, La Jolla, CA) was used as the starting hardware. Labeled PCR product was derived from the amplification of the approximately 200 base pair multiple cloning site of this plasmid with combinations of four different modifications: Alexa Fluor-488 universal primer (Integrated DNA Technologies, Coralville, IA), BODIPY-FL modified reverse primer (BODIPY-FL from Invitrogen, Carlsbad, CA; amine reactive reverse primer from Integrated DNA Technologies), biotin as either a reverse primer (Integrated DNA Technologies) or as biotin-aha-dCTP (Invitrogen), and digoxigenin (DIG) dUTP in a PCR labeling mix (Roche Diagnostics, Indianapolis, IN). Taq polymerase was obtained from New England Biolabs (Ipswich, MA). Resultant labeled DNAs were purified using the QIAquick PCR Purification Kit (Qiagen, Valencia, CA).

To ensure specificity of our detection system, binding proteins – linked to the enzyme horse radish peroxidase (HRP) for chemiluminescent detection or alkaline phosphatase (AP) for colorimetric detection – for the specific labels were tested for cross reactivity using a dot blot. For the blots, 1μl samples of each modified DNA were dotted onto a nylon membrane and permanently bound using a UV crosslinker. Four membranes were set up such that each antibody could be tested against each modified DNA and their respective positive controls. Membranes were incubated in blocking solution (5% BSA, 0.1% Tween-20) for 30 minutes prior to one hour of incubation with the binding protein diluted in blocking solution. Blots were rinsed twice in wash solution (1x TBS, 0.1% Tween) to eliminate excess antibody prior to detection. For the biotin binding protein streptavidin (Roche Diagnostics), detection of binding was visualized via a chemiluminescent substrate for HRP (Pierce Biotech Incorporated, Rockford, IL). For other modifications, antibodies to DIG (Roche Biotech Incorporated), Alexa Fluor-488 (Invitrogen), and BODIPY-FL (Invitrogen) were used. The DIG antibodies are provided with a conjugated AP while the other two modifications were visualized through a secondary anti-rabbit antibody linked to AP. This enzyme was visualized using colorimetric detection with nitro blue tetrazolium (NBT) and 5-bromo-4-chloro-3-indolyl phosphate (BCIP) (Roche Biotech Incorporated), substrates of alkaline phosphatase.

To separate DNA labeled with different modifications, we used magnetic beads linked to binding proteins. The specific modifications of interest bind to the beads which are easily separated from other molecules using a small magnet. We focused initial bead binding experiments on DNA labeled with Alexa Fluor and/or biotin. For capture of biotinylated DNA, streptavidin linked magnetic beads were purchased from Roche. For capture of Alexa Fluor labeled DNA, anti-Alexa Fluor-488 beads were created using Magna-Bind carboxyl derivatized magnetic beads using the manufacturer's protocol (Pierce). Our initial cross-linking of the anti-Alexa Fluor-488 antibody to the beads indicated approximately 65% of the antibody was removed from the solution and thus is presumed to be attached to the beads. Approximately one modification per antibody could bind to the bead, which is more or less as expected given a level of binding of approximately 0.1 pmole Alexa Fluor-488 per 5ul of beads.

For binding, labeled DNA was incubated with the magnetic beads in PBS (130mM NaCl, 2mM KCl, 20mM Na_2HPO_4, 2mM NaH_2PO_4, pH 7.4)) and EDTA (10mM) and placed on a shaker for 30 minutes at room temperature. At the end of this time, the unbound fraction was removed. Initially, to test binding, beads were boiled for 5 minutes in 1x PBS to remove the bound fraction. Visualization of bound and unbound fractions of bead reactions was done through dot blot or Southern blot. For Southern blots, all samples were run on a 10% polyacrylamide Tris borate EDTA Ready Gel (BioRad Corporation, Richmond, CA) and the presence of DNA detected using ethidium bromide. Gels were transferred to a nylon membrane and modifications detected using the appropriate binding protein and reactions as above. We later tried to remove the bound modified DNA fragments using mild conditions so that the DNA remained double-stranded (allowing for rebinding to new beads). We tried a number of conditions including low pH (0.1M glycine pH 2.5), formamide (95-1% at 65°C), Tris EDTA (10mM, 1mM at 70°C), NaOH (0.1N with 1mM EDTA at room temperature), Tris (10mM at 55°C) and SDS (1% in 50mM Tris 10mM EDTA at 65°C for 10 minutes). Visualization and detection were done as above.

3 Implementation and Results

To implement the use of modified nucleotides into a logic problem, we first define the coding of DNA strands. In our case, each modification represents a variable (e.g., "p") that satisfies two conditions. The presence of the modification is taken as the true condition (p) while its absence is taken as the false condition (p'). For example, let's say we randomly assign the DNA modified by Alexa Fluor-488 and biotin with the variables p and q, respectively. If a DNA molecule is labeled with the Alexa Fluor-488 modification, we consider the molecule to be "True" at variable p. If it is labeled with biotin, we consider the molecule to be "True" at variable q. The AND logical operator is satisfied when the DNA molecule contains all variables that satisfy the specified clause (e.g., DNA labeled with both biotin and Alexa Fluor-488 satisfies p AND q). The OR logical operator is satisfied when the DNA contains either one or both variables that satisfy the specified clause (e.g., DNA labeled with either biotin or Alexa Fluor-488 or both modifications satisfies p OR q).

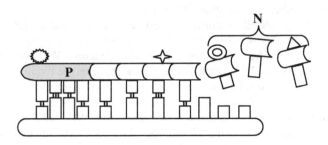

Fig. 1. Encoding strategy to incorporate modifications into DNA strands. Primers (P) and nucleotides (N) containing specific chemical modifications (shapes on strands) can be used in the PCR to create DNA fragments with those chemical modifications as shown. Specific chemical modifications include fluorescent compounds such as Alexa Fluor-488 and Bodipy-FL or non-fluorescent ones such as digoxigenin (DIG) and biotin. These four chemical modifications have successfully been incorporated into DNA fragments (see below).

The writing step (encoding) corresponds to incorporation of the modified nucleotides or primers using PCR (Figure 1). Oligonucleotide primers and nucleotides with specific modifications are readily available commercially (Invitrogen, Carlsbad, CA and Integrated DNA Technologies, Coralville, IA). Reading corresponds to the separation and detection of modified and unmodified molecules. Modified nucleotides are separated using specific binding proteins linked to magnetic beads. The solution to a computational problem is confirmed through detection of the presence or absence of the modification on the DNA via dot blot or Southern blot as described in the methods.

3.1 Step 1 of Implementation: Incorporation of Modified Nucleotides

In the first step toward implementation, we have successfully labeled PCR fragments with 4 different modifications – 2 fluorescent ones, Alexa Fluor-488 and Bodipy-FL (Figure 2), and 2 others, DIG and biotin. The former two were most successfully incorporated using PCR primers with these attached fluorescent moieties as nucleotides with these modifications were not effectively incorporated by Taq

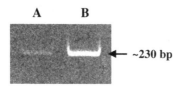

Fig. 2. PCR products incorporating fluorescent modifications. DNA was labeled using modified primers with either BODIPY-FL (A) or Alexa Fluor-488 (B) in the PCR and then separated using 10% acrylamide gel in Tris-borate EDTA buffer. Fluorescent products were directly visualized using UV light.

polymerase. DIG could be incorporated as a dUTP nucleotide derivative while biotin has been used to label PCR products either using a biotinylated primer or using biotinylated dCTP. Because modifications can be incorporated as either primers or nucleotides, it is possible to create DNA with all 16 combinations of the four labels.

3.2 Step 2 of Implementation: Separation of Modified Nucleotides

In the second step, modified nucleotides are separated using specific binding proteins linked to magnetic beads (Figure 3). The specific modifications of interest bind to the

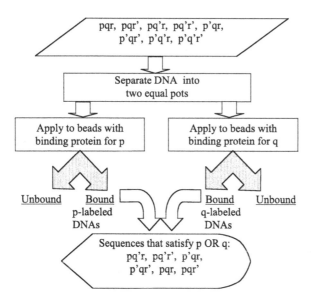

Fig. 3. Steps in computing p OR q. Labeled DNAs ('True') are represented by p, q and r. Unlabeled DNAs ('False') are represented by p', q', r'. Each Boolean variable represents a distinct modification incorporated into the DNA. This figure represents a scheme for the isolation of the labeled DNAs (bound to the binding proteins) although this approach could alternatively be used for retention of unbound material.

beads (bound fraction) and are separated from other molecules using a small magnet. Unlabeled DNAs (unbound fraction) are removed from the tube, leaving only sequences that satisfy the specified clause. Thus, we can isolate either the molecules with the modification or those without it allowing one modification to represent both p and p'.

3.3 Step 3 of Implementation: Detection and Isolation of DNA with Modified Nucleotides

We obtained commercially available binding proteins for each of these four modified nucleotides and tested them for cross reactivity with non-specific modifications. In all cases, the binding proteins recognized only the appropriately labeled PCR products and positive controls (Figure 4). This specificity should allow us to use the binding proteins coupled to magnetic beads as a tool to separate those DNA with modified nucleotides from those without ("True"/"False").

Our next goal was to bind the modified DNA to magnetic beads and work out conditions for removal of the bound material from the beads without denaturing the DNA strands (as shown in Figure 3). We chose to focus initial bead binding experiments on DNA with only 2 modifications – Alexa Fluor-488 and biotin. Both streptavidin and anti-Alexa Fluor linked beads were able to successfully bind labeled PCR products (Figure 5). We had some difficulty, however, with the magnetic beads modified for the binding of Alexa Fluor labeled DNA. Initial cross-linking of the anti-Alexa Fluor antibody to the Magna-Bind magnetic beads indicated that 65% of the antibody was removed from the solution and thus presumed to be attached to the beads. When these beads were washed (involves adding buffer, mixing and pulling

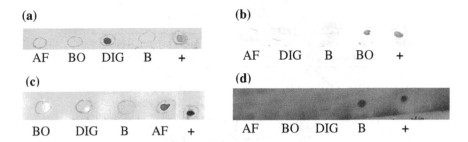

Fig. 4. Dot blots results testing cross-reactivity of binding proteins. PCR products and controls labeled with specific modified nucleotides (AF = Alexa Fluor-488, BO = BODIPY-FL, DIG = digoxigenin, B = biotin, + = positive control) were dotted onto nylon membrane and treated with UV light to crosslink the DNA permanently to the membrane. Membranes were incubated with specific antibodies or binding proteins (a) anti-DIG, b) anti-BODIPY, c) anti-Alex Fluor, and d) streptavidin (binds biotin)) to identify the modified products. Binding proteins were then localized either using alkaline phosphatase (a, b, & c) or horse radish peroxidase as described in the methods. Dark spots on the membranes indicate the presence of antibodies or binding proteins at that location. Binding proteins recognized only specific modifications and not other unrelated compounds.

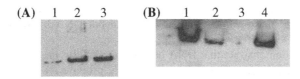

Fig. 5. Southern blot of fractions from binding and release reactions. DNAs were separated using 10% acrylamide gel in Tris-borate EDTA buffer transferred to nylon membranes and then detected using streptavidin-HRP (Panel A) or anti-Alexa Fluor antibodies (Panel B). Panel A. PCR products labeled with biotin (lane 3) were incubated with streptavidin magnetic beads and the unbound fraction saved (lane 1). The bound material was released from the beads by boiling in 1x PBS (lane 2). Panel B. PCR products labeled with Alexa Fluor-488 and biotin (lane 1) were incubated with magnetic beads coated with anti-Alex Fluor antibodies and the unbound material was removed before (lane 2) and after washing (lane 3). The bound material was then removed with an incubation in 1% SDS in 50mM Tris 10mM EDTA at 65°C for 10 minutes (lane 4).

the beads out of solution using a magnet), the washes contained traces of the antibody suggesting that some of the antibody must have been coming off of the beads. We have also noted variability in the amount of Alexa-fluor modified DNA binding to the beads, presumably for this reason.

Removing the bound modified DNA fragments using mild enough conditions such that the DNA stayed double-stranded remained more of a challenge. Many of the standard approaches – high pH or high temperature – cause melting of the DNA which would result in mixing of top and bottom strands. This reorganization of strands may produce different combinations of the modifications than existed in the original sample. Treatment with SDS (1% in 50mM Tris 10mM EDTA at 65°C for 10 minutes) seemed to remove a substantial amount of the bound modified DNA while not causing melting of the strands (see Figure 5B). We tried a number of other treatment conditions which either showed no signal in the removed DNA, multiple bands or higher bands suggesting melting of the DNA fragments (results not shown).

4   Conclusions and Future Work

Using different molecules incorporated into nucleotides, we successfully labeled PCR fragments with 4 different modifications. We found that two of these modifications (Alexa Fluor-488 and BODIPY-FL) were more efficiently incorporated into the PCR product as modified primers. Given that the remaining modifications (biotin and DIG) can be incorporated at any position along the DNA strand, we expect to be able to create all 16 different types of DNA (with and without each of the 4 modified nucleotides) for use in computing problems.

For two of these modifications, biotin and Alexa Fluor-488, we were able to show the ability of binding proteins linked to magnetic beads to isolate labeled DNA fragments. However, we are still working on determining conditions for removing bound DNA from the beads without separating the strands. Treatment with SDS worked well but this approach incurs the added challenge of removing this reagent before rebinding, creating an extra step in the computing process. An ideal approach would be one involving temperature as that would involve no additional purification steps before rebinding the released DNA. We are working, therefore, on temperature conditions where the binding interaction is disrupted but DNA hybridization is not affected [6].

Our experiments demonstrate that DNA can be successfully modified to contain a variety of fluorescent and non-fluorescent labels. We propose an experimental plan for implementation of the aqueous computing approach using these modified molecules. We use DNA for its convenience and ease in incorporating these modifications, however, any molecule can be used as the memory register. In our case, each modification on the DNA represents a variable that satisfies two conditions. The presence of the modification results in the true condition (p) while its absence gives us the false condition (p'). The results presented here allow for the creation of logically consistent molecules at four variables with labeling of only four sites. We expect to be able to scale-up this approach since there are at least 12 fluorescent and 3 non-fluorescent modifications with commercially available binding proteins. However, each binding protein would need to be tested for specificity to their appropriate modification as described in the methods. Any binding protein showing cross reactivity could be further purified using affinity column chromatography to remove the undesired binding protein fraction.

With all of these chemical compounds incorporated, either as modified primers or nucleotides, we would be able to solve 15-variable problems. Thus, this approach has potentially greater computing power than methylation which would require 30 distinct enzymes for the same number of variables. Additionally, modifications used in this approach are not site specific (as are methylases and restriction enzymes), allowing us to label anywhere along the DNA strand. Fluorescent molecules are of particular interest since they can also be visually detected using gel electrophoresis, a spectrophotometer, or fluorescence activated cell sorter. Direct visualization would eliminate the need for chemical detection via dot blot or Southern blot, speeding up computational time. In general, there are millions of compounds to which binding proteins exist. So, in reality, there is no limit to the power of this approach to aqueous computing provided we can work out the conditions for its implementation.

Acknowledgments. The authors would like to acknowledge support from the Air Force for this project, contract number AFOSR FA87500620002.

References

1. Adleman, L.: Molecular computation of solutions of combinatorial problems. Science 266, 1021–1024 (1994)
2. Head, T., Gal, S.: Aqueous computing: Writing into fluid memory. Bulletin of the European Association for Theoretical Computer Science 75, 190–198 (2001)
3. Head, T., Chen, X., Yamamura, M., Gal, S.: Aqueous computing: A survey with an invitation to participate. J. Computer Science & Technology 17, 672–681 (2002)
4. Head, T., Chen, X., Nichols, M.J., Yamamura, M., Gal, S.: Aqueous solutions of algorithmic problems: Emphasizing knights on a 3X3. In: Jonoska, N., Seeman, N.C. (eds.) DNA Computing. LNCS, vol. 2340, pp. 191–202. Springer, Heidelberg (2002)

5. Gal, S., Monteith, N., Shkalim, S., Huang, H., Head, T.: Methylation of DNA may be used as a computational tool: Experimental evidence. In: Mahdavi, K., Culshaw, R., Boucher, J. (eds.) Current Developments in Mathematical Biology, vol. 38, pp. 1–14. World Scientific, New Jersey (2007)
6. Holmberg, A., Blomstergren, A., Nord, O., Lukacs, M., Lundeberg, J., Uhlén, M.: The biotin-streptavidin bond can be reversibly broken using water at elevated temperatures. Electrophoresis 26, 501–510 (2005)

Modeling Non-specific Binding in Gel-Based DNA Computers

Clifford R. Johnson

clifford.johnson@usc.edu

Abstract. In attempting to automate the computation of n-variable 3-CNF SAT problems using DNA, two physical architectures were scrutinized, the "in-line" architecture and the "waste-well" architecture. Computer modeling of the effects of non-specific binding predicted that the in-line version would not work for problems of more than 7 variables. According to the model, the "wrong answer" DNA strands would swamp out the "correct answer" DNA strands in the final computation module. And in fact, the in-line architecture never performed a computation higher than 6 variables.

To perform a 20 variable instance of the 3-CNF SAT problem a manual version of the waste-well architecture was employed. Surprisingly though, after analysis of the modeling results, it appears that through a simple protocol change, the in-line architecture may have been able to perform higher order computations.

1 Introduction

The first molecular computation was performed by Len Adleman at the University of Southern California in 1994. Using DNA molecules to perform the computation, Adleman solved a 7-city, Directed Hamiltonian Path Problem [1]. The molecular implementation used to solve the 7-city DHPP was surprisingly simple. However because of the use of enzymes (ligase) to covalently ligate strands, it was apparent that it would be almost impossible to scale-up the computation, i.e., to solve larger problems, using this paradigm. The making and breaking of covalent bonds using enzymes is notoriously inefficient; 40% reaction completion is considered good. It's messy, difficult, inefficient, error prone. Additionally, the envisioned simplicity of molecular computation disappears - computation schemes begin to look like Rube Goldberg devices. For computations on the order of a 20 variable problem (about 8,000 times more complex than the 7 city problem), a new molecular paradigm was necessary, one that could somehow avoid biology's inherent messiness.

A new paradigm was formulated based on Richard Lipton's method for encoding DNA to represent binary strings [2,3], and was called the "modified sticker model" [4]. It involves no enzymes, and no covalent bond formation or destruction. The computation is performed simply through the hybridization and denaturing of DNA hydrogen bonds. This is the paradigm used to perform the 20 variable 3-CNF SAT problem published in Science [5], which remains at this time, the most complex problem solved using molecules.

M.H. Garzon and H. Yan (Eds.): DNA 13, LNCS 4848, pp. 170–181, 2007.
© Springer-Verlag Berlin Heidelberg 2007

For the most part, molecular computations are performed by hand, at the lab bench. The computation of the instance of the 20 variable 3-CNF SAT problem [5], took 2 to 3 man-weeks to perform by hand. This is labor intensive and error prone. One of the project goals was the automation of the computation process. In trying to automate the computation of 3-CNF SAT problems, the question arose: What *physical* hardware configuration is best? Two different architectures vied for the honor: One was called the "in-line" architecture; the other was called the "waste-well" architecture. Both architectures were actually implemented and tested.

Computer modeling of the effects of non-specific binding (NSB) predicted that the in-line version would not work for problems of more than 7 variables - the wrong "answer" DNA strands would swamp out the "correct answer" DNA strands in the final computation module. In fact, the in-line architecture never performed a computation higher than 6 variables, and the waste well architecture was employed to perform the 20 variable computation. Surprisingly though, after analysis of the modeling results, it appears that through a simple protocol change, the in-line architecture may have been able to perform higher order computations.

2 The 3-CNF SAT Problem

The Satisfiability problem (SAT) is of interest both historically and theoretically. Historically, the SAT problem was the first to be shown to be NP complete. Theoretically, the SAT problem plays a critical role in computer science applications and theory. In practice, the SAT problem is fundamental in solving many application problems in database design, CAD-CAM, robotics, scheduling, integrated circuit design, computer networking, and so on. "Methods to solve the SAT problem play a crucial role in the development of efficient computing systems." [6] SAT problems are a set of computationally intractable NP-complete problems. Problems in Class NP are considered intractable because as the number of variables increases linearly, the computation time increases exponentially. For example, a 100 variable instance of a 3-SAT problem might take IBM's Big Blue, computing at 135 teraFLOPS, 3.2 million centuries to solve, essentially the problem is unsolvable.

The SAT Problem
The goal of the SAT problem is to determine whether there exists a satisfying truth assignment for a given Boolean expression. That is:

Let $U = \{x_1, x_2,..., x_n\}$ be a set of n Boolean variables. A truth assignment for U is a function $T : U \rightarrow \{true, false\}$. Corresponding to each variable x_i are two literals, x_i and $\neg x_i$ (not x_i) that can be assigned to the variable. A literal x_i is true under T iff $T(x_i) = true$; a literal $\neg x_i$ is true under T iff $T(x_i) = false$).

A set of literals surrounded by parentheses is called a clause, and a set of clauses is called a formula.

A satisfying assignment for a formula, φ is called a solution.

The restriction of SAT to instances where all clauses have length k is called k-SAT.

The Conjunctive Normal Form (CNF)

Let φ be a formula. Let C be the set of clauses for that formula. φ is a formula in conjunctive normal form (CNF), implies that a truth assignment $T : U \rightarrow \{true, false\}$ satisfies $c \in C$ iff at least one literal in c is true under T. T satisfies φ iff it satisfies every clause in φ.

Equation 1 shows an instance of a 10 variable 3-CNF SAT problem with 14 clauses. Notice that there are 3 literals per clause separated by the OR symbol v, and that each clause is separated by the AND symbol ^. This is the conjunctive normal form for a formula 3-SAT problem.

$$
\begin{aligned}
\phi = &(X_2 \vee X_4 \vee X_9) \wedge (X_8 \vee \neg X_{10} \vee X_5) \wedge (\neg X_6 \vee \neg X_8 \vee \neg X_{10}) \wedge (X_2 \vee \neg X_4 \vee \neg X_9) \wedge \\
&(\neg X_9 \vee \neg X_3 \vee X_6) \wedge (X_{10} \vee X_5 \vee X_7) \wedge (\neg X_7 \vee X_1 \vee \neg X_2) \wedge (X_2 \vee \neg X_4 \vee X_9) \wedge \\
&(X_3 \vee X_6 \vee \neg X_8) \wedge (\neg X_5 \vee X_7 \vee X_1) \wedge (\neg X_2 \vee \neg X_4 \vee \neg X_9) \wedge (X_2 \vee X_4 \vee \neg X_9) \wedge \\
&(\neg X_1 \vee \neg X_2 \vee X_4) \wedge (X_2 \vee \neg X_4 \vee X_9)
\end{aligned}
\quad (1)
$$

Here ϕ has the unique solution:

$X_1 = F, X_2 = T, X_3 = T, X_4 = F, X_5 = F, X_6 = F, X_7 = F, X_8 = T, X_9 = F, X_{10} = T$.

To solidify these concepts in an informal fashion, think of this as a kind of Agatha Christie murder mystery. Ten professors, named Professor X1, Professor X2, ..., and Professor X10, are invited to dinner. Some of the professors may have been "eliminated" on their way to dinner. We want to know who made it to the dinner, and who didn't. The clauses provide clues. For example, the first clause tells us that either *Professor X2 arrived, OR Professor X4 arrived, OR Professor X9 arrived* for dinner that night. The second clause, for example, tells us that either *Professor X8 arrived, OR Professor X_{10} did not arrive, OR Professor X5 did arrive* for dinner.

If we put all of the clues (clauses) together, we get the solution to the mystery. In the unique solution for ϕ for Equation 1, we see that Professor X1 did *not* arrive to dine, whereas Professor X2 did, and so on. Here is the interesting part. If one were to try to solve Equation 1, without knowing the answer beforehand, it would take a very long time to find the solution, even for this relatively short 10 variable problem. Yet once we are given a solution for ϕ, it is very easy to verify. We just check to see if at least one literal in each clause is true. (This can be seen with Equation 1.) This is the essence of Class NP problems. Problems in Class NP are very, very hard to solve. Yet once a solution is found, it can be *verified* quickly.

3 The Molecular Implementation of the 3-CNF SAT Problem

The implementation paradigm is remarkably straightforward:

1. To represent all possible variable assignments for the chosen n-variable SAT problem, a Lipton encoding [7] for DNA strands is chosen. For each of the n variables $x_1, x_2, \ldots, x_n$, two distinct 15 base value sequences are designed - one representing *true (T)*, $X_k T$, and one

representing *false (F)*, X_kF.   Each of the 2^n truth assignments is represented by a sequence of (n X 15) bases consisting of the concatenation of one value sequence for each variable. In this way all possible assignments are encoded.   DNA molecules with library sequences are termed library strands; a combinatorial pool containing library strands is termed a library.

2.   The probes used for separating the library strands have sequences that are W-C (Watson-Crick) complements of the value sequences.

3.   The *clauses* of an n-var CNF SAT problem are formed with acrylamide gel modules in which the probes for the clause are covalently bonded to the acrylamide gel. For example, for the last clause of Eq. 1, (X_2 v $\neg X_4$ v X_9) the W-C complementary probes for X_2T , X_4F, X_9T are covalently bound to the acrylamide gel. (Figure 1.)

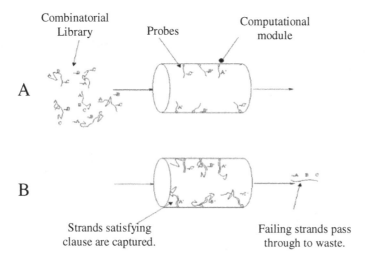

Fig. 1. A and B. A computation. A The combinatorial library enters the gel module which tests the clause (X_2 v $\neg X_4$ v X_9) using covalently bound DNA probes that are W-C complimentary to X_2, $\neg X_4$, and X_9. B Strands that do not satisfy the clause, pass through to waste.

4   The Physical Architectures

The Molecular Algorithm

1.   Under hybridizing conditions (temperature at 15° C), introduce the DNA strand library into the first module via electrophoresis. (This library represents all possible variable assignments.)

2.   Under hybridizing conditions, those strands that satisfy the clause, hybridize to the probes and remain in the gel module. Those strands that do not satisfy the clause pass through.

3.   The gel module is then heated to 65 ° C to release the hybridized (satisfying) strands, which are then passed via electrophoresis to the next cooled module and a new computation.

4.   The strands captured in the final module represent those variable assignments that have successfully satisfied all of the clauses and thus represent a solution.

5.   The final gel module is removed to extract the DNA molecules for PCR amplification and sequencing of the answer.

Note that this molecular algorithm for the SAT computation is massively parallel.

Two different architectures were considered as candidates for automating this algorithm for solving CNF SAT computations using DNA - the in-line architecture and the waste-well architecture.

The In-Line Architecture

The premise of this geometry (Fig. 2) is that as long as the next module down stream is heated, a valid computation can be performed. Fig. 3 diagrams the in-line device set-up during a computation.   In Fig. 3,

Fig. 2. Schematic of the in-line architecture

capture Module 1 is heated to release the combinatorial library, which then moves via electrophoresis to Module 2. The cooled Capture Module 2 captures those strands satisfying its clause while those strands not satisfying the clause pass through to the next consecutive module (Module 3), which is held hot. Theoretically, no strands should remain in the adjacent down-stream module (Module 3) as the temperature is raised to the same release temperature as in Module 1. All non-satisfying strands continue on to Module 4 (which is at room temperature) theoretically clearing out Module 3.   As the Hot-Cold-Hot manifold moves to the right performing computations, all of the non-satisfying strands will eventually empty to the waste reservoir, in theory.

Note that all of the library strands pass through all of the computation modules in this configuration. It is assumed that the heating of down-stream modules is sufficient to release all of the oligonucleotides residing therein and will cause no contamination in the upcoming computation. In short, NSB (non-specific binding) is assumed to be inconsequential.

Though a 6-variable problem was solved using the above set-up, the in-line architecture didn't work for larger n's. Trying to solve a 10-variable problem using the in-line architecture, proved fruitless; and in fact, to solve the 20-variable problem later on, an un-automated version of the waste-well architecture was employed - it was necessary to perform each of the computations by hand.

However, contradicting the assumption that NSB was inconsequential was some experimental evidence that detectable levels of NSB were indeed present in the gels. Computational test runs were performed using ^{32}P labeled library. The library was sent through a series of 24 sample modules, using an in-line construction similar to the one in Fig. 2. Some of these modules acted as capture/release modules; some were simple agarose modules; and some simple acrylamide modules. At the end of the test, all had some level of residual radioactivity as measured with a Beckman Scintillation Counter. The residual radioactivity varied from 0.8% of the total counts to 6.5% of the total counts. When gels were removed from the glass modules (generally the gels slid out very easily), it was determined that about ¼ of the radiation was retained in the glass module, even after squirting distilled water through the gel trough to remove contaminants. The interpretation of these results was: (1) NSB was occurring on the glass surface, and (2) some sort of NSB / oligonucleotide retention was occurring in the gel itself. It is not known what the mechanism is for the retention of oligonucleotides in the gel. Inclusions, micro-fissures, poor gel formation, impurities, or some sort of bonding with the gel, any or all might be responsible for the phenomenon. Some of the radioactivity was probably due to radioactive mononucleotides. However, assuming that NSB is inconsequential is a problematic premise.

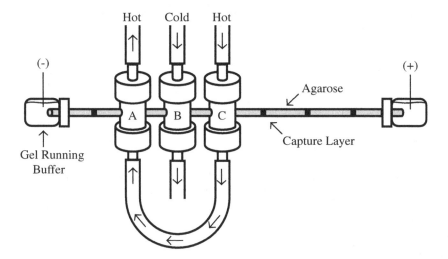

Fig. 3. The In-line Architecture. A 35-cm glass tube loaded with the library module, then with intercalated blank gel modules, and clause modules. The system was fitted with three water jackets (A, B, C). Library strands in the capture layer inside of (A) are released and move into the capture layer inside of (B). There, library strands with subsequences complementary to the probes are captured and retained. The rest of the strands passed into the capture layer inside of (C) but because (C) is kept hot the strands passed through unhindered.

The advantage of the in-line architecture is its simplicity. It is basically a glass tube packed with computation modules at equally spaced intervals, intercalated by gel. To automate the architecture one simply moves a Hot-Cold-Hot manifold down the glass tube (refer to Fig. 3). The disadvantage of the in-line system is that all library strands, "good" strands and "bad," go through every computation module, thus possibly contaminating the modules.

The Waste-Well Architecture

The second geometry (Fig. 4) is called the waste-well geometry. In this architecture, the non-satisfying strands of a computational step avoid passing through every downstream module by going to a waste buffer well, where they are destroyed. Again, strands are released from module 1, which then

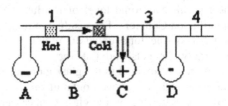

Fig. 4. Schematic of the "waste-well" architecture

pass through module 2. Those strands that satisfy the Module 2 clause are captured and those that do not pass through the module. However, instead of continuing downstream possibly contaminating pristine modules, they are diverted immediately to a waste well, where they are destroyed. This second geometry was specifically conceived to obviate the accumulating effects of NSB. The waste well architecture is not as simple as the in-line architecture but it does preclude the effects of NSB.

This paradigm forms the architecture for the first functional automated molecular computer [8] solving instances of 10 variable 3-CNF SAT problems.

The advantage of the waste-well architecture is that unsuccessful DNA strands go to waste immediately after the computation, thus leaving downstream modules pristine. The disadvantage of the waste-well architecture is that it is complicated to construct [8].

5 The Mathematical Model

A priori, modeling the adsorption (i.e., NSB) of DNA in a gel based system would seem very difficult. First, one would have to determine the dominant forces involved in the binding reaction both to the gel and to the silica. For silica, some studies indicate that three effects, namely: (i) shielded intermolecular electrostatic forces, (ii) dehydration of the DNA and silica surfaces, and (iii) intermolecular hydrogen bond formation in the DNA–silica contact layer, are the dominant contributors to adsorption.[9] For gels, which are mostly fluid, a balance of forces maintains the gel form (sometimes even

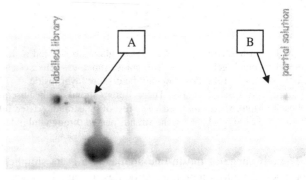

Fig. 5. Composite radioactive image showing prog-ression of a computation

disturbing them infinitesimally can bring on a phase transitions and/or collapsing of the gel).[10]   As mentioned above, inclusions, micro-fissures, poor gel formation, impurities, or some other sort of bonding of DNA with the gel, might contribute to NSB.

These factors would make modeling of NSB a virtual nightmare. However, by using an *output/input* model, and describing the difference of *input - output* as due to NSB, one can arrive at a useful model that seems to be consistent with experimental data. To determine the ratio of molecules left behind in each computational module, that is:

$$(input\text{-}output)/input$$

experimental data is needed.

Fig. 5 was obtained using a Storm phosphor imaging sys-tem. The progression of a computation using lib-rary labeled with ^{32}P was imaged. Arrow A points to the first module in the computation. Here, the heavy residual radio-activity in this module is probably due to radio-active mononucleotides, and not to NSB of the library strands. Thereafter, residual radiation dropped drastically; but there was always a slight amount left. This "slight amount" was deemed to be due to NSB of library strands, and ranged in value from .05% to 1% (barely visible in the above image). Arrow B points to the 7th module with a NSB of about 0.1%. To the right of the module we can just barely see a partial solution progressing through a computation.

Modeling the In-Line Architecture

The model uses the following two assumptions:

1. Non-specific binding takes place in cold modules and has a very low constant of disassociation; i.e., it takes hours instead of seconds for NSB disassociation to occur, even at the elevated temperatures used to denature the probes from the library strands (65°C).

2. Both complimentary and non-complimentary strands bind non-specifically with equal rates.

Both assumptions are reasonable. Assumption 1 ignores NSB in hot modules, yet it is apparent from experimental data that once NSB takes place, it takes hours for those strands to become disassociated.

The model is a set of linear difference equations that take into account binding and dissociation ratios under various conditions. The simulation was run on a spread sheet (Excel).

In the model, the integers *k, n, i* refer to the following:

> *k* refers to the number of variables in the computation;
> *n* refers to the computation step that is in progress;
> *i* refers to the capture module that is in progress;

Let $\mathbf{X}$, $\mathbf{Y}$ be 2^k vectors, the components of which represent percentages of concentrations of each truth assignment strand of the combinatorial DNA.

$\mathbf{X^i}$ - this vector represents the percentage of released strands entering module *i* after having left module *(i – 1)*.

$\mathbf{Y^i}$ - this vector represents the percentage of strands binding in module *i*.

We can represent the state of various quantities of interest for each *i* th module at the *n* th computational step as a series of linear equations:

1. $Y^i_n = C^i\, X^i_n$ $i > n;$ Y^i_n gives the percentage of strands complimentary to the probes of the ith module that actually bind to those probes. C^i is a 2^k x 1 matrix of binding efficiencies.

2. $X^{i+1}_n = [\,1 - C^i\,]\; X^i_n\;;$ X^{i+1}_n gives the percentage of strands that did not bind to the i th module and which will continue on to the (i th + 1) module.

3. $X^{i+1}_{n+1} = R^i\; Y^i_n\;;$ R^i is a 2^k x 1 matrix of release efficiencies; X^{i+1}_{n+1} gives the percentage of strands released from a capture module after it is heated.

4. $E^i_n = H^i\, X^i_n\;;$ E^i represents the percentage of strands that bind non-specifically to the ith module at the n th step. H^i is a 2^k x 1 matrix of non-specific binding efficiencies.

5. $E^{Total} = \sum_{n=1}^{n=k+4} E^i_n - Y^{i=k+4}_{n=k+4}\;;$ E^{Total} gives the total percentage of strands that have bound non-specifically in the final module.

Using Excel for the simulation, we get the following surface graph (Fig. 6) that shows the effects of NSB vs. various binding efficiencies for a 6 variable computation. If we assume that we need at least 10 correct solution strands to every 1 error strand to un-ambiguously PCR amplify the read out, we see that NSB will prevent the correct readout of an answer. In general, we see that with even very small amounts of NSB, the ratio of good strands to contaminating (bad) strands drops drastically.   Binding efficiency - the efficiency with which strands bind to their proper complementary probe - contributes to the problem, but not by very much.

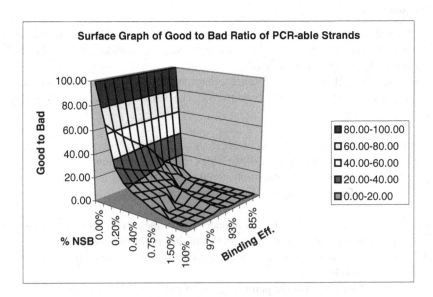

Fig. 6. Error Surface Graph

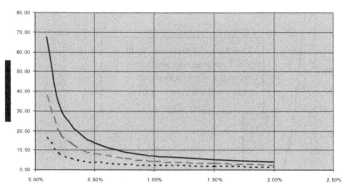

Fig. 7. Binding Efficiency Effect

This is seen more clearly in Fig. 7. In Fig. 7, we compare the ratios of correct answer strands to wrong strands found in the final module for three different binding efficiencies. 100% binding is the ideal, i.e., when all of the strands that should bind to probes in the final module do in fact bind; 95% binding is the efficiency claimed by the technical staff at Mosaic;[1] and 88% binding is the lowest efficiency obtained experimentally in the laboratory. As is seen in Fig. 7, for a 6 variable problem, any rate higher than 0.15% for NSB may cause problems in solution resolution. Fig. 8 extrapolates these results to problems of higher complexity.

Fig. 8 shows the results of the simulation for a NSB rate of 0.1%. The x-axis represents computational complexity, i. e., the number of variables in a 3-CNF SAT computation. From Fig. 8, we see that at the 0.1% rate, the percentage of NSB strands is equal to the percentage of answer strands for a 7 variable SAT problem. This corresponds closely with experimental observation.

Experimental results, phosphor imaging data and computation runs, are consistent with the modeling simulations - that is that the build up of contaminating strands due to NSB in the final module will swamp out the correct answer strands for computations with $n > 7$. So, if the model's premises are true, there seems to be no way to circumvent this build-up, and the in-line architecture seems to be condemned to toy computations of just a few variables.

Or is it?

The surprise lies in the way the final capture module is handled. From the DNA6 paper "Solution of a Satisfiability problem on a gel based DNA computer," [4]

---

[1] Personal communication.

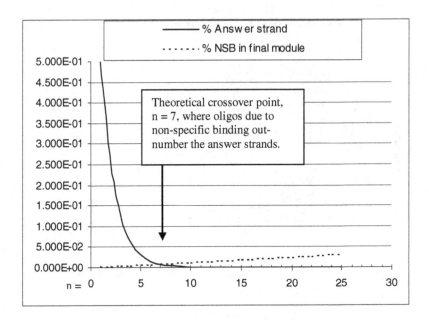

Fig. 8. Swamping Effect of Non-Specific Binding

we see that the gel is extracted from the glass tube and the final capture layer is dissected away. It is then crushed and soaked in 5 ml of water. The captured answer strands are then extracted from the gel by incubating the gel at 65°C for 12 hours.

This procedure allows contaminating strands the time to leach out of the final gel module into solution along with the answer strands. Even though the dissociation time for NSB is on the order of hours, 65°C for 12 hours is long enough for all DNA strands, answer strands as well as error strands, to be eluted from the crushed gel.

A better procedure would be, at the end of the computation, under denaturing conditions (65°C), to elute the answer strands either into a pristine gel module or onto an elution membrane via electrophoresis, for half an hour. This would allow the answer strands cleared out of the final computation module leaving behind the error strands. Here, the long dissociation time for NSB works for us.

6 Conclusion

The in-line architecture is attractive because of its simplicity and the apparent ease of automatability. However, the in-line geometry was plagued by the effects of non-specific binding. We have seen that a simple protocol change would probably have lessened the effects of non-specific binding.

However, many different and poorly understood factors affect the phenomenon of non-specific binding – the type of gel, the buffer, the type of glass, can all affect non-specific binding. To properly characterize non-specific binding would be a lengthy and frustrating undertaking. In fact, the best strategy is probably to employ a geometry that precludes the effects of non-specific binding, that is, to use something like the waste-well architecture that was eventually employed, albeit manually, in the 20 variable computation [5], and as an automated DNA computer in solving 10 variable SAT problems[8].

Acknowledgements

I would like to thank Rebecca A. Anderson for her support on this project.

References

1. Adleman, L.M.: Molecular computation of solutions to combinatorial problems. Science 266, 1021–1024 (1994)
2. Boneh, D., Dunworth, C., Lipton, R.: Breaking DES using a molecular computer. In: Lipton, R.J., Baum, E.B. (eds.) DNA Based Computers: Proceedings of a DIMACS Workshop, April 4, 1995. DIMACS: Series in Discrete Mathematics and Theoretical Computer Science, vol. 27, pp. 37–65. Princeton University. American Mathematical Society, Providence, RI (1996)
3. Lipton, R.: DNA Solution of Hard Computational Problems. Science 268(5210), 542–545 (1995)
4. Braich, R., Johnson, C., Rothemund, P.W.K., Hwang, D., Chelyapov, N., Adleman, L.: Satisfiability Problem on a Gel Based DNA Computer. In: Condon, A., Rozenberg, G. (eds.) DNA 2000. LNCS, vol. 2054, Springer, Heidelberg (2001)
5. Braich, R., Chelyapov, N., Johnson, C., Rothemund, P., Adleman, L.: Solution of a 20-Variable 3-SAT Problem on a DNA Computer. Science 296, 499–502 (2002)
6. Gu, J., Pardalos, P., Du, D. (eds.): Preface, Satisfiability Problem: Theory and Applications. DIMACS Series in Discrete Mathematics and Computer Science, American Mathematical Society, Providence, Rhode Island (1997)
7. Boneh, D., Dunworth, C., Lipton, R.: Breaking DES using a molecular computer. In: Lipton, R.J., Baum, E.B. (eds.) DNA Based Computers: Proceedings of a DIMACS Workshop, April 4, 1995. DIMACS: Series in Discrete Mathematics and Theoretical Computer Science, vol. 27, pp. 37–65. Princeton University. American Mathematical Society, Providence, RI (1996)
8. Johnson, C.: Automating the DNA Computer. In: Mao, C., Yokomori, T. (eds.) DNA Computing. LNCS, vol. 4287, Springer, Heidelberg (2006)
9. Melzak, K.A., Sherwood, C.S., Turner, R.F.B., Haynes, C.A.: Driving forces for DNA adsorption to silica in perchlorate solutions. J. of Colloid and Interface Science (181), 635–644 (1996)
10. Tanaka, T.: Gels. Scientific American 244(1), 124–138 (1981)

Stepwise Assembly of DNA Tile on Surfaces

Kotaro Somei[1], Shohei Kaneda[2], Teruo Fujii[2], and Satoshi Murata[1]

[1] Interdisciplinary Graduate School of Science and Engineering,
Tokyo Institute of Technology, Yokohama, 226-8502 Japan
`somei@mrt.dis.titech.ac.jp, murata@dis.titech.ac.jp`
[2] Institute of Industrial Science,
The University of Tokyo, Tokyo, 153-8505 Japan
`{shk,tfujii}@iis.u-tokyo.ac.jp`

Abstract. A method of solid-phase self-assembly for building DNA nanostructure in a microfluidic device is proposed in this paper. In this method, pre-assembled DNA lattice is anchored on solid surface, which gives nuclei for further growth of the lattice. Flushing out the solution around the nuclei by flow and replace it by different solution enables us stepwise self-assembly in a single chamber at the constant temperature. The results of experiment to verify feasibility of the proposed method will be shown.

Keywords: DNA tile, self-assembly, stepwise assembly, microfluidic device.

1   Introduction

The production technology of the nanometer order is divided roughly into two categories: top-down, and bottom-up. The top-down production technology, typified by semiconductor processing technology, achieves the resolution of less than a hundred nanometer based on lithography technology. This technology is applied in the MEMS/NEMS, enabling the fabrication of microsensors, micromotors and other micromachines. However, top-down processing technology has an essential drawback; the size of manufacturing facilities greatly increases as the processing method evolves to be more sophisticated. On the other hand, the bottom-up technology realizes nanoscale objects made of atoms, molecules and nanoparticles, not by using external apparatus but by designing interaction among them. This kind of technique, where those components coalesce into complicated nanostructures by self-organization, is now drawing increased attention. Especially, active researches are done on the self-assembly of biomolecules such as DNA and proteins, and on the self-organization using polymer. The bottom-up technology does not require a huge plant for production while making minute processing possible; however there still remain numerous problems with practical applications such as low reliability during the assembly process.

The DNA nanotechnology was initiated in 1982 by Seeman when he proposed self-assembled nanostructures made of DNA molecules [1]. The key in this technology is immobilization of Holliday junction (crossover) to make well-defined DNA structures. Winfree and Seeman utilized one of such structures called DX (double

M.H. Garzon and H. Yan (Eds.): DNA 13, LNCS 4848, pp. 182–190, 2008.
© Springer-Verlag Berlin Heidelberg 2008

crossover) tile to realize a patterned lattice made of these tiles [2]. This method allows us to construct not only simple pattern such as periodic stripes or barcodes, but also the complex algorithmic pattern such as Sierpinski's fractal [3]. However, it is very difficult to obtain perfect DNA lattices in one-pot reaction. The growth process of DNA lattice is strongly influenced by the concentration of monomer tiles as well as the temperature of the solution [4]. As the reaction progresses, decreased concentration of monomer tiles in the tube is inevitable which results in erroneous assembly.

Reif proposed another method of DNA tile self-assembly called the hierarchical assembly procedure [5]. In this method, instead of mixing all kinds of DNA in a single step, several equimolar DNA solutions are combined at an appropriate temperature for the self-assembly of one specific subcomponent. Each subcomponent is independently assembled, and then mixed with solution of another subcomponent at higher temperature to build higher order components and so on. In this manner, they built fully addressable DNA lattice made of larger DNA tiles called 4×4 tile [6]. Advantage of this method is that it requires less kinds of orthogonal sequence of DNA than that of one-pot self-assembly.

We have proposed a method of DNA tile assembly by using microfluidic device [7,8]. We expect that the microfluidic device is advantageous to obtain large high-quality DNA lattice, because it provides a reaction chamber in which the concentration of each DNA component can be kept optimal concentration for the crystals' growth. For this purpose, we have designed and fabricated a special microfluidic device utilizing capillary pump and open reaction chamber that enables real-time, direct AFM observation [7]. We also confirm the flow in the microchannel enhance the hybridization efficiency between immobilized DNA on the wall and DNA molecule in the solution [8].

In this paper, we propose a stepwise self-assembly on surfaces for building DNA nanostructure in a microfluidic device. In our method, pre-assembled DNA lattices are anchored on the microfluidic channel, which serves as nuclei for further growth of the DNA lattices. Flushing out the solution around the nuclei and switching solutions by flow enables us to realize stepwise self-assembly whose can produce the results similar to those of the hierarchical assembly procedure, even in a single reaction chamber. This method also enables us to build a layered structure of DNA lattice at a constant temperature. In the following sections, we show the detail of the concept of the stepwise self-assembly on surfaces and the results of experiment to verify feasibility of the proposed method.

2   Concept of Stepwise Self-assembly on Surfaces

Schematic of the stepwise self-assembly on surfaces is given in Fig.1. This method is suitable for building layered DNA tile lattices with reduced number of orthogonal sticky ends [7]. Note that our method is not limited to this but also effective in building nano-objects with structured hierarchy.

For ease of understanding, a simple example is used hereafter. Three kinds of solution are prepared,

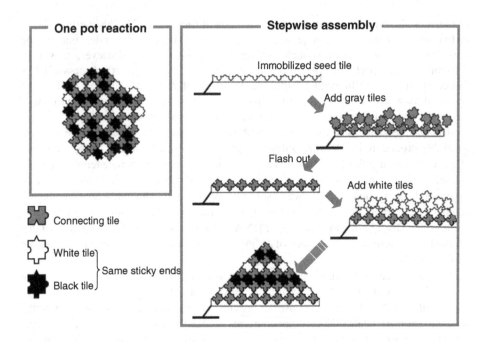

Fig. 1. Stepwise assembly on surfaces

each of which containing a single type of DNA tiles. They are shown as black, gray, and white tiles in the Figure. The incision of four edges represents the sequence of sticky end. Each tile can only be connected to the particular tiles with complementary sticky ends. Here, we assume that the white tiles and black tiles have the identically shaped sticky ends to reduce the number of different sticky ends. Moreover, we designed so that both the white and the black tiles are allowed to attach themselves only to gray tiles, limiting the possibility to the combination of either gray-and-white or gray-and-black.

One-pot reaction, where all three types of the tiles are simultaneously thrown into a tube, results in randomly patterned lattice (Fig.1 left). On the other hand, the solid-phase self-assembly enables us to build well-defined pattern in stepwise fashion. First, pre-assembled seed lattices (initial nuclei) are anchored on a surface by hybridization between the immobilized DNA and the seed lattice. Next, a solution containing only gray tiles is applied. After the reaction, any unconnected DNA tile is washed off with the buffer flow. Then, another solution containing either black or white tiles is applied so the new tiles can hybridize with the previous layer of gray tiles. Further application of the solution containing gray tiles will build a new foundation for yet another layer of black or white tiles. By repeating this process, arbitrary layered pattern can be made from only three kinds of tiles. Also note that the whole process can be done under the same temperature, thus we do not have to change the length of sticky ends for each stage of assembly.

3 Pre-assembly of Nuclei and Their Anchoring on a Gold Surface

In order to initialize lattice growth in the solid-phase, we need crystal nuclei (seed lattices). They must have well-defined lattice structure for the further growth. For this purpose, we prepared a DNA tile lattice made of two columns of DNA tiles (called 2-column DNA lattice, hereafter) (Fig.2.A).

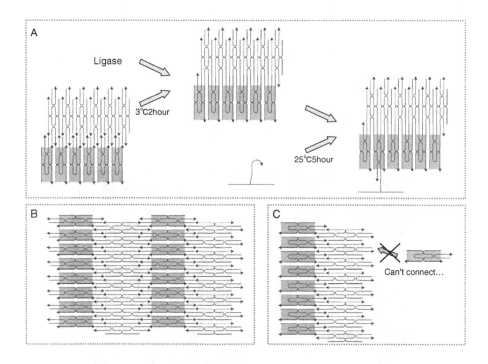

Fig. 2. Pre-assembly of 2-column lattice

3.1 2-Column DNA Lattice

A simple DNA tile set, consisting of two DX tiles is used to make a 2-column DNA lattice. Original tile set generates a large periodic lattice with alternating row of two kinds of tiles, however the shape of the lattice cannot be defined (Fig.2.B). We modified the original tile set to obtain a 2-columned lattice (Fig. 2.C). Here, the strands that comprise lower sticky ends for the gray tile is removed from the original set. We used the same sequence as in the literature [2] for all the strands. The solution must be kept at low temperature (3°C), because this lattice is formed by only one matching sticky ends. Also we have to stabilize the structure by ligation. Then the solution containing the 2-columin lattices is applied on the gold surface, and anchored by an immobilized ssDNA (Fig.2.A).

The pre-assembly of the 2-column DNA lattice was evaluated by gel electrophoresis. The experimental protocols are as follows: 1) A DNA solution for the 2-column lattice is heated up to 95°C (ssDNA 1 μM, 1×TAE, $MgCl_2$ 12.5 mM). 2) It is slowly cooled down to room temperature in water bath (Styrofoam container filled with hot water). It takes overnight. 3) Ligase (T4 DNA Liagase; Takara) is applied to the solution and kept at 10 °C for 2 hours. The resultant solution was evaluated by gel electrophoresis (15%PAGE, 250V, 45 min, room temperature) (Table.1 and Fig.3). Lane 1 is a reference experiment where a full DNA lattice is formed. Most ingredients of the solution remained on the top of gel because of their size. The 2-colum lattice is in Lane 3. The long smear region between full lattice and a band of monomer tile indicates it

Table 1. Experimental condition

	Sample	Ligase
Lane 1	Full DNA lattice	O
Lane 2	Full DNA lattice	X
Lane 3	2-column DNA lattice	O
Lane 4	2-column DNA lattice	X
Lane 5	White tile	O
Lane 6	White tile	X

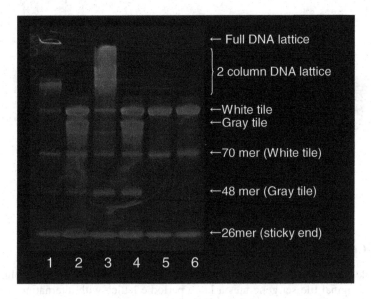

Fig. 3. Gel electrophoresis

forms aggregated structure of various sizes. Ligation process was omitted for lanes 2 and 4. Here, we did not observe any larger structure than a monomer tile. It implies that the 2-column lattices are broken during the electrophoresis.

3.2 Anchoring of 2-Column Lattice

The pre-assembled 2-column lattice is anchored on a solid surface by the following protocols: 1) A 36-base ssDNA (5'-TCA CTC TAC CGC ACC AGA ATG GAG ATT

TTT TTT TTT-SH-3') is put onto gold surface patterned on a glass substrate (72mm x 50 mm, Matsunami). This strand was immobilized by Au-SH bonding (DNA: 50 μM, $MgCl_2$: 200 mM). 2) The surface was rinsed with buffer (1×TAE, $MgCl_2$ 12.5 mM). 3) Solution of 2-column lattice (1μM of each ssDNA, 1×TAE, $MgCl_2$ 12.5 mM) was applied on the surface. The lattice is anchored by hybridization with the immobilized ssDNA (4 hours). One of the sticky end for upper (light gray) tile was modified with FITC for evaluation by fluorescence. 4) The surface was rinsed again with buffer.

Anchoring of the 2-colimn lattice is confirmed by fluorescence intensity (Table.2, Fig.4, 5). As control experiments, another ssDNA with FITC that do not match with the immobilized strand were also tested. The correct combination of the immobilized

Table 2. Experimental condition

	Immobilization	Sample
No.1	O	2-column lattice
No.2	X	2-column lattice
No.3	O	Mismatched ssDNA
No.4	X	Mismatched ssDNA
No.5	O	None

No.1   No.2   No.3   No.4   No.5

Fig. 4. Image of fluorescence microscope (excitation wavelength: 488nm, Emission wavelength: 530nm)

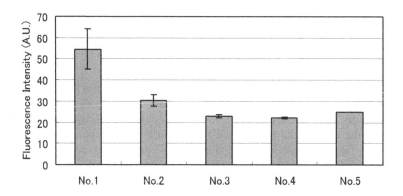

Fig. 5. Fluorescent intensity of each sample

strand and the 2-column lattice showed the highest intensity. Intensity of other cases were at the same level of background (No.5).

4 Stepwise Self-assembly of DNA Tile on Surfaces

We examined the stepwise self-assembly initiated by the anchored pre-assembled 2-column DNA lattice. In order to demonstrate the method, two kinds of DNA tiles (gray and white tile in Fig.6) are prepared. Only the white tile was modified by FITC, and also, only the gray tiles can associate with the 2-column lattice. The lattice can grow only when applying solutions in the order of "gray, wash, white, wash, gray, wash ...." Fig.6.A illustrates first two steps of such "correct" sequence for the growth. By contrast, Fig.6.B illustrates "wrong" sequence (white, wash, gray, wash, ...). In this case, white tiles cannot associate with the 2-column lattice, thus no further growth occurs. In other words, the growth of DNA lattice must be dependent on the order of solution applied to the nuclei.

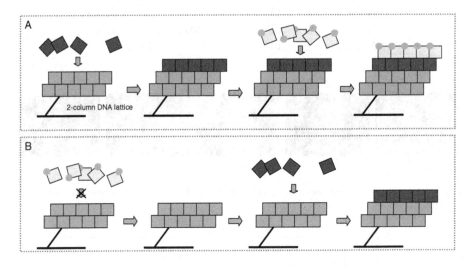

Fig. 6. Experimental scheme of stepwise assembly

This was examined by the following protocols: 1) A 2-column DNA is anchored by the same process described in section 3.2. 2) 2 hours later, its surface was washed by buffer and was dried quickly by N_2 blower. 3) A solution of monomer DNA tiles was applied on the surface and left for two hours to allow the association of the tiles. 4) Then the surface was washed and dried again. This process was repeated according to the specific order of application.

Four kinds of different sequence were compared (Table.3). Fig.7 and 8 show the results of fluorescence intensity measurement. Only No.1 has high intensity, which implies that DNA lattice grows from 2-column DNA lattice only when the monomer solution is applied in the correct order.

Table 3. Experimental condition

	1_{st} Step	2_{nd} Step	3_{rd} Step
No.1	2-column DNA lattice	Gray tile	White tile
No.2	2-column DNA lattice	White tile	Gray tile
No.3	2-column DNA lattice	Gray tile	None
No.4	2-column DNA lattice	White tile	None

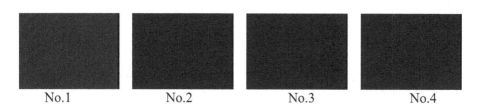

No.1   No.2   No.3   No.4

Fig. 7. Image of fluorescence microscope

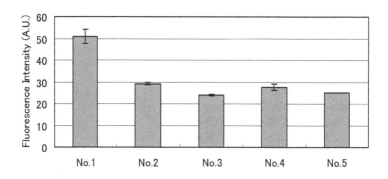

Fig. 8. Fluorescent intensity

5   Conclusion

In this paper, a method of stepwise self-assembly of DNA tile on surfaces is proposed. First, a 2-column DNA lattice as a seed structure is made and anchored on a surface by hybridization. The formation of the 2-colum lattice and its anchoring on the surface is confirmed by fluorescence microscope. Next, the stepwise self-assembly around the anchored seed lattice is evaluated by applying solutions in different order. We verified that only the correct sequence allows the lattice to grow. Although our experiment was simple, it was designed to demonstrate the feasibility and potential of fabricating complicated DNA nanostructure by the solid-phase self-assembly.

All the experiments shown here were done on a glass plate. We have to verify that the same process is valid in the microfluidic device. According to our preliminary experiment, there are hardly any problems except for the absorption of fluorescent particle in PDMS matrix of the microfluidic device. We think this is solved by some straightforward approach. For the future work, we have to observe the surface with AFM to directly examine the product. Improving yield for multi-step assembly is also an important issue.

Acknowledgement

This work is supported by Ministry of Education, Culture, Sports, and Science and Technology of Japan under Grant-in-Aid for Scientific Research on Priority Areas, No. 170590001, 2006.

References

[1] Seeman, N.C.: Nucleic Acid Junctions and Lattices. Journal of Theoretical Biology 99, 237–247 (1982)
[2] Winfree, E., Liu, F., Wenzler, L.A., Seeman, N.C.: Design and self-assembly of two-dimensional DNA crystals. Nature 394(6693), 539–544 (1998)
[3] Rothemund, P., Papadakis, N., Winfree, E.: Algorithmic Self-Assembly of DNA Sierpinski Triangles. PLoS Biology 2(12), 424 (2004)
[4] Winfree, E.: Algorithmic Self-Assembly of DNA. Ph.D Thesis, California Institute of Technology (1998)
[5] Reif, J.: Local Parallel Biomolecular Computation. In: Rubin, H., Wood, D.H. (eds.) DIMACS Series in Discrete Mathematics and Theoretical Computer Science, vol. 48, pp. 217–254 (1999)
[6] Park, S., Pistol, C., Ahn, S., Reif, J., Lebeck, A., Dwyer, C., LaBean, T.: Finite-size, Fully-Addressable DNA Tile Lattices Formed by Hierarchical Assembly Procedures. Angew. Chem, Int. Ed. 45, 735–739 (2006)
[7] Somei, K., Kaneda, S., Fujii, T., Murata, S.: A Microfluidic Device for DNA Tile Self-Assembly. In: Carbone, A., Pierce, N.A. (eds.) DNA Computing. LNCS, vol. 3892, pp. 325–335. Springer, Heidelberg (2006)
[8] Somei, K., Kaneda, S., Fujii, T., Murata, S.: Hybridization in a Microfluidic Device for DNA Tile Self-Assembly. In: Proc. Foundations of Nanoscience, Self-Assembled Architectures and Devices (FNANO 2006), pp. 148–152 (2006)

An Interface for a Computing Model Using Methylation to Allow Precise Population Control by Quantitative Monitoring

Ken Komiya, Noriko Hirayama, and Masayuki Yamamura

Department of Computational Intelligence and Systems Science, Interdisciplinary Graduate
School of Science and Engineering, Tokyo Institute of Technology
{komiya,my}@dis.titech.ac.jp

Abstract. We developed an interface to enable feedback control for a methylation-based computing model, in which a bit string is represented by the methylated and unmethylated status of the specific locations on a DNA molecule. On construction of a reaction system for the computational purpose, it is problematic that an open loop system without feedback control is easy to lose the molecular variety required for computation. It is, thus, important for the methylation-based computing to achieve quantitative sensing for feedback control. Difference in methylation status can be converted into the sequence variation by the bisulfite reaction. As a consequence, distribution between methylated and unmethylated DNA molecules could be quantitatively monitored by combining the polymerase chain reaction (PCR) using methylation specific primers with quantitative PCR. In the present study, we experimentally investigated the feasibility of the proposed interface for controlling the population of a library of DNA registers that have distinct methylated patterns representing different bits. Result indicated that, quantitative measurement of population was successfully performed by discriminative amplification using the methylation-specific primer. This interface, which allows us to generate a homogenous or biased library as expectedly, would be useful for molecular evolutionary computation and molecular learning.

1 Introduction

Development of the technique to quantitatively monitor the population of molecules allows precise population control through computation. In many DNA-based computing models, it is assumed that modifications on a variety of DNA molecules encoding distinct information are equivalently and completely performed by reactions implemented as computational procedure. However, it seems impossible to achieve error-free modifications using biological reactions. As an alternative, an interface allowing feedback control is effective to achieve precise and robust computation, and thus, essential for making DNA-based computing practical. In early studies of DNA-based computing, several solutions to combinatorial problems were demonstrated by taking advantage of hybridization between complementary sequences of DNA molecules. Molecules that encoded the correct solution were successfully selected out of a library of molecules that encoded candidate solutions [1,2]. In those exercises, whether the population of

M.H. Garzon and H. Yan (Eds.): DNA 13, LNCS 4848, pp. 191–200, 2008.

candidate solutions had the unexpected bias did not matter since molecules of the excessive amount to the problem size were contained in a test tube. Towards molecular solution to large scale problems or molecular implementation of evolutionary computation [3] and learning [4], construction of a tractable reaction system, provided with an interface for quantitative monitoring to avoid the unintended bias in the molecular library, is expected as a breakthrough.

In the present study, we developed an interface to enable feedback control for a methylation-based computing model, in which a bit string is represented by the methylated and unmethylated status of the specific locations on a DNA molecule. Difference in methylation status of DNA molecules can be converted into the sequence variation by the bisulfite reaction [5]. As a consequence, distribution between methylated and unmethylatad DNA molecules could be quantitatively monitored by combining the PCR using methylation specific primers [6] with quantitative PCR. We experimentally investigated the feasibility, necessity and effectiveness of the proposed interface for generation of the random library of DNA registers that have distinct methylated patterns representing different bits, each of the expected amount.

2   Methylation Computing

2.1   Methylation-Based Aqueous Computing

Methylation here refers to the reaction performed by methyltransferase, or methylase in short, to add a methyl group to a cytosine or adenine on a double-stranded DNA molecule. It is known that, many bacteria have their own pair of restriction and methylation enzymes sharing a short recognition sequence [7] and methylation plays a role to keep their inherent DNA from self-attacking by the restriction enzyme acting as a protection system against the foreign DNA. In eukaryotes, methylation takes part in regulation of gene expression. Methylation pattern on DNA is succeeded as epigenetic information by hemi-methyltransferase that methylates a hemi-methylated DNA, that is, an unmethylated strand of a double-stranded DNA molecule is methylated according to another methylated strand [8]. Inspired by the strategy of nature, in which genetic information is stored in double-stranded DNA molecules and only necessary parts are processed by cooperative enzymes on demand, computation using double-stranded DNA molecules and sequence specific enzymes, such as restriction enzymes and methylases, would be promising. Although hybridization by single-stranded DNA molecules with their complements to transform into the double-stranded form is utilized for the central process in many DNA-based computing, it might be difficult to achieve precise control of that process in competition with formation of a number of sub-optimal structures [9].

The use of methylation for aqueous computing [10] and Boolean logic [11] was, so far, proposed. Aqueous computing is a characteristic computing model to implement content addressable parallel processing, using molecules dissolved in water as nano-scale registers. In an aqueous solution, the location of a vast number of molecules is randomized. When a solution is partitioned into a chosen number of portions, every portion contains the same variety of molecules. Different modifications on molecules can be performed in parallel in each portion, and then, partitioned portions can be united again into a single solution. In aqueous computing, each molecule has identifiable "stations" and computation is performed by altering station settings according to the constraints. In contrast to the conventional paradigm in DNA-based computing, called Adleman-Lipton's paradigm, in which computation begins with the initial step to generate a library of molecules encoding all candidate solutions, aqueous computation begins with a single variety of molecules, and thus prevents us from sequence design that is often harder than the given problem to be solved, and keeps the number of molecular varieties as small as possible through computation. Many NP-complete algorithmic problem families can theoretically be solved by a single aqueous algorithm [12]. Among many potential reactions proposed to implement aqueous computing [12,13,14,15], the use of methylation as writing operation has the advantages. Erase operation could be performed by PCR since DNA polymerase can not reproduce the methylation status of the template. Copy operation could also be achieved by combinatorial use of DNA polymerase with hemi-methyltransferase. Thus, a rewritable and amplifiable DNA-RAM would be realized. Restriction enzymes, that can not digest their specific cutting sites only when those sites are methylated, provide readout of methylation status. A solution to 3-variable, 4-clause satisfiability problem was attempted [16], using multi-cloning site of a plasmid, methylases and restriction enzymes. Under an assumption that every reaction can be performed at the sufficient efficiency, many problems including the NP-complete can be solved only by methylation and digestion. However, fine control of methylation and digestion reactions for faultless discrimination has not yet been established.

On construction of a reaction system for the computational purpose, it is commonly problematic that an "open loop" system without feedback control is easy to lose the molecular variety required for computation. It is, thus, important for the methylation-based aqueous computing to develop an interface allowing feedback control by quantitative monitoring.

2.2 Interface for Quantification of Methylation

The bisulfite reaction is the method to determine the methylation status of cytosine residues in a DNA molecule by converting intact cytosines in a single-stranded DNA molecule to uracils whereas 5-methylcytosines are unreactive [5] (Fig. 1A). After the conversion, the resulted DNA is subjected to the reaction for discriminating the sequence variation derived from methylation. By PCR with the primer that is complementary to the methylated region, and thus has mismatches to the unmethylated DNA and with Pol I-type DNA polymerase that lacks the 3' to 5' exonuclease activity for proof-reading, methylation-specific PCR can be performed [6]. Quantitative PCR (qPCR) is the method to determine the relative amount of target DNA by comparing the cycle number at which the exponential increase of fluorescence intensity begins to be detected. qPCR using SYBR Green, which can do without expensive fluorescent probes to be carefully designed, is tractable. For achieving an interface to quantitatively monitor the methylation status, we employed qPCR using SYBR Green with primers, each having one single mismatch at its 3' end with either the methylated or unmethylated station of DNA register. (Fig. 1B).

3   Materials and Methods

3.1   Preparation of a DNA Register

We prepared by PCR amplification a short double-stranded DNA molecule, **R4S**, of 80 base pairs (bp) as a DNA register of high density, having four stations to be altered by methylation. The sequence of **R4S** was generated by simply putting recognition sequences of HhaI and HpaII methylases and restriction enzymes between the sequences commonly used in molecular biology study as primers for T7 promotor and M13, involving recognition sequences of HaeIII and AluI in them. All oligonucleotides used in the experiment were commercially synthesized and purified by Sigma Genosys (Fig. 1C).

PCR reaction for preparation was performed in a 50-μl solution containing PrimeSTAR HS DNA polymerase (Takara Bio) buffer, 0.2mM dNTP, primers **P1** and **P2** (50 pmol each), 1.25 units PrimeSTAR HS DNA polymerase, and 3.9 ng **R4S**. Reaction mixture was incubated at 98°C for 2min, then subjected to the thermalcycle condition of 30 cycles (98°C for 10 sec, 55°C for 5 sec, 72°C for 10 sec) followed by an incubation at 72°C for 2min. PCR product was purified by 16% polyacrylamide gel electrophoresis (PAGE).

3.2   Methylation Specific Amplification

For the confirmation of discriminative amplification using the methylation-specific or non-methylation-specific primer that has only one single mismatch at its 3' end, we performed realtime PCR either with the control oligonucleotide, **methyl-R4S** or **non-methyl-R4S**, of the sequences same as the bisulfite-converted **R4S** after methylation on all four stations or before methylation, respectively. PCR reaction was performed at a 20-μl scale in the rTaq DNA polymerase (Toyobo) buffer containing 2.5 mM magnesium chloride, 0.2mM dNTP, 5 pmol the methylation-specific or non-methylation-specific primer, 5 pmol the reverse primer **P2 bisulfite**, 2 units rTaq DNA polymerase, 1X SYBR Green I (Molecular Probes) and 40 pg the control oligonucleotide. Reaction mixture was incubated at 95°C for 2min, then subjected to the thermalcycle condition of over 30 cycles (95°C for 30 sec, 40°C for 30 sec, 72°C for 30 sec) followed by an incubation at 72°C for 2min. Fluorescence intensity was measured at every annealing step with Mx3005P (Stratagene).

3.3   Quantification of Methylation

We first optimized digestion reaction to determine the least amounts of restriction enzymes required for complete digestion of unmethylated DNA, then optimized methylation reaction to determine the least amounts of methylases required for complete blocking of digestion. All methylases and HhaI, HaeIII, AluI restriction enzymes were purchased from New England Biolabs. HpaII and HindIII restriction enzymes from Takara Bio. Digestion reaction was performed at a 25-μl scale with 25 ng methylated or unmethylated **R4S** in the buffer recommended by the supplier, NEBuffer 4 with BSA for HhaI, NEBuffer 2 for HaeIII and AluI, L buffer for HpaII, and M buffer for HindIII, respectively. Reaction mixture was incubated at 37°C for 1 hour. Methylation reaction was performed

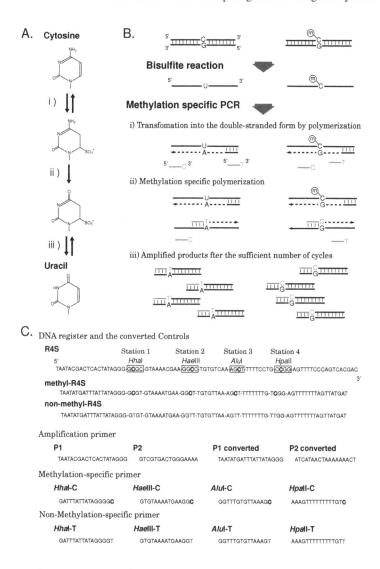

Fig. 1. A. Bisulfite reaction Cytosine is converted to uracil by i) sulphonation, ii) deamination, and iii) desulphonataion. 5-methyl cytosine remains unconverted. **B. methylation specific PCR** The black horizontal lines, vertical lines and dashed arrows indicate DNA strands, base parings and extensions by DNA polymerase, respectively. The letter, m in a circle indicates the methylation at the 5th carbon of cytosine. The gray horizontal lines with the 3' terminal base, C and T represent the methylation-specific and non-methylation-specific primers, respectively. Only the primers perfectly matching with the template can hybridize upon PCR amplification. **C. Sequences used in this study** Recognition sequences of methylases to implement stations are indicated by rectangles. Cytosines in the sequence of DNA register (**R4S**) to be methylated, of the methylation control (**methyl-R4S**), and at the 3' end of the methylation-specific primers are indicated by the bold letters. Thymines of the non-methylation control (**non-methyl-R4S**) and at the 3' end of the non-methylation-specific primers are indicated by the gray bold letters.

at a 10-μl scale with 50 ng **R4S** and 80 μM S-adenosylmethionine in the Methylase Buffer 1 (50 mM Tris-HCl (pH 7.5), 10 mM EDTA, 5 mM 2-mercaptoethanol) for HhaI, AluI and HpaII, or the Methylase Buffer 2 (50 mM NaCl, 50 mM Tris-HCl (pH 8.5), 10 mM DTT) for HaeIII. Reaction mixture was incubated at 37°C for 1 hour.

The bisulfite reaction is ordinarily performed for the conversion of long genomic DNA. The short DNA register of this study required some modifications of the common protocol to avoid the damage and loss of DNA. After the partial methylation reaction (details in reaction conditions are described in Sec. 4.2), we performed bisulfite reaction following the instruction of MethylEasy DNA Bisulphite Modification Kit (Human Genetic Signatures) except for the incubation time reduced to 3 hours for the bisulfite reaction step and the temperature lowered to -20°C for the isopropanol precipitation step. Finally, we quantified the amounts of the methylated and unmethylated DNA registers for HhaI and HeIII stations to determine the population by methylation specific qPCR. The experimental protocol was same as Sec. 3.2 and the control oligonucleotide, **methyl-R4S** or **non-methyl-R4S** (each 0.4 to 400 pg) were also amplified to generate the standard curves for quantification. The amounts of DNA were calculated using the software MxPro (Stratagene).

4   Results

4.1   Methylation Specific Amplification

Fig. 2 illustrates the resulted amplification plots for 8 specific primers. Clear discrimination was confirmed for methylation status of the stations implemented by HhaI and HaeIII recognition sites, but not for AluI. For the station of HpaII, discrimination was achieved, though the efficiency of amplification was low probably due to the short length of the amplified product unsuitable for qPCR.

4.2   Quantification of Methylation

Fig. 3A illustrates the results of optimized methylation and digestion reactions. Although complete digestion by AluI restriction enzyme and perfect blocking of HaeIII-digestion by HaeIII methylase could not be achieved by increasing the amount of enzyme and/or incubation time, the alternative use of HindIII restriction enzyme whose recognition sequence involves that of AluI and repeated methylation reactions by HaeIII methylase, respectively, allowed almost digital discrimination between the methylated and unmethylated status, confirmed by visual inspection of PAGE analysis. Complete writing and readout operation on all of four stations appeared to be achieved. The determined least amounts of HhaI, HaeIII, HindIII and HpaII required for the complete digestion were 4.0, 4.4, 12 and 4.0 units/25 ng DNA, respectively. The determined least amounts of HhaI, HaeIII, AluI and HpaII methylases required for the perfect blocking were 1.5, 5.5 (3 times), 2.4 and 1.0 units/50 ng DNA, respectively.

Fig. 3C illustrates the amplification plots obtained by qPCR for the stations of HhaI and HaeIII, following the partial methylation reaction. For HaeIII station, methylation was performed only one time with 5.5 units HaeIII methylase. For other stations, methylation was performed under an incubation at 37°C for 45 min with 30% amount of

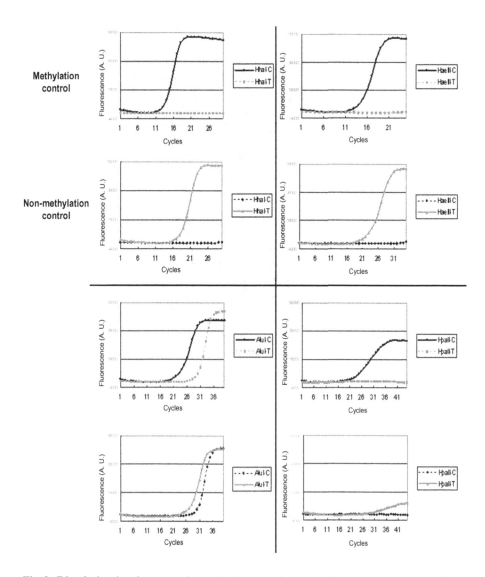

Fig. 2. Discrimination between the methylated and unmethylated status by methylation specific PCR The black and gray curves are the amplification plots of methylation specific PCR with the methylation-specific and non-methylation-specific primers, respectively. The curve represented by the bold or dashed line indicates the matched (methylation control with the methylation-specific primer or non-methylation control with the non-methylation-specific primer) or mismatched (methylation control with the non-methylation-specific primer or non-methylation control with the methylation-specific primer) template-primer pair, respectively.

the respective methylase required for perfect blocking. The calculated values of the methylated population were 37.1% and 81.7% for the stations of HhaI and HaeIII, respectively.

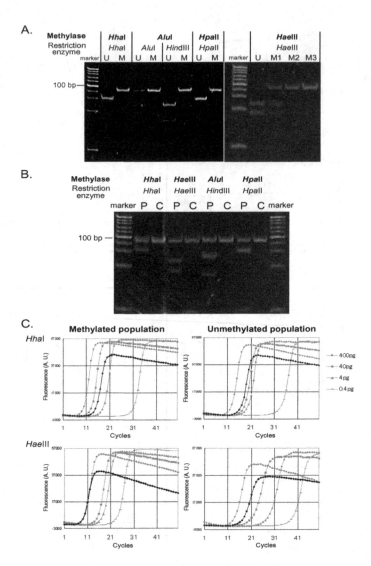

Fig. 3. A. Digital discrimination between the methylated and unmethylated status U and M denotes the lanes of the unmethylated and methylated DNA registers, subjected to digestion, respectively. M1, M2 and M3 denotes the methylation repeated for 1, 2 and 3 times. Comparing to the 20-bp ladder marker, bands of uncut DNA appeared at the position of around 80 bp and bands of cut DNA appeared at the appropriate positions. **B. Confirmation of partial methylation** P and C denotes the lanes of DNA registers, digested after methylation reaction under the condition for partial and complete methylation, respectively. Partial methylation was confirmed by the incomplete digestion in the lanes P. **C. Quantification of methylation status** The gray and black curves in the left panels are the amplification plots of the methylation control as standards and the partially methylated samples, respectively, with the methylation-specific primer. Those in the right panels are the amplification plots of the non-methylation control as standards and the partially methylated samples, respectively, with the non-methylation-specific primer.

5   Discussion

The experimental results presented in the current work provide a clear validation of the feasibility of an interface for quantitative monitoring of methylation status. This interface allows feedback control of the computation implemented by methylation, and thus enables precise population control through computation. For molecular solution to combinatorial problems at a large scale, generation of a homogenous library is important, that is, molecules encoding all distinct candidate solutions in a test tube should be given the equal chance to be processed. Molecular learning and evolutionary computation, as an optimization process under some constraints, would take advantages of the ability of biomolecules that had evolved in nature by optimizing their properties according to the environment, and thus, be a promising direction to expand DNA-based computing. For evolutionary computation, how to maintain the diversity of a library is an issue. Similarly to that techniques to generate truly random numbers have been devised for making a library with the expected population in conventional evolutionary computation *in silico* [17], molecular evolutionary computation *in vitro* also requires establishing the technique to construct a homogenous or biased molecular library as expectedly. For the molecular learning in which the population of molecules is changed according to the difference between the amount of those consistent with training examples and those inconsistent [4], quantitative sensing is necessitated at each round of training. When the proposed interface is applied to molecular learning and evolutionary computation, parallel operation in a 1.5-ml test tube containing a library of molecules beyond the size of 10^{18} could surpass conventional computers and retrieve the attraction of DNA-based computing. Moreover, a distinctive feature that outcome of computation is an aqueous solution containing molecules adapted to a certain constraint, might lead to engineering of an intelligent aqueous solution or a functional molecule set for biological and diagnosis purposes. For the next step of the present study, the functionality of feedback control with a quantitative interface in methylation-based computing should be investigated in the experiment.

Acknowledgements

This research was partially supported by the Ministry of Education, Culture, Sports, Science and Technology of Japan, Grant-in-Aid for Scientific Research on Priority Areas (14085101) and Young Scientists (B) (18700298). We thank to Tom Head and Susannah Gal at Binghamton Univ. and Daisuke Kiga at TITech for their kind advisements.

References

1. Adleman, L.: Molecular computation of solutions to combinatorial problems. Science 266, 1021–1024 (1994)
2. Ouyang, Q., Kaplan, P.D., Liu, S., Libchaber, A.: DNA solution of the maximal clique problem. Science 278, 446–449 (1997)
3. Zhang, B.-T., Jang, H.-Y.: Molecular programming: Evolving genetic programs in a test tube. In: 2005 Genetic and Evolutionary Computation Conference, Proceedings, vol. 2, pp. 1761–1768 (2005)

4. Sakakibara, Y.: Population computation and majority inference in test tube. In: DNA Computing. 7th Int'l Workshop on DNA-Based Computers, pp. 82–91 (2002)
5. Hayatsu, H., Wataya, Y., Kai, K., Iida, S.: Reaction of sodium bisulfite with uracil, cytosine, and their derivatives. Biochemistry 9, 2858–2866 (1970)
6. Herman, J.G., Graff, J.R., Myohanen, S., Nelkin, B.D., Baylin, S.B.: Methylation-specific PCR: A novel PCR assay for methylation status of CpG islands. Proc. Nat. Acad. Sci. USA 93, 9821–9826 (1996)
7. Roberts, R., Vincze, T., Posfai, J., Macelis, D.: REBASE - restriction enzymes and DNA methyltransferases. Nuc. Acics Res. 33, D230–D232 (2005), http://rebase.neb.com
8. Bacolla, A., Pradhan, S., Robert, R.J., Wells, R.D.: Recombinant human DNA (cytosine-5) methyltransferase. ii. steady-state kinetics reveal allosteric activation by methylated DNA. J. Biol. Chem. 274, 33011–33019 (1999)
9. Dimitrov, R.A., Zuker, M.: Prediction of hybridization and melting for double-stranded nucleic acids. Biophys. J. 87, 215–226 (2004)
10. Head, T.: Writing by methylation proposed for aqueous computing. In: Mitrana, C.M.-V.V. (ed.) Where Mathematics, Computer Science, Linguistics and Biology Meet, pp. 353–360. Kluwer Academic Publishers, Dordrecht (2001)
11. Dimitrova, N., Gal, S.: Methylogic: Implementaion of Boolean logic using DNA methylation. In: DNA Computing. 12th Int'l Workshop on DNA-Based Computers, pp. 404–417 (2006)
12. Head, T., Yamamura, M., Gal, S.: Aqueous computing: writing on molecules. In: 1999 Congress on Evolutionary Computation, Proceedings, pp. 1006–1010 (1999)
13. Head, T., Rozenberg, G., Bladergroen, R.S., Breek, C.K.D., Lommerse, P.H.M., Spaink, H.: Computing with DNA by operating on plasmid. BioSystems 57, 87–93 (2000)
14. Yamamura, M., Hiroto, Y., Matoba, T.: Another realization of aqueous computing with peptide nucleic acid. In: DNA Computing. 7th Int'l Workshop on DNA-Based Computers, pp. 213–222 (2002)
15. Takahashi, N., Kameda, A., Yamamoto, M., Ohuchi, A.: Aqueous computing with DNA hairpin-based RAM. In: DNA Computing. 10th Int'l Workshop on DNA-Based Computers, pp. 355–364 (2005)
16. Gal, S., Monteith, N., Shkalim, S., Huang, H., Head, T.: Methylation of DNA may be useful as a computational tool: Experimental evidence. In: Mahdavi, K., Culshaw, R., Boucher, J. (eds.) Current Developments in Mathematical Biology. Series in Knots and Everything, vol. 38, pp. 1–14. World Scientific (2007)
17. Kita, H., Yamamura, M.: A functional specialization hypothesis for designing genetic algorithms. Proceedings of 1999 IEEE International Conference on Systems, Man and Cybernetics III, 579–584 (1999)

Hardware Acceleration for Thermodynamic Constrained DNA Code Generation

Qinru Qiu[1], Prakash Mukre[1], Morgan Bishop[2], Daniel Burns[2], and Qing Wu[1]

[1] Department of Electrical and Computer Engineering, Binghamton University,
Binghamton, NY 13902
[2] Air Force Research Laboratory, Rome Site, 26 Electronic Parkway, Rome, NY 13441
qqiu@binghamton.edu, pmukre1@binghamton.edu,
Morgan.Bishop@rl.af.mil, Daniel.Burns@rl.af.mil,
qwu@binghamton.edu

Abstract. Reliable DNA computing requires a large pool of oligonucleotides that do not cross-hybridize. In this paper, we present a transformed algorithm to calculate the maximum weight of the 2-stem common subsequence of two DNA oligonucleotides. The result is the key part of the Gibbs free energy of the DNA cross-hybridized duplexes based on the nearest-neighbor model. The transformed algorithm preserves the physical data locality and hence is suitable for implementation using a systolic array. A novel hybrid architecture that consists of a general purpose microprocessor and a hardware accelerator for accelerating the discovery of DNA under thermodynamic constraints is designed, implemented and tested. Experimental results show that the hardware system provides more than 250X speed-up compared to a software only implementation.

1 Introduction

A single DNA strand (i.e. oligonucleotides) is a sequence of four possible nucleotides denoted as A, C, G and T. Short DNA sequences can be synthesized easily and be used for different applications, including high density information storage [2], molecular computation of hard combinatorial problems [1], and molecular barcodes to identify individual modules in complex chemical libraries [3]. These applications rely on the specific hybridization between DNA code words and their Watson-Crick complements. The key to success in DNA computing is the availability of a large collection of DNA code word pairs that do not cross-hybridize.

The capability of hybridization between two oligonucleotides is determined by the base sequences of the hybridizing oligonucleotides, the location of potential mismatches, the concentrations of the molar strand, the temperature of the reaction and the length of the sequences [4]. The *melting temperature* (T_m) is a parameter that characterizes these factors [4]. It is defined as the temperature at which 50% of the DNA molecules have been separated to single strands. Another closely related measure of the relative stability of a DNA duplex is its Gibbs free energy, denoted as ΔG^O. The nearest-neighbor (NN) model [7][10] was proven to be an effective and accurate estimation of the free energy. In [12], the concept of *t-stem block insertion-deletion*

M.H. Garzon and H. Yan (Eds.): DNA 13, LNCS 4848, pp. 201–210, 2008.
© Springer-Verlag Berlin Heidelberg 2008

codes was introduced that captures the key aspects of the nearest neighbor model. In the same reference, a dynamic programming algorithm is presented to calculate the maximum weight of the t-stem common subsequence.

Search methods for DNA codes are extremely time-consuming [5], and this has limited research on DNA codeword design, especially for codes of length greater than about 12-14 bases. For example, the largest known DNA codeword library, which has been generated based on the edit distance constraint with length 16 and edit distance 10, consists of 132 pairs, and composing such codes takes several days on a cluster of 10 G5 processors, with no guarantee of optimality.

In [8], we presented a novel accelerator for the composition of reverse complement, edit distance, DNA codes of length 16. It incorporates a hardware GA, hardware edit distance calculation, and hardware exhaustive search which extends an initial codeword library by doing a final scan across the entire universe of possible code words. The proposed architecture consists of a host PC, a hardware accelerator implemented in reconfigurable logic on a *field programmable gate array* (FPGA) and a software program running in a host PC that controls and communicates with the hardware accelerator. The proposed architecture uses a modified genetic algorithm that uses a locally exhaustive, mutation-only heuristic tuned for speed. The architecture reduces the search time from 6+ days (on 10 Pentium processors) to 1.5 hours, achieving an effective 1000X speed-up, and it produces locally optimum codes.

The edit distance metric only provides a first order approximation of the free energy of binding of DNA duplexes. To improve the quality of DNA codes, more accurate metrics based on the thermodynamics of binding of DNA duplexes must be considered. This paper focuses on implementing the nearest-neighbor based free energy calculation on a reconfigurable hardware accelerator. We present a transformed algorithm to calculate the maximum weight of the 2-stem common subsequence of two DNA oligonucleotides. The result is the key part of the Gibbs free energy of the DNA cross-hybridized (CH) duplexes based on the nearest-neighbor model. The transformed algorithm preserves the physical data locality and hence is suitable for implementation using a systolic array. A new hardware accelerator for accelerating the discovery of locally optimum DNA codes with thermodynamic constraints is described. At this writing the proposed architecture provides more than 250X speed-up compared to a software only implementation.

The remainder of this paper is organized as follows: Section 2 describes the transformed algorithm and its hardware implementation using a 2D systolic array. Section 3 presents our formulation of the problem, and the solution technique in hardware GA. Section 4 provides a performance comparison between the software version and the hardware version of the codeword search. Section 5 presents final conclusions.

2   Calculation of NN Free Energy Using 2D Systolic Array

The thermodynamics of binding of nucleic acids has been widely studied and reported in the literature. The nearest-neighbor (NN) model [10] was proven to be an effective and accurate estimate of the thermodynamic binding energy. The NN model assumes that stability of a DNA duplex depends on the identity and orientation of neighboring base pairs. There are 10 possible NN pairs: AA/TT, AT/TA, TA/AT, CA/GT, GT/CA,

CT/GA, GA/CT, CG/GC, GC/CG, and GG/CC. Based on the NN model, the total free energy change of a DNA duplex at temperature T can be calculated by the following equation:

$$\Delta G^{o}_{T}(total) = \Delta G^{o}_{T,initiation} + \Delta G^{o}_{T,symmetry} + \sum_{i \in Watson-Crick\ NNs} G^{o}_{T,stack}(i) + \Delta G^{o}_{T,AT\ Terminal} \tag{1}$$

where $\Delta G^{o}_{T,initiation}$ is the initiation energy, $\Delta G^{o}_{T,symmetry}$ is a parameter that reflects whether the duplex is self-complementary, $\Delta G^{o}_{T,AT\ Terminal}$ is a parameter that accounts for the differences between duplexes with terminal AT versus terminal GC, and $\Delta G^{o}_{T,stack}(i)$ gives the thermodynamic energy of Watson-Crick NN duplex i, which is determined by the structure of the primary sequence of the DNA duplex. This work focuses on accelerating the calculation of NN free energy using reconfigurable hardware and applies it to hardware based DNA code word search.

We developed a dynamic programming algorithm to calculate the NN free energy based on the technique presented in [12]. Given a CH duplex $x : y'$, where y' is the Watson-Crick complement of y, we define 3 matrices. They include a *suffix matrix* (s) which stores the longest common suffix between x and y, a *weighted suffix matrix* (ws) which stores the accumulated weight of each common stem-2 and an energy matrix (e) which stores the accumulated free energy of the possible NNs. The value of the ijth entry of these matrices can be calculated using the following equations.

$$s_{ij} = \begin{cases} s_{i-1,j-1} + 1 & \text{if } x[i] = y[j] \\ 0 & \text{otherwise} \end{cases} \tag{2}$$

$$ws_{i,j} = \begin{cases} ws_{i-1,j-1} + w(x[i-1], x[i]) & \text{if } x[i] = y[j] \ \& \ x[i-1] = y[i-1] \\ 0 & \text{otherwise} \end{cases} \tag{3}$$

$$e_{ij} = \begin{cases} \max(ws_{i,j} - ws_{i-1,j-1} + e_{i-2,j-2}, ws_{i,j} - ws_{i-2,j-2} + e_{i-3,j-3}, \\ \quad ....., ws_{i,j} - ws_{i-s_{ij},j-s_{ij}} + e_{i-3,j-3}, \ e_{i,j-1}, e_{i-1,j}) & \text{if } x[i] = y[j] \\ \max(e_{i-1,j-1}, e_{i,j-1}, e_{i-1,j}) & \text{otherwise} \end{cases} \tag{4}$$

The parameter $w(a[i-1], a[i])$ is the stack-pair free energy between nearest-neighbor base pairs $a[i-1]$ and $a[i]$. The bottom right entry of the e matrix gives the NN free energy of $x : y'$.

Systolic array processing has been widely used in parallel computing to enhance computational performance. The general systolic architecture has $N \times N$ connected processors, as shown in Figure 1 (b). Each processor performs an elementary calculation. The processor $P(i,j)$ reads data from its up stream neighbors $P(i-1,j)$, $P(i,j-1)$ and $P(i-1, j-1)$, and propagates the results to its down stream neighbors $P(i+1,j)$, $P(i,j+1)$ and $P(i+1, j+1)$. After an initialization, or latency period that fills the pipeline, the array generates one result per 2 clock periods.

Equations (2)~(4) cannot be directly mapped to a 2D systolic array architecture because to calculate e_{ij} we need the value of $ws_{i-d, j-d}$ ($e_{i-d, j-d}$), $1 \le d \le s_{ij}$. The variable e_{ij} is calculated by processor $P(i,j)$. The variables $ws_{i-d, j-d}$ and $e_{i-d, j-d}$ are calculated by processor $P(i-d, j-d)$. If the calculation of e_{ij} is performed at clock period t, then the calculations of $ws_{i-d, j-d}$ and $e_{i-d, j-d}$ for the same DNA duplex are performed at clock period $t - 2d$. Because cells in the systolic array will register the new input and update their results every 2 clock periods, it is not possible for us to access the values of $ws_{i-d, j-d}$ and $e_{i-d, j-d}$ at clock period t if d is greater than 1. One way to handle this problem is to store the values of $ws_{i-d, j-d}$ and $e_{i-d, j-d}$ in memory or in registers. Because the maximum value of s_{ij} can be as high as the length of the DNA strand, which in our case is 16, this solution would require duplication of each cell in the systolic array 16 times. This is not practical as it significantly increases the hardware cost.

In this work, we use function transformation to simplify the hardware design. We define a *minimum weighted suffix matrix* (**min_ws**) which stores the minimum value of the difference between $ws_{i-d, j-d}$ and $e_{i-d-1, j-d-1}$, where $1 \le d \le s_{ij}$. The ijth entry of **min_ws** can be calculated as

$$min\_ws_{ij} = \begin{cases} min(min\_ws_{i-1, j-1}, ws_{ij} - e_{i-1, j-1}) & \text{if } x[i] = y[j] \\ 1{,}000{,}000 & \text{otherwise} \end{cases} \tag{5}$$

when $x[i] \ne y[j]$, min_ws_{ij} will be set to an extremely large number, otherwise, it is the minimum between $min\_ws_{i-1,j-1}$ and $ws_{ij}-e_{i-1,j-1}$. The calculation of e_{ij} and ws_{ij} is transformed into the following equations.

$$ws_{i,j} = \begin{cases} ws_{i-1, j-1} + w(x[i-1], x[i]) & \text{if } x[i] = y[j] \ \& \ min\_ws_{i-1, j-1} \ne 1{,}000{,}000 \\ 0 & \text{otherwise} \end{cases} \tag{6}$$

$$e_{ij} = \begin{cases} max(ws_{i,j} - min\_ws_{i-1, j-1}, e_{i, j-1}, e_{i-1, j}) & \text{if } x[i] = y[j] \\ max(e_{i-1, j-1}, e_{i, j-1}, e_{i-1, j}) & \text{otherwise} \end{cases} \tag{7}$$

Equations (5)~(7) are equivalent to equations (2)~(4), however, only information from adjacent cells is needed in the calculation, hence, they can be implemented using the systolic array architecture.

The hardware design of the 2D systolic array can be derived directly from equations (5)~(7). The systolic array is an $n \times n$ array of identical cells. Each cell in the array has 7 inputs, among which the inputs $e_{i-1,j}$ and $x[i-1, j]$ are from the cell that is located above, the inputs $e_{i,j-1}$ and $y[i, j-1]$ are from the cell that is located to the left, and the inputs $e_{i-1,j-1}$, $ws_{i-1,j-1}$ and $min\_ws_{i-1,j-1}$ are from the cell that is located to the upper left. Each cell performs the computations that are described in equations (5)~(7). For cell (i,j), the outputs $x_{i,j}$ and $y_{i,j}$ are equal to the inputs $x_{i-1,j}$ and $y_{i,j-1}$. Figure 1 (a) gives the structure of each cell, including its input/output connections and the computation implemented. The variables $x_{i,j}$ and $y_{i,j}$ are represented as 2 bit binary numbers with A=00, C=01, G=10, and T=11. The variables $e_{i,j}$, $ws_{i,j}$ and $min\_ws_{i,j}$ are represented as 14 bit signed integer numbers.

The overall architecture of the 2D systolic array as well as the data dependency and timing information are shown in Figure 1 (b). In order to prevent ripple through operation, the cells in the even columns and even rows or odd columns and odd rows are synchronous to each other and perform computations in the same clock period. The rest of the cells are also synchronous to each other but perform the computation in the next clock period. Streams of operands enter a set of shift registers along the edges of the array that synchronize the presentation of bases in the operands with the results of calculations that propagate through the array diagonally.

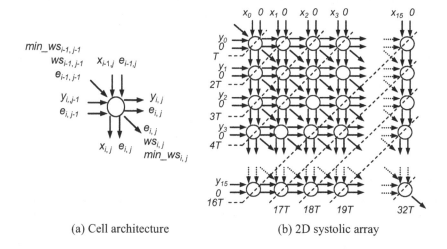

(a) Cell architecture   (b) 2D systolic array

Fig. 1. 2D systolic array for maximum weighted 2-stem common subsequence

3   Problem Formulation and Solution Technique

We consider each DNA codeword as a sequence of length n in which each symbol is an element of an alphabet of 4 elements. Let $G(x:y)$ denote the nearest neighbor free energy of duplex $x:y'$. In this work, we focus on searching for a set of DNA codeword pairs S, where S consists of a set of DNA strands of length n and their reverse complement strands e.g. $\{(s_1, s_1'), (s_2, s_2'), \ldots\}$, where (s_1, s_1') denotes a strand and its Watson-Crick complement. The problem can be formulated as the following constrained optimization problem:

$$\max |S| \tag{8}$$

such that
$$g - range \leq \max\left(G(s_1 : s_1'), G(s_1': s_1)\right) \leq g, \qquad (9)$$

$$g - range \leq \max_{s_2 \in S, s_2 \neq s_1} \left(G(s_1 : s_2), G(s_1 : s_2'), G(s_1': s_2), G(s_1': s_2')\right) \leq g \qquad (10)$$

where g and $range$ are user defined threshold called *CH upper bound* and *CH range*. Equation (8) indicates that our objective is to maximize the size of the DNA codeword library. Constraints (9)~(10) specify that the NN free energy of any CH duplexes must be lower than or equal to g but greater than or equal to g-$range$. The range was initially introduced because we thought that adding the code words that are too far away from the rest of the library would restrict future growth of the library. Therefore, we only add code words that are "just good enough". Later in the experiments we found that the range has little impact on library size, however, it has a significant impact on the convergence speed of the GA.

The optimization problem is solved using a genetic algorithm. A genetic algorithm (GA) is a stochastic search technique based on the mechanism of natural selection and recombination. Potential solutions, which are also called *individuals*, are evolved from generation to generation, with *selection*, *mating*, and *mutation* operators that provide an effective combination for exploring the global search space.

Given a codeword library S, the fitness of each individual d reflects how well the corresponding codeword fits into the current codeword library. Two values define the fitness, the *reject_num* and *max_match*. The *reject_num* is the number of codewords in the library which do not satisfy the condition (9)~(10) and
$$max\_match = \max_{s_2 \in S, s_2 \neq s_1} \left(G(s_1 : s_2), G(s_1 : s_2'), G(s_1': s_2), G(s_1': s_2')\right).$$

A traditional GA mutation function might randomly pick an individual in the population, randomly pick a pair of bits in the individual representing one of its 16 bases, and randomly change the base to one of the 3 other bases in the set of 4 possible bases. In the proposed algorithm, however, we randomly select an individual, but then exhaustively check all of the 48 possible base changes. This is an attempt to speed beneficial evolution of the population by minimizing the overhead that would be associated with randomly picking this individual again and again in order to test those mutations. We also specify that if none of the 48 mutations were beneficial, a random individual will be generated to replace the individual. For more details about the genetic algorithm and its hardware implementation, refer to [8]. In this work, we extend the architecture of the hardware GA presented in [8] to incorporate the consideration of nearest-neighbor free energy. The 2D systolic array that is presented in section 4 is used as the fitness evaluation module and the main state machine controller of the GA is modified so that it checks constraints (9)~(10).

4   Experimental Results and Discussions

A hardware accelerator that uses a stochastic GA to build DNA codeword libraries of codeword length 16 has been designed, implemented, and tested. The design was implemented on the reconfigurable computing platform that is composed of a desktop computer and an Annapolis WildStar–Pro FPGA board [9]. The FPGA board is plugged into the PCI-X slot of the host system. The WildStar-Pro uses one XC2VP70 FPGA that has 74,448 programmable logic cells. The hardware accelerator uses about 80% of the logic resources. It runs at a 45 MHz clock frequency. A hardware based code extender that uses exhaustive search to complete the codeword library generated from GA was also designed and implemented. All the code word libraries that have been found are verified using the online tool SynDCode[11]. Since the GA is a sto-chastic algorithm, all results reported are the average of 5 runs.

The first set of experiments compares the performance of the hardware-based and the software-only DNA codeword search. Two versions of each search algorithm were implemented. They are denoted as "deterministic search" (DS) and "randomized search" (RS). A population size of 16 was used for both versions. The population for DS was initialized using 16 sequential internal values from 0x000003F0 to 0x000003FF, which correspond to DNA codewords 3'AATTTAAAAAAAAAAAA'5 through 3'TTTTTAAAAAAAAAAAA'5, while the population for RS was initialized randomly. When a new codeword is found, or when none of the mutated codewords has lower fitness than the original individual, a new individual is generated to replace the original one. In DS, a counter is used to generate the new individual. The counter is initialized to 0x000006D6. In RS, the new individual is generated randomly. We found that random search is more effective than the deterministic search. However, in order to compare the speed of hardware-based implementation and software-based implementation, we must ensure that the two systems perform exactly the same com-putation tasks. This is achievable only with a deterministic algorithm. All experi-ments were run with $g = 8.5$ and $range = 1.0$, and were terminated after 300 code word pairs were found.

Figure 2 shows the time required to build large thermodynamically constrained DNA code word libraries, for software running on a single processor workstation, and for the hardware accelerator. The lower curves indicate faster speed. As we can see, the software-based deterministic search has the lowest performance, while the hard-ware-based random search has the highest performance. The hardware-based deter-ministic search provides approximately 240X speed-up compared to the software-only version while the hardware-based random search provides approximately 260X

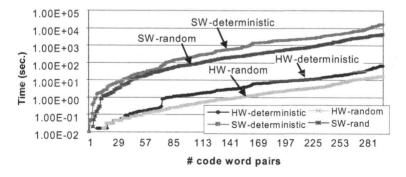

Fig. 2. Comparison between hardware-based and software-based implementation

speed-up compared to the software-only version. Compared to deterministic search, random search provides approximately 3.7X and 4X speed-ups using software-only and hardware-based implementations respectively. The plot also shows that the curves for software-only implementation and the hardware-based implementation are almost parallel to each other, which indicates that they both have the same complexity. Therefore, the performance gain that has been achieved by using hardware acceleration is a constant ratio.

The second set of experiments evaluates the impact of CH range on the speed and quality of the code word search. Figure 3 (a) gives the time to find 400 code word pairs for different CH ranges. In the next experiment, we ran the GA until it converged (i.e. could not find any new code words for 10 minutes), and then used exhaustive search to complete the codeword library. Figure 3 (b) shows the size of the final library. As we can see, the GA converges faster when the range is set to an appropriate value. For example, compared to range = 0.5, the runtime of GA is 26% and 24% longer at range= 0.05 and 3.0 respectively. Contrary to our original belief, the distance range does not have significant impact on library size. The size of final locally optimum libraries found with the addition of ES differ by only 3%. Exhaustive search usually finishes within 2 hours, depending on the number of words not found by GA.

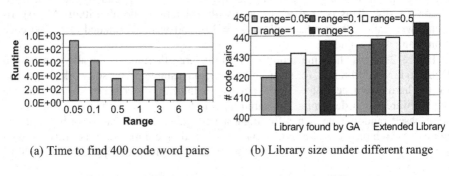

(a) Time to find 400 code word pairs   (b) Library size under different range

Fig. 3. Impact of different ranges on the search speed and library size

The third set of experiments compared the search speed for different CH upper bounds (g). We varied the CH upper bound from 6.5 to 10.0 and ran GA-based code word search. Figure 4 (a) shows the number of code word pairs found in 5 minutes for CH upper bounds from 5 to 8.0 while Figure 4 (b) shows the runtime required to find 300 code word pairs for CH upper bound from 8.5 to 10. The results indicate that the time to find 300 code words increases exponentially as CH upper bound increases.

The significance of the hardware accelerator is that for the first time it enables us to evaluate different code word search algorithms and explore the lower bound of optimal code word library size in a reasonable amount of time. For example, without the hardware accelerator, each experiment in our second set would have taken more than 20 days.

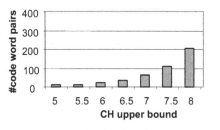

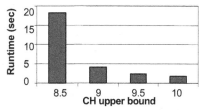

(a) # code word pairs found in 5 minutes.   (b) Time to find 300 code word pairs.

Fig. 4. Code word search under different CH upper bound

While it is true that the hardware accelerator does not explicitly consider constraints preventing bulges or internal loops, the free energy metric checking in a 2D systolic does impose those constraints implicitly by covering all sliding of the mers against each other. We believe that it should be possible to extend this work to include other secondary constraints commonly used in DNA code design, such as CG content, disallowing specific sequences, and checking all concatenations of two library words against each other (i.e. 32 mers vs 32 mers) in future hardware versions. Interestingly, scaling up to 32 mer x 32 mer checking may or may not result in longer checking times. The challenge of using hardware to calculate the free energy of DNA codewords of length 32 is that it may require more programmable hardware resources than any present single chip FPGA can provide. Possible solutions are to implement a large systolic array using multiple connected FPGAs and perform all computations in parallel, or implement a small systolic array on one FPGA and time-multiplex the computation, or await larger future generation FGPAs. While the first two solutions are feasible today, compared to the first solution, the second solution has lower cost but also lower performance. Careful tradeoff decisions must be made based on the available resources, and the given cost and performance requirements. It is also noted that DNA code design problem is only slightly different than the tag-antitag and probe set design problem faced in composing diagnostic micro arrays, where mers of length 25-60 must be checked in many alignments against longer mers drawn from large and potentially multiple genomes. Hardware accelerators similar to our own should be adaptable to that problem. Finally, DNA codes designed in-silicon for both problems must be checked by fabrication and wet chemistry experiments run under use conditions to verify their true utility

5   Conclusions

In this work, we propose a novel systolic array architecture to calculate the nearest-neighbor free energy of DNA duplexes that is based on a transformed version of a dynamic programming approach. A single chip FPGA hardware accelerator has been developed that builds large, locally optimum libraries of DNA codewords with GA and exhaustive search, both based on thermodynamic energy constraints. The present version, run at 45 MHz clock frequency, provided more than a 250X speedup over a software only approach running on a 2.5 GHz Pentium processor.

References

[1] Adleman, L.M.: Molecular Computation of Solutions to Combinatorial Problems. Science 266, 1021–1024 (1994)

[2] Mansuripur, M., Khulbe, P.K., Kuebler, S.M., Perry, J.W., Giridhar, M.S., Peyghambarian, N.: Information Storage and Retrieval using Macromolecules as Storage Media. In: Optical Data Storage (2003)

[3] Brenner, S., Lerner, R.A.: Encoded Combinatorial Chemistry. Natl. Acad. Sci. USA 89, 5381–5383 (1992)

[4] Deaton, R., Garzon, M.: Thermodynamic Constraints on DNA-based Computing. In: Computing with Bio-Molecules: Theory and Experiments, Springer, Heidelberg

[5] Brenneman, A., Condon, A.: Strand Design for Biomolecular Computation. Theoretical Computer Science 287, 39–58 (2002)

[6] Tanaka, F., Kameda, A., Yamamoto, M., Ohuchi, A.: Design of Nucleic Acid Sequences for DNA Computing based on a Thermodynamic Approach. Nucleic Acids Research 33(3), 903–911 (2005)

[7] Santalucia, J.: A Unified View of polymer, dumbbell, and oligonucleotide DNA nearest neighbor thermodynamics. In: Natl. Acad. Sci. Biochemistry, pp. 1460–1465 (1998)

[8] Qiu, Q., Burns, D., Wu, Q., Mukre, P.: Hybrid Architecture for Accelerating DNA Codeword Library Searching. In: IEEE Symposium on Computational Intelligence in Bioinformatics and Computational Biology (2007)

[9] Annapolis Micro System, http://www.annapmicro.com/

[10] SantaLucia Jr., J., Hicks, D.: The thermodynamics of DNA Structural Motifs. Annu. Rev. Biophys. Biomol. Struct. 33, 415–440 (2004)

[11] Bishop, M.A., Macula1, A.J., Renz, T.E.: SynDCode: Cooperative DNA Code Generating Tool. In: 3rd Annual Conference of Foundations of Nanoscience (2006)

[12] D'yachkov, A.G., Macula, A.J., Pogozelski, W.K., Renz, T.E., Rykov, V.V., Torney, D.C.: A Weighted Insertion-Deletion Stacked Pair Thermodynamic Metric for DNA Codes. In: Ferretti, C., Mauri, G., Zandron, C. (eds.) DNA Computing. LNCS, vol. 3384, pp. 90–103. Springer, Heidelberg (2005)

Hardware and Software Architecture for Implementing Membrane Systems: A Case of Study to Transition P Systems

Abraham Gutiérrez, Luís Fernández, Fernando Arroyo, and Santiago Alonso

Natural Computing Group - Universidad Politécnica de Madrid
28031 Madrid, Spain
{abraham, setillo, farroyo, salonso}@eui.upm.es

Abstract. Membrane Systems are computation models inspired in some basic features of biological membranes. Many variants of such computing devices have already been investigated. Most of them are computationally universal, i.e., equal in capacity to Turing machines. Some variant of these systems are able to trade space for time and solve, by making use of an exponential space, intractable problems in a feasible time. This work presents a software architecture that completes the generic hardware prototype based on microcontrollers presented in a previous work. This parallel hardware/software architecture is based on a low cost universal membrane hardware component that allows to efficiently run any kind of membrane systems. This solution was less enclosed and floppier than the hardware specifically designed, and cheaper than those based on clusters of PC's.

Keywords: Natural Computing, Transition P System, Hardware implementations, Software architecture.

1 Introduction

Membrane computing is a new computational model based on the membrane structure of living cells [1]. It must be stressed that they are not intended to model the functioning of biological membranes. Rather, they explore the computational character of various features of membranes that can be used for modeling new computational paradigms inspired by Nature. This model has become, during last years, a powerful framework for developing new ideas in theoretical computation and connecting the Biology with Computer Science.

The membrane structure of a P System is a hierarchical arrangement of membranes, embedded in a skin membrane, the one which separates the system from its environment. A membrane without any membrane inside is called elementary. Each membrane defines a region that contains a *multiset* of objects, and a set of evolution rules. The objects are represented by symbols from a given alphabet. The objects can pass through membranes, the membranes can change their permeability, they can dissolve, or they can divide. These features are used in defining transitions between different system configurations, and these sequences of transitions are used to define computations. Membrane systems are synchronous, in the sense that a global

M.H. Garzon and H. Yan (Eds.): DNA 13, LNCS 4848, pp. 211–220, 2008.

clock is assumed, i.e., the same clock holds for all regions of the system. At each time unit, a transformation of a system configuration takes place by applying rules in all regions, in a nondeterministic and maximally parallel manner. This means that the objects to evolve and the rules governing this evolution are chosen in a nondeterministic way.

Nowadays, membrane systems have been sufficiently characterized from a theoretical point of view. Their computational power has been settled – many variants are computationally complete. However, the way in which these models can be implemented is an open problem today. As usually happens, implementation of these systems has been attacked from two different approaches: software and hardware models. An overview of membrane computing software can be found in [3]. It can be found several membrane systems simulators very elaborated [2]. However, the hardware model seems to be the most appropriate -apart from the biological implementation– able to obtain the massive parallelization membrane systems claim.

The main research lines in hardware model design are, hardware ad-hoc and simulation over local networks using cluster of microprocessors:

- The hardware specifically designed has the advantage of being a massively parallel solution [5][6][7]. Their weak point resides in the lack of flexibility that presents, because this type of solutions only allows the simulation of a specific kind of membrane systems (for each membrane system a specific hardware is needed). They are also very enclosed solutions because they can be applied only to a very little range of problems (reduced number of objects in the alphabet and small number of evolution rules).
- The solutions based on cluster of microprocessors [4] and local networks have as main advantage the use of very common and well-known architectures. They are floppy systems also because a change at software level allows the simulation of any kind of membrane systems. Its main problem, as Ciobanu recognizes [4], is caused by the network congestion. Although there are new studies that solve these problems [11], the best simulation times are reached always with few units, so the obtained solutions have a low degree of parallelism.

Our solution, based on the use of microcontrollers, tries to be an intermediate point among the previously exposed research lines:

- Microcontroller architecture is as flexible as cluster of microprocessors and less enclosed and floppier than the hardware specifically designed.
- Microcontroller architecture has more level of potential parallelism than cluster of microprocessors but does not have intrinsic parallel nature of the hardware specifically designed.
- Finally, it is the cheapest architecture (8 US$ approximately per basic unit).

Due to the balance between flexibility, parallelism and cost reached by this architecture it's very possible to obtain better results than the previous works did.

This work is structured as follows: firstly presents basic units of the hardware prototype. Secondly, algorithms for applying evolution rules in Transition P systems are characterized. After that, next sections are devoted to present the software architecture developed and how the algorithms are adapted to the hardware prototype. Finally, the obtained results and some conclusions are presented.

2 Hardware Prototype

The designed prototype expects to simulate the operation of just one membrane, and not the whole system. It makes use of three basic components: a microcontroller that carries out the processing unit features, an EEPROM memory that carries out the storage unit features and a communications bus. For a complete description of this prototype see [12].

2.1 Processing Unit: PIC16F88 Microcontroller

The microcontroller stores and run the program that simulates the behaviour of the membrane. The possibility of change this program, without modifying the associate hardware, provides a great flexibility to the solution allowing the simulation of any kind of membrane system.

The election of PIC (Peripheral Interface Controller) 16F88, comes determined by its low cost ($1.90), that allows us to think about solutions with a high grade of potential parallelism and also because of being the half rank microcontroller that provides the minimum necessary requirements: a enough processing speed (20Mhz, 200 nanosecond instruction execution), a suitable word wide (8bits), a possibility of manipulating data with a sufficient precision (8bits) and a easy–to-program instruction set (only 35 single word instructions).

2.2 Storage Unit: 24LC1025 Eeprom

The memory module defines the compartment that delimits the membrane and it is the container of its main components: the evolution rules and the multisets. The large store capacity of this memory module makes the solution less enclosed than the hardware specifically designed, being valid for a bigger range of problems.

The solution presented in this prototype uses a 24LC1025 memory module, also manufactured by Microchip Inc. Technology, that gathers a reduced cost ($4,50) and a great storage capacity (128 Kbytes). It uses a low consumption CMOS technology (3mA in writing process), with a maximum writing cycle of 5ms and is guaranteed for more than a million of write operations. It is also compatible with I2C protocol, supporting the standard mode of 100 KHz and the fast mode of 400 KHz, which allows a suitable speed for the kind of application that we are developing.

2.3 Communication Bus: I2C Bus

The communication bus allows the effective communication among the membranes, facilitating the distribution of the information for the whole system. It is also the responsible for content exchange between the memory module and the microcontroller.

The prototype uses the I2C bus (Inter Integrated Circuit Bus) of Philips Semiconductors implemented in MSSP module (Master Synchronous Serial Port) of the own microcontroller. I2C defines a synchronous, bidirectional protocol, of master-slave type that uses a serial bus, formed by two threads, to which several devices can be connected by a very simple hardware. Each wired device to the bus is recognized by an only address that differentiates it of the rest of the wired components. One of the advantages from the usage of this kind of bus is its simplicity which allows us to connect new components, for example, we can simply expand the storage capacity of our solution until 512 Kbytes adding other three memory modules (24LC1025) to the same bus, that they would be accessed making use of the available address lines.

2.4 General Structure of the Circuit

The following figure shows us the general structure of the circuit used for the implementation of universal membrane hardware component; the connection between the microcontroller and the memory module through the I2C bus:

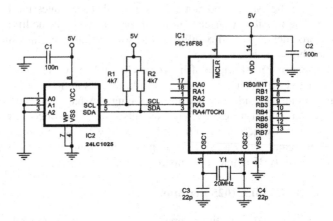

Fig. 1. General circuit diagram

3 Software Architecture

The proposed software solution tries to define a generic architecture that allows its adaptation to any kind of membrane system, in the simplest possible way. It arranges the developed features in three layers: a low layer that provides the basic routines for hardware performance; an intermediate layer composed by the modules that provide the main routines to manipulate membrane systems; and a higher layer, or implementation layer, in which the algorithms for different models of membrane systems are defined.

3.1 Data Structures

There are five main structures in our architecture: *multiset* structure, *evolution rules* structure, *rules table* structure, *applicable rules bitmap* structure and *priority rules bitmap* structure.

Each region of a membrane can potentially host an unlimited number of objects, represented by the symbols from a given alphabet. In our case, these objects are not implemented individually but only their *multiset* is represented using the *Parikh vector*: let *{a1, ..., an}* be an arbitrary alphabet; the number of occurrences of a symbol a_i in x is denoted by $| x |_{ai}$; the *Parikh vector* associated with x with respect to

$a_1, ..., a_n$ is $(|\, x\, |_{a1}\, , ..., \, |\, x\, |_{an}) = (m_1, ...,m_n)$. In our implementation, it is possible to choose between 8-bit and 16-bit per object, which allows to store up to $27 = 128$ or up to $215 = 32768$ instances of an object (the remaining bit is being used to detect a capacity overflow). This value limits the maximum number of object instances that can be stored, but it is easy to change this restriction if it is necessary.

The byte's order reflects the order of the objects within the alphabet and consequently, the byte position directly indicates which symbol's multiplicity is being stored. There are no restrictions for the number of objects of the alphabet, just the maximum capacity of the memory. To indicate the end of the *multiset* a special character is used.

To represent the *evolution rules* we use the same type of vector, indicating the end of each rule by a special character too. When the membrane is first loaded in the EEPROM memory, a memory map is created with information about the beginning address of each evolution rule in our memory. This table is initialized only at the beginning of the simulation and do not change afterwards (the *P System's evolution rules* do not change throughout a computation, except when membrane is dissolved).

Furthermore, after the application of a rule inside the membrane, the program constantly computes whether a rule can be applied or not. The routine takes the membrane rules as inputs and generates a bitmap array, the *applicable rules bitmap*, indicating the set of applicable rules. This bitmap is later used to select applicable rules.

The last main structure is the *priority rules bitmap*. It consists in a byte array per evolution rule indicating their priority regarding the rest of the rules. This structure is used to select the rule to be active.

3.2 Hardware Abstraction Layer (HAL)

This is an abstraction layer, implemented with a set of routines libraries, between the physical hardware of the prototype and the specific software that runs on higher layers. Its function is to hide differences in hardware from the simulation software, so it not needs to be changed to run on systems with different hardware. A HAL allows routines from intermediate layer to communicate with lower level components, such as directly with hardware. This allows portability of the implemented code to a variety of microcontrollers, with different memory modules or even different communication buses. This layer is composed by four main modules:

- The *microcontroller.inc* module takes charge of providing the necessary subroutines for the initialization and basic operation of the microcontroller. Emphasize among others, the routines that work with the inner memory of the microcontroller (*flash* memory for general purpose and *eeprom* memory for main program).
- The *memory.inc* module takes charge of providing the necessary subroutines for the initialization and basic operation of the memory modules, as well as the management operations (load and recovery) of their main contents: *multisets* and *evolution rules*. To facilitate individualized accesses to different contents a random access reading system has been configured.

- The **bus.inc** module takes charge of providing the necessary subroutines for the initialization and basic operation of the I2C bus, as well as the communication operations between different membranes and different components in the prototype. The defined routines are specifically adjusted to the requirements imposed by the information manipulation. The bus fast mode support (400 KHz) has been used.
- The **synchronization.inc** module takes charge of providing the necessary subroutines for the correct synchronization of the different components. It also includes subroutines that allow implementing the necessary synchronization barrier in the membrane evolution steps. Special care has been take it with the reading and writing operations in the storage device (maximum writing cycle of 5ms) or in the microcontroller enabled areas.

3.3 Membrane Basic Runtime Layer (MBRL)

This layer defines a set of routine libraries that provide a common interface to the top layer (MIL). They make use of routines implemented in the low layer (HAL). This interface is made up of generic functions that allow carrying out the basic operations present in most of the membrane systems. Four main modules compose this layer:

- The **multiset.inc** module takes charge of providing the necessary subroutines for storage, retrieval and management of *multiset* defined in a membrane. We can find routines like: *load_multiset, restore_multiset, write_multiset, read_multiset...*
- The **rules.inc** module takes charge of providing the necessary subroutines for storage, retrieval and management of evolution rules defined in a membrane. We can find routines like: *load_rules, restore_rules, next_rule, applicable_rules, select_ramdon_rule...*
- The **priority.inc** module takes charge of providing the necessary subroutines for storage, retrieval and management of priority map defined among different evolution rules of a membrane. We can find routines like: *load_priorities_vector, restore_priorities_vector, set_rule_priority, get_rule_priority...*
- The **communication.inc** module takes charge of providing the necessary subroutines for information communication and message passing among the different membranes that compose a system. It processes the transfer commands present in the evolution rules (in_j, out_j, *here*). It is also the module in charge of implementing the operations associated to membrane permeability. We can find routines like: *send_to_region, dissolve_region, pause_computation, resume_computation...*

3.4 Membrane Implementation Layer (MIL)

In this layer are located the programs that implements the different kinds of membrane systems. In our case we have three programs that implements the P System algorithms described in the next section. These programs make use of routine libraries present in the intermediate architecture level (MBRL). New implementations may need the adaptation of some of the MBRL libraries.

The following image shows the structure of the proposed software architecture:

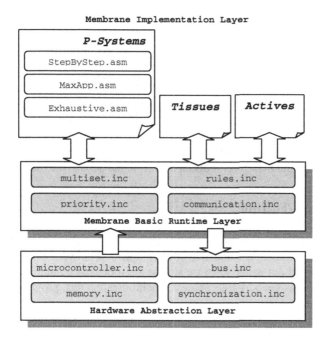

Fig. 2. Software Architecture Layers

4 Implemented Transition P System

In this paper we implement three algorithms. The first one is widely used on simulation tools, and the other two are new algorithms based on applicability benchmarks for evolution rules in which, a certain degree of parallelism is obtained – at a given iteration step of the algorithm, a given evolution rule is applied several times over the *multiset* of objects–. A total parallelism cannot be achieved but these new algorithms go one step ahead in two different senses: firstly, implementations are closed to the theoretical model and to biology, and secondly, due to parallel application of evolution rules by chunks, it will be more efficient. For a complete description of these algorithms see [8][11].

4.1 Step by Step Algorithm

In literature about simulation tools of membrane systems [9], it is widely used an algorithm in which rules are applied one by one in a set of micro-steps of evolution, in particular, during the process of calculating the *multiset* of evolution rules. This algorithm involves the random selection of one of the active evolution rules and the modification of the *multiset* of objects until the set of active evolution rules became empty.

4.2 Maximal Applicability Algorithm

This algorithm is based on considering the *maximal applicability benchmark* of evolution rules over a *multiset* of objects [8]. Process is as follows: once an evolution rule has been selected in a non deterministic manner, the rule is applied a random number of times between 1 and the maximal applicability benchmark, per iteration. It is expected that this higher consume of objects will accelerate the end of execution.

4.3 Elimination Rules Algorithm

Process is as follows: all rule, different from X, in the active rules set is applied. The number of times that is applied will be a random number between 0 and its *maximal applicability*. This way, each one has the possibility of being applied. Rule X is applied an equal number of times to its *maximal applicability*. This way the rule is not any more active and therefore, it disappears from the active rules set.

5 Hardware Testing

Before construction phase different kinds of tests have been carried out over individualized components and over the complete prototype. So many simulations have been carried out using software simulation tools and emulation test boards. The testing process has gone by four phases: synchronization test, program test, simulation test and emulation test.

5.1 Synchronization Test

One of the most important elements for a good operation of the proposed solution is the communication bus, mainly, if we think of a board composed by hundred of components like those presented in this work. To obtain a great productivity, the fast mode (400 KHz) of the I2C bus it has been used. This implies the adaptation of all signals and necessary delays for all the components. To make this adjustment the communication has been proven for each one component with the bus. Their behaviors have been defined using a signal analyzer.

5.2 Programs Test

For the simulation programs and the library routines we used the MicroChip MPLAB IDE integrated environment (integrated toolset for the development of embedded applications employing Microchip's PIC®). The programming tool was macro assembler MPASM, that generates relocatable object files for the MPLINK object linker, Intel® standard HEX files, MAP files to detail memory usage and symbol reference, absolute LST files that contain source lines and generated machine code and COFF files for debugging.

5.3 Simulation Test

For the complete software simulation of the prototype the Proteus IDE integrated environment has been used. Its correct operation for the different algorithms has been verified. In this tool all obtained factors in previous tests have been considered (synchronization analysis and program fine tuning).

5.4 Emulation Test: In Circuit Debugger

Once stopped the software simulation tests, the prototype has been proven on a tests board. This way, the real operation of the prototype is verified. For it MPLAB ICD2 trainer has been used.

6 Conclusions

In this work we present a parallel hardware/software architecture based on a low cost universal membrane hardware component that allows to run efficiently any kind of membrane systems. The solution is based on generic hardware such as microcontrollers. So, the resulting architecture is less enclosed and floppier than the hardware specifically designed. Moreover, it has a higher parallelism degree than cluster of microprocessors and it is the cheapest solution.

In particular, as a test of their flexibility, in this work three different algorithms for application of evolution rules in *Transition P Systems* (the basic variant of membrane systems) are implemented. Moreover adding the necessary software components, algorithms for active membranes P-systems, tissues-like P-systems or many variants of the proposed model could be implemented.

In addition, the propose architecture does not impose functionality or arrangement restrictions to their basic units. It can be adapted without problems to different software architectures [8], like: *Parallel architecture oriented to membranes*, *Parallel architecture oriented to processors* and *Parallel architecture oriented to rules*. It can be also adapted to different hardware architectures [11] like: *Peer to Peer architecture*, *Master-Slave architecture* and the new *Hierarchic architectures*.

References

1. Păun, Gh.: Computing with Membranes. Journal of Computer and System Sciences, 61 (2000), Turku Center of Computer Science-TUCS Report nº 208 (1998)
2. Păun, Gh.: Membrane computing. Basic ideas, results, applications. In: Pre-Proceedings of First International Workshop on Theory and Application of P Systems, Timisoara, Romania, pp. 1–8 (September 26-27, 2005)
3. Ciobanu, G., Pérez-Jiménez, M., Paun, G.: Applications of Membrane Computing. Natural Computing Series. Springer, Heidelberg (2006)
4. Ciobanu, G., Wenyuan, G.: A P system running on a cluster of computers, Membrane Computing. In: Martín-Vide, C., Mauri, G., Păun, G., Rozenberg, G., Salomaa, A. (eds.) Membrane Computing. LNCS, vol. 2933, pp. 123–150. Springer, Heidelberg (2004)
5. Petreska, B., Teuscher, C.: A reconfigurable hardware membrane system. In: Alhazov, A., Martín-Vide, C., Paun, G. (eds.) Preproceedings of the Workshop on Membrane Computing, Tarragona, pp. 343–355 (July 2003)
6. Fernández, L., Martínez, V.J., Arroyo, F., Mingo, L.F: A Hardware Circuit for Selecting Active Rules in Transition P Systems. In: Workshop on Theory and Applications of P Systems. Timisoara (Rumanía) (September 2005)
7. Martínez, V., Fernández, L., Arroyo, F., Gutiérrez, A.: A hardware circuit for the application of active rules in a Transition P Systems region. In: Fourth International Conference Information Research and Applications, Bulgaria, Varna (June 20-25, 2006)

8. Fernandez, L., Arroyo, F.: Parallel Software Architectures Analysis for P-Systems Implementation. In: NCA 2006, Natural Computing and Applications Workshop. 8th International Symposium on Symbolic and Numeric Algorithms for Scientific Computing Timisoara, Romania (September 26-29, 2006)
9. Arroyo, F., Luengo, C., Baranda, A.V., Mingo, L.F.: A software simulation of transition P systems in Haskell. In: Păun, G., Rozenberg, G., Salomaa, A., Zandron, C. (eds.) Membrane Computing. LNCS, vol. 2597, pp. 19–32. Springer, Heidelberg (2003)
10. Fernández, L., Arroyo, F., García, I., Gutiérrez, A.: Parallel software architectures for implementing P systems. In: AROB 12th 2007, XII International Symposium on Artificial Life and Robotics, Oita, JAPAN (January 25-27, 2007)
11. Tejedor, A., Fernández, L., Arroyo, F., Bravo, G.: An architecture for attacking the bottleneck communication in P System. In: AROB 12th 2007, XII International Symposium on Artificial Life and Robotics, Oita, JAPAN (January 25-27, 2007)
12. Gutiérrez, A., Fernández, L., Arroyo, F., Martínez, V.: Design of a hardware architecture based on microcontrollers for the implementation of membrane systems. In: SYNASC 2006, 8th International Symposium on Symbolic and Numeric Algorithms for Scientific Computing Timisoara, Romania (September 26-29, 2006)

Towards a Robust Biocomputing Solution of Intractable Problems

Marc García-Arnau[1,*], Daniel Manrique[1], Alfonso Rodríguez-Patón[1], and Petr Sosík[1,2]

[1] Departamento Inteligencia Artificial, Universidad Politécnica de Madrid (UPM)
Boadilla del Monte s/n – 28660 Madrid, Spain
mgarciaarnau@alumnos.upm.es
[2] Institute of Computer Science, Faculty of Philosophy and Science, Silesian University
Bezručovo nám. 13, 74601 Opava, Czech Republic

Abstract. An incremental approach to construction of biomolecular algorithms solving intractable problems is presented. The core idea is to build gradually the space of candidate solutions and remove invalid solutions as soon as possible. We demonstrate two examples of this strategy: a P system with replication and inhibitors for solving the Maximum Clique Problem for a graph, and an incremental DNA algorithm for the same problem inspired by the membrane solution. The DNA implementation is based on the parallel filtering DNA model featuring error-resistance of the employed operations. The algorithm is compared with two standard papers that addressed the same problem and its DNA implementation in the past. The comparison is carried out on the basis of a series of computational and physical parameters. The incremental algorithm features a dramatically lower cost in terms of time, the number and size of DNA strands, together with a high error-resistance. A probabilistic analysis shows that physical parameters (volume of the DNA pool, concentration of the solution-encoding strands) and error-resistance of the algorithm should allow to process *in vitro* instances of graphs with hundreds to thousands of vertices.

Keywords: Membrane Computing, DNA Computing, NP-Complete problem, Maximum Clique Problem, Incremental Strategy.

1 Introduction

One of the major sources of popularity of biomolecular computing in 1990's was the promise to solve computationally intractable problems in polynomial (often linear) time due to the massive parallelism, minute energy consumption and nanoscale dimensions of biomolecular computing elements. After the pioneering experiment published in [1], an explosion of publications appeared, many of them proposing molecular solutions of NP-complete problems. Soon another abstract biocomputing model emerged: membrane systems (also known as P systems), inspired by the structure and behaviour of living cells [20]. For an overview of both fields see [22], [3], [21] or online resources.

However, although it has been theoretically possible to solve NP-complete problems in polynomial time, researchers have not managed to limit the exponential

---

* Corresponding author.

M.H. Garzon and H. Yan (Eds.): DNA 13, LNCS 4848, pp. 221–230, 2008.

growth of some resources involved in problem solving, such as the number of molecules. Another serious issue is the unreliability of biological operations, leading to errors exponentially growing with the number of iterations. Therefore, a great deal of effort in molecular computing was put into optimizing these issues. We mention several possible strategies: the adaptation of especially space-efficient classical algorithms to a DNA scenario [18]. The creation of constructive algorithms designed to optimize the number of DNA strands [4]. The reduction of the number of biological operations needed and the resulting risk of error [15]. The application of dynamic DNA programming techniques [6], or the creation of computational models based on a destructive strategy over RNA strands as in [9]. These and other similar papers suggest a compromise between space optimization and algorithm time, without overlooking the number and complexity of the operations involved. Despite this effort, the biggest known instance of the SAT problem solved yet *in vitro* by a biocomputing machinery has 20 variables [23].

In this paper we propose an incremental strategy for biomolecular solutions of intractable problems. The core idea is rather simple: to build gradually the space of candidate solutions and remove invalid solutions as soon as possible. We demonstrate two biocomputing algorithms based on this strategy for solving the Maximum Clique Problem for a graph: a P system with replication and inhibitors, and an incremental DNA algorithm based on the parallel filtering model. Mathematical analysis of the algorithm, including its error-resistance and scalability, proves superior parameters when compared to standard approaches.

The article is organized as it follows. Section 2 reviews the Maximum Clique Problem and proposes a P system with an incremental strategy to solve the problem. In Section 3 we propose an implementation of the incremental strategy using a DNA algorithm. The efficiency of the proposed solution is compared in Section 4 with that of two standard papers that tackled the same problem earlier in [19] and [11]. A series of computational and physical parameters are defined to make this comparison. Section 5 sets out final remarks

2 Incremental Strategy in P Systems

2.1 The Maximum Clique Problem and Related Work

Let $G = (V, E)$ be a graph with n nodes. A clique in G is a subset $V' \subset V$ such that each two vertices in V' are connected by an arc in E. The Maximum Clique Problem then involves finding the biggest subset V' of totally connected nodes in the graph.

Over the last decade, many papers have tackled this problem from the natural computing paradigm point of view. For example, a theoretical proposal of a DNA solution was presented in [2]. However, a DNA bioalgorithm to solve this problem was implemented for the first time in [19]. Later many others have addressed the same problem from different viewpoints. For example, the work of [5] focuses on the relation between evolutionary computing and DNA computing. Later, a parallel algorithm using fluids displacement in a three-dimensional microfluidic system was presented in [8]. Almost simultaneously another microflow reactor was used in [17] to solve the same problem using a brute force strategy. An aqueous algorithm was proposed in

[11] to solve the same instance of the problem as in [19]. Finally, the work of [24] is an example of use of the DNA computing sticker model to get all the cliques of size K of a graph.

2.2 A P System with Replicated Rewriting and Inhibitors

P systems with replicated rewriting are defined as membrane systems whose basic objects are not symbols but structured elements, like strings. These systems contain multisets of strings that are processed using replicated rewriting rules, that is, rules that, apart from modifying strings, can increase the number of their copies. Each such rule consists of $n \geq 1$ subrules. When a rule is applied to a string, the string is first replicated into n copies and then each subrule is applied to one copy [14].

Another idea is the use of promoters or inhibitors in P systems with a clear biological inspiration [7], [12]. A rule modelling a biological reaction can or cannot take place in the presence of certain enzymatic proteins. In our case, for each rule R_i of the P system, we define a set of inhibitors U_i containing those objects a_j in presence of which the rule cannot be applied.

Consider a graph G and the complementary graph $G' = (V, E')$ containing all edges missing in the original graph G. Then only those sets of vertices that do not contain nodes linked by an arc in E' are valid cliques. Then E' is what we will call the *constraints set*, which will be implemented by sets of inhibitors in our P system.

Hence, the P system proposed here uses the information from the constraint set E' to gradually build only those cliques of G that are valid, thereby minimizing the number of strings in the system. Given a graph $G = (\{a_1,...,a_n\}, E)$, we formally define this P system as a construct:

$$\Pi = (V, \mu, M_1, ..., M_n, R_1, ..., R_n), \text{ where:}$$

$$V = \{a_i, d \mid 1 \leq i \leq n\}$$
$$\mu = [_n ... [_2 [_1]_1]_2 ...]_n$$
$$M_1 = \{d\}; \qquad M_i = \{\lambda\}, \quad \text{with } 2 \leq i \leq n$$
$$R_i = \{d \rightarrow (da_i, out)_{\neg U_i} \parallel (d, out)\} \text{ with } U_i = \{a_j \in V \mid \{a_i, a_j\} \in E'\}, \ 1 \leq i \leq n$$

In the above description, V is an alphabet consisting of an auxiliary symbol d and symbols $a_1,...,$ a_n representing the nodes of the graph. Furthermore, Π has a membrane structure μ composed of n embedded membranes. Initially, the innermost membrane 1 contains the string with a single symbol d, whereas the other membranes are empty ($M_i = \{\lambda\}$, $2 \leq i \leq n$). Each membrane contains a set of rules R_i with replication. Each rule is of the form $X \rightarrow (v_1, tar_1)_{\neg U_1} \parallel (v_2, tar_2)_{\neg U_2} \parallel ... \parallel (v_k, tar_k)_{\neg U_k}$, where

$X \in V$, $v_i \in V^*$, $U_i \subset V^*$ and $tar_i \in \{out, here\} \cup \{in_j \mid 1 \leq j \leq n\}$, for $1 \leq i \leq k$.

The sets U_i are finite and their elements are called *inhibitor strings*. A subrule $X \rightarrow (v_i, tar_i)_{\neg U_i}$ is applicable to a string w only if w contains X and does not contain any inhibitor string of U_i. If a rule of the above form is applied to a string w, then for each its applicable subrule a separate copy of w is created, to which the subrule is applied. This application means that one occurrence of X in w is rewritten by v_i and the resulting string is sent to the membrane indicated by tar_i. In this manner, the number of the processed strings can increase during the computation. At each step, each string in each membrane i to which at least one rule in R_i is applicable, must be processed by one such rule. Other strings pass unchanged into the next step.

In the P system described above, each set R_i contains a single rule composed of a pair of subrules $r_{i,a}: d \rightarrow (da_i, out)_{\neg U_i}$ and $r_{i,b}: d \rightarrow (d, out)$. The system starts to work by applying the two rewriting subrules of membrane 1 to the string d. At the first step, two strings are created and sent to membrane 2. The process is repeated in each membrane applying their respective rules R_i so that, step by step, all the valid cliques of the graph are generated. Note that the subrules $r_{i,b}$ are always applicable to all strings, whereas the subrules $r_{i,a}$ are only applicable to strings which do not contain any of the symbols a_j representing vertices linked by an arc in G' to the node a_i. At step n the system outputs a set of strings corresponding to all valid cliques for the graph G, the biggest of which is the solution to the Maximum Clique Problem.

Therefore, the P system algorithm presented above generates simultaneously all the valid cliques of the graph G. It could be improved to separate the solutions representing maximal cliques. However, when implemented in the DNA framework, one can use tools specific for DNA computing for this separation. Hence we leave a solution of the separation problem for the next sections.

3 A Robust DNA Algorithm for the Maximum Clique Problem

The computational benefits of using DNA molecules come from their chemical properties, mainly from the Watson-Crick complementarity principle. Also the massive parallelism of performed operations is of crucial importance, as well as an extremely low energy consumption and volume compared to the classical silicon technology. DNA strands are formed by repeating four types of nitrogen bases {A, C, G, T} attached to a sugar-phosphate backbone. Chemical properties of DNA guarantee that adenine (A) can only pair with thymine (T), whereas guanine (G) can only pair with cytosine (C).

Since the pioneering Adleman's DNA computational model, several other DNA computing models have been presented. Recently, in the monograph [3] and other publications, authors focus on the reliability and error-resistance of the elementary lab operations used for DNA computing. For example, properties of so-called *parallel filtering* model, first presented in [2], are thoroughly examined in [3]. Therefore, we adopt this model as a basis of our implementation of the P system described in Section 2.2. However, to preserve advantages of our incremental solution construction, we enrich the parallel filtering model with the operation *Append* implemented by the Parallel Overlap Assembly (POA). Also, we replace the operation *Copy* with a similar operation *Replicate*. While *Copy* was designed just to split the content of a test tube into k other tubes of equal volumes, *Replicate* actually replicates the content of a source tube using the PCR operation. To summarize, the following operations on DNA single-stranded molecules (oligos) are used:

1. *Union* (t_1, t_2, t_3): Mixes multisets of strings (i.e. tubes) t_1 and t_2 to form their union in t_3.
2. *Replicate* (t_1, t_2, t_3): Replicates the contents of a tube t_1 into two tubes t_2 and t_3 such that each of t_2 and t_3 contains (at least) one copy of each strand originally present in t_1, which is emptied.
3. *Remove* $(t, \{a_i\})$: Given a tube t and a set of strings $\{a_i\}$, we remove from t all the strings containing some a_i as a substring.
4. *Append* (t_1, a): Given a tube t_1, append the string a to the end of all the strings the tube contains. If t_1 is empty, then the result of the operation is just $\{a\}$.
5. *MeasureStrings* (t): Measure the strings of tube t to identify the longest one(s).

Following the design instructions in [19], the use of 20-base long oligos (20-*mer*) to encode each node a_i of the problem domain is considered. Therefore, a clique consisting of k vertices will be represented as a catenation of k 20-mers. Each of these 20-mers must contain a restriction enzyme binding site, for example GATC for the enzyme *Mbo*I. The same enzyme is used for all the 20-mers. For a possible implementation of the operations listed above we refer to [3]. We only note that the operation *Append* (t_1, a) is implemented using POA. For each element a_i we need exactly $i-1$ *append operators* of the form $5'-\overline{a_i}\,\overline{a_j}-3'$, $1 \leq j < i$. Now we present the DNA algorithm working in linear time with respect to the number of nodes $|V|$, which solves the Maximum Clique Problem.

$\forall \ (a_i \in V, \ i = 1,2,...,n) \ begin$

$\quad \{$

$\qquad Replicate\,(t_1, t_2, t_3)$ // *This step generates tubes* t_2 *and* t_3 *and empties tube* t_1.

$\qquad Remove\,(t_2, \{a_j \mid \{a_i, a_j\} \in E', 1 \leq j < i\})$ // *Removal of invalid cliques.*

$\qquad Append\,(t_2, a_i)$ // *This step adds node* a_i *to all the valid cliques.*

$\qquad Union\,(t_2, t_3, t_1)$ // *This step merges the two tubes to reconstruct* t_1.

$\quad \}$

$\quad MeasureStrings\,(t_1)$ // *This step measures the DNA strings in* t_1. *The biggest one*
$\qquad\qquad\qquad\qquad\qquad$ *is the clique that solves the problem.*

Contrary to the algorithms used in [1], [19], [24], [3] and many other authors, the initial test tube does not contain the pool of all the possible candidate solutions of the problem. Instead, it contains a certain amount (say, $p = 20$) of copies of an initial 20-mer (primer binding site) representing empty cliques. This amounts the number of copies of each future solution strand to ensure that, despite potential errors, the computation would proceed correctly. The strings representing the problem solution (valid cliques for the input graph) are built gradually, as in the P system described above.

4   Comparison with Standard Approaches

In this section, we compare our incremental algorithm with standard strategies used in [19] and [11] to solve the Maximum Clique Problem. The former one proposes the use of a brute force strategy. The cliques are represented as binary strings of n pairs of elements (*position, value*). The *position* component indicates the respective node, whereas the *value* component indicating whether or not the node is in the clique. The cases *yes* (*not*) are represented by an empty (10 *bp* long) strand, respectively. Once the space of all possible problem solutions (2^n possible cliques) has been generated, a series of restriction enzymes are applied to break the strands encoding all the invalid cliques. Hence, these strands are not multiplied in the successive PCR cycles. As a result, gel electrophoresis can be applied in the last step to measure the strands in the set and identify the optimum problem solution.

On the contrary, [11]. proposes the use of plasmids (circular DNA molecules) as a binary data medium. Information about the clique is encoded in a 175 *bp* subsegment of the plasmid, called MCS (Multiple Cloning Site). That subsegment is further divided into *n* regions called stations. Each station represents a node of the graph and is associated with a particular restriction enzyme. The model has an operation *Reset(k)* which sets the value of station *k* to 0, increasing its size by 4 *bp*. Initially, all the stations are set to 1, indicating that all the nodes are present in the clique. The initial amount of molecules corresponds to the size of the solution space (2^n) and remains constant during the solution. Cyclically, a step is carried out for every arc $\{a_i, a_j\}$ that is in the complementary graph G'. The solution space is split into two subsets t_1 and t_2 in each step. The molecules in t_1 are subject to the *Reset(a_i)* operation and the molecules in t_2 to the *Reset(a_j)* operation. Finally, the contents of t_1 and t_2 are poured into a single set. At the end of the algorithm, the number of 1's in the molecules is counted to determine the maximum clique that solves the problem.

We have chosen the following set of computational and physical parameters for the intended comparison of the three algorithms.

1. *No. of iterations:* How many abstract iterations (due to the used model) the algorithm requires.
2. *No. of different molecules per iteration:* The number of different molecules (not copies of the same molecule) the algorithm generates prior to the first iteration and then after each iteration.
3. *Solution ratio after the last iteration*: Calculated as the number of molecules encoding a solution to the problem / total number of molecules. Reflects the state after the last iteration.
4. *Minimum solution ratio*: Similar as the previous parameter, but measured as a minimum over all the iterations including the initial state prior to the first iteration.
5. *No. of operations:* Total number of principal biological operations carried out by the algorithm. We have counted the number of *Cut* operations using restriction enzymes for the brute force algorithm, each *Reset* on a station for the aqueous algorithm, and each *Remove* and *Append* for the incremental algorithm.
6. *Mean string size per iteration:* The weighted mean size of the present molecules taken *prior* to the first iteration and then *after* each iteration. Measured in terms of number of bases or base pairs (*bp*). For the aqueous algorithm, the size refers to the MSC region of the plasmid.
7. *Total mean string size:* Arithmetic mean of the values in the previous parameter.
8. *No. of restriction enzymes:* The total amount of restriction enzymes required by the algorithm.

We analyse the performance of the three algorithms in the case of random Erdös – Renyi graphs [13]. Let $G = (V, E)$ be a random graph with *n* nodes and let *p* be the density of G, i.e., a fixed probability that two randomly chosen nodes are connected by an edge. Let the nodes be indexed $0, 1, \ldots, n - 1$. We fix the following notation:

$\langle card(E) \rangle = pn(n-1)/2$, average number of edges of graph G

$q = \langle card(E') \rangle = (1-p)n(n-1)/2$, average no. of edges of the complementary graph

$\langle N_c \rangle = \binom{n}{c} p^{c(c-1)/2}$, average number of cliques of size c, $1 \le c \le n(n-1)/2$

$s = \sum_{c=1}^{n(n-1)/2} \langle N_c \rangle$, average total number of all cliques in G

$m = n - (1 - (1 - p)^n)/p$, avg. no. of nodes connected to a node with a lower index

$t = n(1 - p^{n-1})$, average number of nodes of G incidental with at least one edge in E'

$c_{\max} = [2\log_b(n) - 2\log_b \log_b(n) + 2\log_b(e/2) + 1]$, asymptotic clique number of G, where $b = 1/p$

$r = q(1 - (1 - c_{\max}/n)^2)$, avg. no. of edges in E' incidental with a clique of size $c_{\max}$

The formulas for $\langle N_c \rangle$ and $c_{\max}$ given above can be found in [16] or [13]. We use the notation $[x]$ to denote the integer closest to a real value x. The other formulas are derived by rather straightforward application of elements of the probability theory and graph theory. Then the parameters of the three algorithms are the following:

Table 1. General comparison between the three DNA algorithms studied in Section 4

	Brute force algorithm	Aqueous algorithm	Incremental algorithm
No. of iterations	q	q	n
No. of different molecules per iteration	min $s+1$ max 2^n	min 1 max 2^n	min 1 max s
Solution ratio after the last iteration	$\dfrac{1}{s+1}$	2^{-r}	$\dfrac{1}{s}$
Minimum solution ratio	2^{-n}	2^{-r}	$\dfrac{1}{s}$
No. of operations	$2q$	$2q$	$m+n$
Mean string size per iteration[1]	min $20(n+1) + 5n$ max $20(n+1)+10n$	min $25(n+1)$ max $25(n+1)+4n$	min 20 max $20 + 20\,c_{\max}$
No. of restriction enzymes[2]	t	t	1

[1] upper bounds could eventually be further improved
[2] a single restriction enzyme is used in the *Incremental algorithm*

Most of the formulas in Table 1 follow rather easily by careful inspection of the three described algorithms. The solution ratio for the *aqueous algorithm* may deserve an explanation. Each step of the algorithm corresponding to an edge incidental with a maximal size clique halves the number of molecules which will encode this clique after the last iteration.

As a test case we assume a rather dense graph with $n = 1000$ nodes and the edge probability $p = 1/3$. Results of the calculation are summarized in Table 2. The size of the chosen graph is dramatically higher than that of the largest problem cases which have been up-to-date successfully solved in biocomputing labs. The more surprising is the observation that even in this case the number of molecules of the incremental algorithm reaches values easily manageable in a test tube, and that concentrations of solution-encoding molecules are quite probably detectable.

Table 2. Comparison between algorithms for the case of a graph with $n = 1000$ and $p = 1/3$

	Brute force algorithm	Aqueous algorithm	Incremental algorithm
No. of iterations	333000	333000	1000
No. of different molecules per iteration	min 3.3 e +9 max 1.07 e +301	min 1 max 1.07 e +301	min 1 max 3.3 e +9
Solution ratio after the last iteration	3.0 e −10	4.96 e −2194	3.0 e −10
Minimum solution ratio	9.33 e −302	4.96 e −2194	3.0 e −10
No. of operations	666000	666000	1997
Mean string size per iteration	min 25020 max 30020	min 25025 max 29025	min 20 max 240
No. of restriction enzymes	1000	1000	1

5 Conclusion

We have presented an incremental strategy of biomolecular computing which is based on a subsequent building of the solution space and eliminating non-perspective partial solutions. This is in contrast with traditional approaches which generate initially a large pool of candidate solutions and then filter out invalid elements. Two variants of a biomolecular algorithm based on the incremental strategy for solving the Maximum Clique Problem have been presented. Finally, we have compared our algorithm with two other standard DNA approaches that solved this problem earlier.

A series of computational and physical parameters have been used to carry out the comparison. The incremental strategy features a number of advantages in terms of efficiency and error-resistance. The comparison results focusing on the efficiency are summarized in Table 1 and 2. It follows that even for large problem cases (thousands of vertices) the incremental strategy can provide a tractable solution measured in the number different molecules, of steps and the size of the DNA pool.

Concerning the error-resistance of the incremental strategy, its key property is the need for step-by-step replication of partial solutions. Therefore, one cycle of the PCR and POA operations are a part of each iteration of the algorithm. The PCR is considered one of the sources of errors in DNA computing, due to possible mutations, and also due to the fact that it can quickly multiply residual molecules encoding invalid solutions which should have been destroyed during the *Remove* operation. Hence, one could expect rather high error rate when the number of iterations grows. Assume, e.g., the probability of failure of the operation *Remove* on an individual molecule $p_E = 0.005$ (i.e., 0.5%). A probabilistic analysis (not included for space limitations) explains, rather surprisingly, why even for large graphs with as much as $n = 500$ nodes and $k = 100$ copies of each molecule in the test tube, the algorithm would report, on average, 0.0047 false cliques larger that the actual maximal clique, implying a very low probability of error.

All the results presented in this paper are based on a theoretical model. However, the operations required to implement our algorithm have already been carried out in laboratory many times and their features are known. In any case, another project to test in the lab an in-vitro implementation of the proposed algorithm is under consideration.

Acknowledgements

This research has been partially funded by the Spanish Ministry of Science and Education under projects TIN2006-15595 and DEP2005-00232-C03-03, by the Ramón y Cajal Program of the Spanish Ministry of Science and Technology, and by the Czech Science Foundation, grant No. 201/06/0567.

References

1. Adleman, L.M.: Molecular computation of solutions to combinatorial problems. Science 266, 1021–1024 (1994)
2. Amos, M., Gibbons, A., Hodgson, D.: Error-resistant implementation of DNA computations. In: Proceedings of the Second Annual Meeting on DNA Based Computers. Princeton University, pp. 87–101 (1996)
3. Amos, M.: Theoretical and experimental DNA computation. Springer, Heidelberg (2005)
4. Bach, E., Condon, A., Glaser, E., Tanguay, C.: DNA models and algorithms for NP-complete problems. In: Proceedings of the 11th IEEE Conference on Computational Complexity, pp. 290–300. IEEE Computer Society Press, Los Alamitos (1996)
5. Bäck, T., Kok, J.N., Rozenberg, G.: Evolutionary computation as a paradigm for DNA-based computing. In: Landweber, L., Winfree, E., Lipton, R., Freeland, S. (eds.) Proceedings: DIMACS Workshop on Evolution as Computation, Princeton, NJ, pp. 67–88 (1999)
6. Baum, E.B., Boneh, D.: Running dynamic programming algorithms on a DNA computer. In: Proceedings of the Second Annual Meeting on DNA Based Computers, Princeton University, pp. 141–147 (1996)

7. Bottoni, P., Martin-Vide, C., Păun, G., Rozenberg, G.: Membrane systems with promoters/inhibitors. Acta Informatica 38, 695–720 (2002)
8. Chiu, D.T., Pezzoli, E., Wu, H., Stroock, A.D., Whitesides, G.M.: Using three-dimensional microfluidic networks for solving computationally hard problems. In: Proceedings of the National Academy of Sciences of USA, vol. 98, pp. 2961–2966 (2001)
9. Cukras, A.R., Faulhammer, D., Lipton, R.J., Landweber, L.F.: Chess games: a model for RNA based computation. Biosystems 52, 35–45 (1999)
10. Graham, R.L., Knuth, D.E., Patashnik, O.: Concrete Mathematics. Addison-Wesley, Reading (1992)
11. Head, T., Yamamura, M., Gal, S.: Aqueous computing: Writing on molecules. In: Proceedings of Congress on Evolutionary Computation, IEEE Service Center, Piscataway, pp. 1006–1010 (1999)
12. Ionescu, M., Sburlan, D.: On P systems with promoters/inhibitors. Journal of Universal Computer Science 10, 581–599 (2004)
13. Janson, S., Luczak, T., Rucinski, A.: Random Graphs. John Wiley & Sons, Chichester (2000)
14. Krishna, S.N., Rama, R.: P Systems with replicated rewriting. Journal of Automata, Languages and Combinatorics 6, 345–350 (2001)
15. Manca, V., Zandron, C.: A clause string DNA algorithm for SAT. In: Jonoska, N., Seeman, N.C. (eds.) DNA Computing. LNCS, vol. 2340, pp. 172–181. Springer, Heidelberg (2002)
16. Matula, D.W.: On the complete subgraphs of a random graph. In: Bose, R.C., et al. (eds.) Proc. 2nd Chapel Hill Conf. Combinatorial Math. and its Applications, pp. 356–369. Univ. North Carolina, Chapel Hill (1970)
17. McCaskill, J.S.: Optically programming DNA computing in microflow reactors. Biosystems 59, 125–138 (2001)
18. Ogihara, M.: Breadth first search 3-SAT algorithms for DNA computers. Technical Report 629, University of Rochester, NY (1996)
19. Ouyang, Q., Kaplan, P.D., Liu, S., Libchaber, A.: DNA solution of the maximal clique problem. Science 278, 446–449 (1997)
20. Păun, G.: Computing with membranes. Journal of Computer and System Sciences 61, 108–143 (2000)
21. Păun, G.: Membrane Computing. In: An Introduction, Springer, Heidelberg (2002)
22. Păun, G., Rozenberg, G., Salomaa, A.: DNA Computing. In: New Computing Paradigms, Springer, Heidelberg (1998)
23. Ravinderjit, B., Chelyapov, N., Johnson, C., Rothemund, P., Adleman, L.: Solution of a 20 variable 3-SAT problem on a molecular computer. Science 296, 499–502 (2002)
24. Zimmermann, K.H.: Efficient DNA sticker algorithms for NP-complete graph problems. Computer Physics Communications 144, 297–309 (2002)

Discrete Simulations of Biochemical Dynamics

Vincenzo Manca

Department of Computer Science, University of Verona, Italy
vincenzo.manca@univr.it

Abstract. Metabolic P systems, shortly MP systems, are a special class of P systems, introduced for expressing biological metabolism. Their dynamics are computed by *metabolic algorithms* which transform populations of objects according to a *mass partition* principle, based on suitable generalizations of chemical laws. The definition of MP system is given and a new kind of regulation mechanism is outlined, for the construction of computational models from experimental data of given metabolic processes.

Keywords: P Systems, Metabolism, Discrete Biological Models.

1 Introduction

In [7] a discrete perspective was introduced in the analysis of metabolic processes, which was then developed in papers [2,9,4,8,10,11], and which is focused on the notion of *Metabolic* P systems, shortly MP systems. Here, we outline the possibility of deducing an MP model, for a given metabolic process, from a suitable macroscopic observation of its behavior along a certain number of steps.

MP systems are a special type of P systems [12] which were proven to effectively model the dynamics of several biochemical processes: the Belousov-Zhabotinsky reaction (Brusselator) the Lotka-Volterra dynamics, the SIR (Susceptible-Infected-Recovered epidemic) [1], the leukocyte selective recruitment in the immunological response [5,1], the Protein Kinase C activation [2], circadian rhythms, mitotic cycles [8], [6][1].

The perspective introduced by MP systems can be synthesized by a principle which replaces the mass action principle. We call it the *mass partition principle* because, according to it, the system is observed along a discrete sequence of steps, and at each step, all the matter of any kind of substance, consumed in the time interval between two consecutive steps, is partitioned among all the reactions which need it for producing their products. If we are able to determine the amount of reactants that any reaction takes in that step, according to the stoichiometry of the reactions (which we assume to know), we can perfectly establish the amount of substances consumed and produced between two steps, therefore all the dynamics can be discovered. As a consequence of mass partition principle, two important aspects follow. In MP system rules act on

---

[1] The package Psim, developed in Java within the research group on *Natural Computing* led by the author, at the Department of Computer Science of the University of Verona (Italy), provides representations and dynamics generations of MP systems (Psim is available from the site of the *Center for BioMedical Computing*, at the University of Verona: www.cbmc.it).

M.H. Garzon and H. Yan (Eds.): DNA 13, LNCS 4848, pp. 231–235, 2008.

object populations, rather than on single objects. Moreover, dynamics is deterministic at population level, but nothing can be said about the dynamical evolution of single objects.

2 Metabolic P Systems

MP systems are deterministic P systems where the transition to the next state (after some specified interval of time) is calculated according to a *mass partition strategy*, that is, the available matter of each substance is partitioned among all reactions which need to consume it. A special class of MP systems was proved to be stongly related to differential models [4]. The notion of MP system we consider here is based on those given in [8,10,11].

Let us consider a set X of substances and a set of R of reactions over them, as pairs of strings, represented in the arrow notation according to which any rule $r \in R$ is identified by $\alpha_r \rightarrow \beta_r$ with α_r, β_r strings over X (α_r represents the reactants of r, while β_r represents the products of r, for example, $aab \rightarrow cd$ is a reaction where two *molecules* of a with a *molecule* of b react by producing a *molecule* of c and a molecule of d).

For a string γ and a symbol x we denote by $|\gamma|_x$ the number of occurrences of the symbol x in γ, while $|\gamma|$ is the length of γ. Then, the *stoichiometric matrix* A_R correspondent to a set R of reactions over a set X of substances is defined by setting $A_R = (A_R(x,r) \mid x \in X, r \in R)$ and, for every $x \in X$ and $r \in R$, $A_R(x,r) = |\beta_r|_x - |\alpha_r|_x$. Moreover, we define $R_\alpha(x) = \{r \in R \mid |\alpha_r|_x > 0\}$. Two reactions r_1, r_2 are *competing* if $r_1, r_2 \in R_\alpha(x)$ for some substance $x \in X$. We call *regulator*, of a reaction r, any reactant of r or any reactant of a reaction which is competing with r.

Definition 1 (MP System). *An MP system is a construct*

$$M = (X, R, Q, U, \nu, \sigma, \tau, q_0, \Phi)$$

where:

- $X = \{x_1, \ldots, x_n\}$ *is a finite set of substances (the types of molecules);*
- $R = \{r_1, \ldots, r_m\}$ *is a finite set of reactions over X;*
- Q *is the set of states, that is, the functions $q : X \rightarrow \mathbb{R}$ from substances to real numbers. The state q of the instant i can be identified as a vector $(x_1[i], x_2[i], \ldots, x_n[i])$ of real numbers, constituted by the values which are assigned, by q, to the elements of X.*
- $U = \{u_1, \ldots, u_m\}$ *is the set of reaction units, where, for each rule r, u_r is a function from states to real numbers (the amount of molar quantity consumed/produced by the rule r in correspondence to any occurrence of reactant/product occurring in it);*
- ν *is a natural number which specifies the number of molecules of a (conventional) mole of M, as population unit of M;*
- σ *is a function which assigns to each $x \in X$, the mass $\sigma(x)$ of a mole of x (with respect to some measure unit);*

- τ is the temporal interval between two consecutive states;
- $q_0 \in Q$ is the initial state, also denoted by $X[0] = (x_1[0], x_2[0], \ldots, x_n[0])$;
- Φ is a set of regulation maps.

The temporal evolution of an MP system M is calculated by means of the following system of autonomous first-order difference equations (1) (2), called metabolic algorithm, where $X[i]$ and $U[i]$ are the vectors of substance quantities and reaction units at step i, A_R is the stoichiometric matrix of dimension $n \times m$ corresponding to the reactions of R (n is the number of different substances and m the number of reactions), Φ is the vector of functions (as many as the reactions), and $\times, +$ are the usual matrix product and vector sum:

$$X[i+1] \;=\; (A_R \times U[i]\,) + X[i] \tag{1}$$
$$U[i] \;=\; \Phi(X[i]) \tag{2}$$

The parameters τ, ν, μ have no role in the mathematical definition of dynamics. Nevertheless, they are essential for giving a determinate physical meaning to the numerical values, according to a specific time/mass measure scale.

3   Metabolic Algorithms and Log-Gain Regulation

Given a real metabolic system that we can observe for a certain number of steps, is it possible to determine an MP system which could predict, within an acceptable approximation, the future behaviour of the given system? We will show how this task could be achieved. In fact, in some cases, we can determine, in a systematic way, an MP system which is an adequate model of some observed metabolic dynamics.

In order to discover the reaction units at each step, we introduce the notion of *log-gain regulation*. In fact, it seems to be perfectly natural that a proportion should exist among the relative variation of substances and the relative variation of the reaction unit of r. The relative variation of a substance x is defined as the ratio $\Delta(x)/x$. In differential notation (with respect to the time variable), this ratio is related to $\frac{dx}{dt}/x$, and from elementary calculus we know that it is the same as $\frac{d(\lg x)}{dt}$. This equation explains the term "log-gain" for expressing relative variations. In this way, we can derive the values of the reaction units at any observation time, therefore, these parameters determine the dynamics of MP systems. More precisely, we set the following principle.

Principle 2 (Log Gain Regulation). *For $i \geq 0$ let $Lg(u_r[i]) = (u_r[i+1] - u_r[i])/u_r[i]$ be the log-gain of the reaction unit u_r at the step i, and let $Lg(x[i]) = (x[i+1] - x[i])/x[i]$ be the log-gain of the substance x at the same step, then $Lg(u_r[i])$ is a linear combination of the log-gains of the regulators of r:*

$$Lg(u_r[i]) = \sum_{x \in X} p_{r,x} Lg(x[i]) + p_r \tag{3}$$

$p_{r,x}$ with $x \in X$ are the log-gain parameters. If the parameter $p_{r,x} \neq 0$, then x is a regulator of r. The parameter p_r is called the log-gain offset of the rule r.

Given the dynamics of a system that we *observe* for a sufficient number of steps, we want to know, with a sufficient precision, the (molar) quantities of all different kinds of molecules, for a sequence of steps. Let us denote these quantities with the sequence, for $i = 0, \ldots k$, of vectors:

$$X[i] = (x_1[i], x_2[i], \ldots, x_n[i])$$

Moreover, we assume to know the structure of the system, that is, kinds of substances, reactions, time unit, molar unit, and initial state. We want to predict the vectors $X[i]$ for steps $i > k$, which follows the observation steps. We solve the problem when we discover the regulation maps Φ.

Let us consider the system of equations $LG[i] + \Delta S[i + 1]$, obtained by putting together system (3) at step i with system (1) of Definition 1 at step $i + 1$. We call it **observation module**. This system of equations has $n + m$ equations. The variables of this system are the reaction units and the log-gain parameters (and offsets). In general, the number of these variables: $u_1[i + 1], u_2[i + 1], \ldots, u_m[i + 1], \ldots$ is greater than the number of equations. Moreover, in order to discover the dynamics underlying the passage of the MP system, from one step to its next step, it is enough to know. at any step. the value of reaction units. Despite the difference between the number of equations and the number of variables, the following theorem holds, as a consequence of log-gain principle (we omit the proof).

Theorem 3. *The system $LG[i] + \Delta S[i + 1]$ has one and only one solution.*

Let us assume to know $U[0]$ (in fact, there are some methods for determining it [11]). The value $X[0]$ is known because it corresponds to the initial state of the system. Therefore, if we solve this system for $i = 0$, that is, $LG[0] + \Delta S[1]$, we get the value of $U[1]$. So, if vectors $X[i]$ for $i = 1, \ldots, k$, are given by observation, we can apply the same procedure, again for $i = 1, 2, \ldots, k$, and get $U[2], U[3], \ldots, U[k + 1]$.

Now assume that these vectors depend on the substance quantities with some polynomial dependence of a given degree, say a third degree, then we can use some standard interpolation tools for finding the functional dependence of vector U with respect to the substance quantities. The resulting polynomials are some approximations of the regulation functions Φ we are searching for, and our task was completed. In fact, now we can use the metabolic system (1) (2) of Definition 1 for computing the evolution of the given MP systems in all the steps i for $i > k$.

We applied this method to many metabolic systems (e.g. Lotka-Volterra, Brusselator, and Mitotic Cycles) and we were able to reconstruct, almost exactly their dynamics. But this procedure assumes the knowledge of $U[0]$. Actually, there are several possibilities under investigation. However, we discovered experimentally a very interesting fact, which deserves a more subtle theoretical investigation. If we consider the system $\Delta S[0]$ and choose as $U[0]$ one of its infinite solutions (imposing some additional very natural constraints), then in many cases, we found that, independently from the chosen value of $U[0]$, after a small number of steps, say $k = 3$ steps, our procedure will generate, with a great approximation, the same vectors $U[i + k]$, for all $i > 0$. This means that the data collected in the observation steps are sufficient to determine the functions which, on the basis of substance quantities, regulate the dynamics of the system.

Numerical elaborations of our simulations were performed by MATLAB® standard operators (*backslash* operator for square matrix left division or in the least squares sense solution) and interpolation was performed by polynomials of third degree. Specific observation strategies were adopted, by using about one hundred steps. In almost all cases, the observed dynamics were correctly reconstructed. This means that the regulation functions, deduced according to the outlined method, provided MP systems with the same dynamics of the observed systems. In conclusion, in the case of natural systems, from suitable observations, we could discover, with good approximation, the underlying dynamical regulation maps, and consequently, reliable computational models of their dynamic. However, applications of our method to more complex dynamics and deeper theoretical analyses of the simulation results will be topics for further research.

References

1. Bianco, L., Fontana, F., Franco, G., Manca, V.: P systems for biological dynamics, pp. 81–126. in [3]
2. Bianco, L., Fontana, F., Manca, V.: P systems with reaction maps. International Journal of Foundations of Computer Science 17(1), 27–48 (2006)
3. Ciobanu, G., Păun, G., Pérez-Jiménez, M.J. (eds.): Applications of Membrane Computing. Springer, Heidelberg (2006)
4. Fontana, F., Manca, V.: Discrete solutions of differential equations by metabolic P systems. Theoretical Computer Science 372, 165–182 (2007)
5. Franco, G., Manca, V.: A membrane system for the leukocyte selective recruitment. In: Martín-Vide, C., Mauri, G., Păun, G., Rozenberg, G., Salomaa, A. (eds.) Membrane Computing. LNCS, vol. 2933, pp. 180–189. Springer, Heidelberg (2004)
6. Franco, G., Guzzi, P.H., Mazza, T., Manca, V.: Mitotic Oscillators as MP Graphs. In: Hoogeboom, H.J., Păun, G., Rozenberg, G., Salomaa, A. (eds.) WMC 2006. LNCS, vol. 4361, pp. 382–394. Springer, Heidelberg (2006)
7. Manca, V., Bianco, L., Fontana, F.: Evolutions and oscillations of P systems: Applications to biological phenomena. In: Mauri, G., Păun, G., Pérez-Jiménez, M.J., Rozenberg, G., Salomaa, A. (eds.) WMC 2004. LNCS, vol. 3365, pp. 63–84. Springer, Heidelberg (2005)
8. Manca, V., Bianco, L.: Biological networks in metabolic P systems. BioSystems (to appear) doi:10.1016/j.biosystems.2006.11.009
9. Manca, V.: MP systems approaches to biochemical dynamics: Biological rhythms and oscillations. In: Hoogeboom, H.J., Păun, G., Rozenberg, G., Salomaa, A. (eds.) WMC 2006. LNCS, vol. 4361, pp. 86–99. Springer, Heidelberg (2006)
10. Manca, V.: Metabolic P Systems for Biochemical Dynamics. Progress in Natural Sciences (Invited Paper) 17(4), 384–391 (2007)
11. Manca, V.: The Metabolic Algorithm for P Systems: Principles and Applications. Theoretical Computer Science (to appear, 2007)
12. Păun, G.: Membrane Computing. An Introduction. Springer, Heidelberg (2002)

DNA Splicing Systems

An Ordinary Differential Equations Model and Simulation

Elizabeth Goode and William DeLorbe

Mathematics Department, Towson University, Towson MD 21252, USA

Abstract. DNA splicing in the test tube may generate DNA molecules other than well-formed ones having two blunt ends. We introduce an example splicing system that generates all possible molecular types. We use differential equations to model the system, and Mathematica to simulate its dynamics. We find that most simulation results match our predictions, and acknowledge that a more comprehensive program is needed for further investigations. This is the first model and simulation of which we are aware that specifically treats the fact that several molecular types in addition to well-formed molecules may be present in a splicing system, even at equilibrium.

Keywords: DNA, splicing systems, limit languages, differential eqs.

1 Introduction

In [4] Tom Head introduced *splicing systems* to model the generative power of double-stranded DNA (dsDNA) in the presence of appropriate enzymes that cut and paste dsDNA. Researchers have explored many extensions of the basic theory. One of the most important results assumes a finite initial number of different dsDNA sequences and a finite number of enzymes. In this case the *splicing language* generated by the system is a regular language in the Chomsky hierarchy of formal languages. (See [1], [12], and [13].) Other formulations of splicing yield systems that perform universal computation. (See [5], [10], [8], [9], and [14].)

In discussions several years ago Head noted that in the laboratory splicing can generate molecules that are *transient* in the sense that they are used up as the splicing operation "runs to completion." In light of this, the molecules of interest are those left in the test tube at equilibrium. In [3] these persistent molecules are called *limit words* of the system, and the corresponding formal language is the *limit language*. In its standard formulation, the splicing language is defined as the set of well-formed molecules (i.e., those having two blunt ends) that are created anytime during the evolution of the system, and the limit language is those well-formed molecules present at equilibrium. In the test tube, however, splicing produces molecules that are not well-formed, i.e. molecules that have one or two sticky ends and/or molecules that are circular. Results concerning circular molecules appear in [6], [11] and [15]. Pixton's cut-and-paste splicing systems

M.H. Garzon and H. Yan (Eds.): DNA 13, LNCS 4848, pp. 236–245, 2008.

introduced in [13] and cutting/recombination systems presented by Freund et. al. in [2] specifically treat sticky ends. When this type of splicing model is applied as done in Examples 2.6 and 2.7 in [3], both the splicing and the limit languages are interpreted to include molecules that are not well-formed. A cut-and-paste example of a limit language with circular strings appears in [3].

In [3] Pixton introduced the first differential equations model of the limit behavior of a splicing system. The system generated only well-formed molecules during iterated cutting and pasting, thus the full power of the cut-and-paste model was not required. A dynamical systems model of splicing that treats all molecule types has not appeared before, nor has a computer simulation of any splicing dynamics been given to date. In this work we model a system that generates all the molecule types and give a computer simulation of its dynamics.

Sect. 2 contains a brief explanation of the cut-and-paste approach to splicing. In Sect. 3 we introduce our example system, and Sect. 4 contains the differential equations that model its dynamics. In Sect. 4 we also verify that the amount of dsDNA stays constant throughout the reaction. In Sect. 5 appear results from the Mathematica simulation and a discussion of how they compare with our predictions about how the system should behave. Over time the example system generates progressively longer molecules. Our program models molecules up to a fixed length, therefore we see only the initial system dynamics. We need a more comprehensive simulation program for further investigation.

2   Cut-and-Paste Splicing

A splicing system (σ, I) is a mathematical model of the generative behavior of dsDNA in the presence of certain enzymes. Biochemically speaking, I models an initial set of well-formed dsDNA, and σ models a test-tube environment in which possibly several restriction enzymes (site-specific cutting enzymes) and a ligase (a pasting enzyme) act simultaneously. The splicing language $\sigma^*(I) = L$ models the set of all molecules that appear anytime during iterated splicing. The limit language $L_\infty \subset L$ represents the set of molecules that persist when the system reaches equilibrium. In either language, a circular word w is denoted by $\hat{}w$.

We use Pixton's cut-and-paste model of splicing from [13] because it clearly shows the 3-step splicing operation (2 cuts and 1 paste). The model has special symbols called "end markers" that encode the blunt ends as well as the sticky ends generated during cutting. A cutting rule $\alpha z \beta$ encodes the cutting action $xzy \implies x\alpha + \beta y$, and a pasting rule $\alpha w \beta$ corresponds to the pasting action $x\alpha + \beta y \implies xwy$. Thus a splicing rule $(u_1, v_1; u_2, v_2)$ in the standard splicing notation seen in [3] is represented by two cutting rules $\alpha_1 u_1 v_1 \beta_1$ and $\alpha_2 u_2 v_2 \beta_2$ and one pasting rule $\alpha_1 u_1 v_2 \beta_2$. In this case α_1 encodes the sticky end on the "left" half after cutting between u_1 and v_1 and β_2 encodes the sticky end on the "right" half after cutting between u_2 and v_2. The pasting rule reconstitutes $u_1 v_2$ when the sticky ends α_1 and β_2 are reattached.

3   The Example Splicing System

Example 1 (The Model System). The initial set is $I = \{\rho(uxu)^{10}\rho\} = \{N^{10}\}$. The rule set in σ has one rule that is used for both cutting and pasting: $\gamma uu\gamma$. The words in $L = \sigma^*(I)$ and L_∞ are the same. Verification of (1) is left to the reader.

$$L = \rho\,(uxu)(uxu)^+\rho + \rho(uxu)^*ux\gamma + \gamma xu(uxu)^*\rho + \gamma x(uux)^*\gamma + \hat{\ }(uux)^+. \quad (1)$$

Explanation of Ex. 1: The initial set contains many copies of one strand, denoted N^{10}, that is the concatenation of ten copies of a dsDNA sequence F having two sticky ends. One can think of an F-unit as a sequence uxu, where the suffix u concatenated with the prefix u forms a restriction (cut) site uu. We shall assume this F-unit reads as its own reverse complement so its orientation in space will not impact the reading of its sequence. Although F-units have sticky overhangs, we shall assume the N^{10} molecules are well-formed and have blunt ends, indicated by ρ, that cannot ligate together. Such molecules can be manufactured by "filling in" the necessary nucleotides to blunt the sticky ends of N^{10} and dephosphorylating these ends. We ignore these small end deviations. Because the initial blunt ends are dephosphorylated and new ones are not created during iterated splicing, blunt-end ligation never occurs.

The system has a ligase that catalyzes the covalent bonds necessary to paste any two molecules whose complementary sticky ends come into close proximity with one another and hybridize. There is one restriction enzyme that cuts only at site uu, and uu occurs only between copies of F. We shall assume the cut produces sticky overhangs that are reverse complements of one another. Such sticky ends and an F-unit that is its own reverse complement impact the counting involved when writing the differential equations. Specifically:

1) when molecule Z is cut with the restriction enzyme to generate Z_1 and Z_2, not only can Z_1 and Z_2 rehybridize and ligate together to reform Z, but Z_1 can rotate $180°$ in space and ligate to another copy of Z_1, and
2) circular molecules can form because a dsDNA fragment can circularize if it has two complementary sticky ends.

Another implication is stated in the following lemma:

Lemma 1. *All molecules with matching ends (both blunt or both sticky) generated by the system of Ex. 1 have palindromic sequences.*

Proof. A word in L with sticky ends has the form $\gamma x(uux)^*\gamma = \gamma(xuu)^*x\gamma$. A word with blunt ends has the form $\rho(uxu)(uxu)^+\rho = \rho\,(uxu)^+(uxu)\rho$. $\qquad\square$

There are four basic types of molecules that appear in the system. An initial N-type molecule can be cut between copies of the F-unit to produce two new molecules, called A-type molecules, each having one blunt end and one sticky overhang. An A-type molecule can be cut between F-units to produce a shorter A-type molecule and a B-type molecule having sticky overhangs on both ends. (See Fig. 1.) The B-type molecule can ligate to itself to form a circular C-type molecule. Creation of N, A and B-type molecules can occur in other ways once the system generates all four molecular types.

Molecules are characterized by length as well as type. An N^k molecule is the concatenation of k copies of the F-unit. Both of its ends inherit ρ, the blunt end filling performed on the initial set of N^{10} molecules. We denote the set of all N^k-type molecules by N_k. We will abuse this notation and let N_k also denote the cardinality of set N_k. This will be unambiguous in context. We use similar notation for the other molecule types, i.e., there are A^k, B^k and C^k-type molecules in sets A_k, B_k and C_k having cardinality A_k, B_k and C_k, respectively.

Fig. 1. The N, B, and A-type molecules appear left to right. The C-type is not shown.

4   The Ordinary Differential Equations Model

In this section we introduce the ordinary differential equations that describe the change in the number of each type of molecule in one time step, i.e. in one small time interval denoted by Δt. We assume only one action occur on a given strand of DNA in a given time step. For example, a molecule could be cut in a given time step, or one paste with another molecule, but not both in the same time step. We use parameters α and β, where α is the probability that a given pair of molecules X and Y will paste to yield XY in a unit of time, and β is the probability that a cut operation will occur at a given site on a given molecule X in a unit of time. Note α and β will vary with the concentrations of the reactants.

We let Z denote the total number of Z-type molecules of any length for $Z \in \{N, A, B, C\}$, i.e., $Z = \sum_k Z_k$. We now derive (2) and (3).

N_k changes as N^k molecules are created and destroyed. Consider first that an N^k molecule has $k - 1$ cutting sites, so the probability it will be cut in one time unit is $\beta(k - 1)$. The net change in N_k due to this is $-\beta(k - 1)N_k\Delta t$. An A^i molecule and an A^j molecule can paste together to form an N^k molecule if $i + j = k$. At first glance it appears this reaction generates $\alpha \sum_{i+j=k} A_i A_j \Delta t$ new N^k molecules in each step. However, the paste of a particular pair of A^i and A^j molecules is counted twice because an $A^i A^j$ molecule rotated in space and is thus also counted an $A^j A^i$ molecule. The change in N_k due to pasting A molecules is therefore $\frac{1}{2}\alpha \sum_{i+j=k} A_i A_j \Delta t$. The rate of change N'_k appears in (2).

Next we find A'_k. An A^k molecule can form if short A and B-type molecules paste together. The paste can occur two different ways since B molecules can rotate $180°$. We sum A_i and B_j when $i + j = k$ to find a $2\alpha \sum_{i+j=k} A_i B_j \Delta t$ contribution to A_k in one time step. A^k molecules also form if an N^m molecule with $m > k$ is cut to form an A^k and an A^{m-k} molecule. The cut can (generally) occur in two places in N^m for fixed m, yielding $2\beta N_m$ new A^k molecules. Summing over all $m > k$ yields a $2\beta \sum_{m>k} N_m \Delta t$ change in A_k. Last, A^k's form if an A^m molecule for $m > k$ is cut to produce an A^k and a B^{m-k} molecule. For fixed m this can occur at one site in an A^m molecule, contributing βA_m to A_k in one time step. The corresponding net change in A_k is $\beta \sum_{m>k} A_m \Delta t$.

An A^k molecule is lost when it pastes with another A molecule to produce a longer A molecule. This contributes a $-\alpha A A_k \Delta t$ net change to A_k. Also, an A^k molecule can paste with a B molecule in two ways (recall B's 180° rotation), each with probability $\alpha \Delta t$. The resulting net change in A_k is $-2\alpha B A_k \Delta t$. Lastly, an A^k molecule can be cut to produce shorter B and A molecules. Each A^k has $k-1$ cutting sites so there are $(k-1)A_k$ possible cuts that together contribute a net change of $-\beta(k-1)A_k \Delta t$ to A_k. The A'_k equation appears in (3).

For brevity we omit the derivations of B'_k and C'_k appearing in (4) and (5).

$$N'_k = \frac{1}{2}\alpha \sum_{i+j=k} A_i A_j - \beta(k-1)N_k \tag{2}$$

$$A'_k = 2\alpha \sum_{i+j=k} A_i B_j + 2\beta \sum_{m>k} N_m + \beta \sum_{m>k} A_m - \alpha A A_k - 2\alpha B A_k - \beta(k-1)A_k \tag{3}$$

$$B'_k = 2\alpha \sum_{i+j=k} B_i B_j + \beta(k)C_k + \beta \sum_{m>k} A_m + 2\beta \sum_{m>k} B_m$$
$$- \alpha B_k - 2\alpha A B_k - 4\alpha B B_k - \beta(k-1)B_k \tag{4}$$

$$C'_k = \alpha B_k - \beta(k)C_k \tag{5}$$

Lemma 2. *The amount of material in the test-tube remains constant over time during the reaction according to this model, i.e. the total number of F-units remains unchanged from one time step to the next.*

Proof. Let variables X, P, Q, and R denote the numbers of F-units stored in molecules of type N, A, B and C, respectively, at time t. By definition we have:

$$X(t) = \sum_k kN_k, \quad P(t) = \sum_k kA_k, \quad Q(t) = \sum_k kB_k, \quad \text{and} \quad R(t) = \sum_k kC_k. \tag{6}$$

We use (2) – (6) to find the terms in X', P', Q' and R'.
The two terms in X' are

$$\frac{1}{2}\alpha \sum_k \sum_{i+j=k} kA_i A_j = \frac{1}{2}\alpha \sum_{i,j}(i+j)A_i A_j = \alpha PA, \quad \text{and} \tag{7}$$

$$-\beta \sum_k k(k-1)N_k = -\beta \sum_m m(m-1)N_m. \tag{8}$$

The six terms in P' are

$$2\alpha \sum_{k} \sum_{i+j=k} kA_iB_j = 2\alpha QA + 2\alpha PB, \tag{9}$$

$$2\beta \sum_{k} \sum_{m>k} kN_m = \beta \sum_{m} \left(2\sum_{m>k} k \right) N_m = \beta \sum_{m} m(m-1)N_m, \tag{10}$$

$$\beta \sum_{k} \sum_{m>k} kA_m = \frac{1}{2}\beta \sum_{m} m(m-1)A_m, \tag{11}$$

$$-\alpha \sum_{k} kAA_k = -\alpha PA, \tag{12}$$

$$-2\alpha \sum_{k} kBA_k = -2\alpha PB, \text{ and} \tag{13}$$

$$-\beta \sum_{k} k(k-1)A_k = -\beta \sum_{m} m(m-1)A_m. \tag{14}$$

The eight terms in Q' are

$$2\alpha \sum_{k} \sum_{i+j=k} kB_iB_j = 4\alpha QB, \tag{15}$$

$$\beta \sum_{k} k^2 C_k = \beta kR, \tag{16}$$

$$2\beta \sum_{k} \sum_{m>k} kB_m = \beta \sum_{m} m(m-1)B_m, \tag{17}$$

$$\beta \sum_{k} \sum_{m>k} kA_m = \frac{1}{2}\beta \sum_{m} m(m-1)A_m, \tag{18}$$

$$-2\alpha \sum_{k} kAB_k = -2\alpha QA, \tag{19}$$

$$-4\alpha \sum_{k} kBB_k = -4\alpha QB, \tag{20}$$

$$-\alpha \sum_{k} kB_k = -\alpha Q, \text{ and} \tag{21}$$

$$-\beta \sum_{k} k(k-1)B_k = -\beta \sum_{m} m(m-1)B_m. \tag{22}$$

Lastly, the two terms in R' are

$$\alpha \sum_{k} kB_k = \alpha Q, \text{ and} \tag{23}$$

$$-\beta \sum_{k} k^2 C_k = -\beta kR. \tag{24}$$

The total number of F-units present at time t in the system is given by

$$M(t) = X(t) + P(t) + Q(t) + R(t). \tag{25}$$

To prove the lemma we verify that $M' = X' + P' + Q' + R' = 0$. We must check the cancelations that occur between the terms in (7) – (24).

X' terms αPA in (7) and $-\beta \sum_m m(m-1)N_m$ in (8) cancel with P' terms $-\alpha PA$ in (12) and $\beta \sum_m m(m-1)N_m$ in (10). Four terms for P' remain. Working backwards, $-\beta \sum_m m(m-1)A_m$ in (14) cancels with term $\frac{1}{2}\beta \sum_m m(m-1)A_m$ that appears twice – once in (11) and once in (18). Then $-2\alpha PB$ in (13) and $2\alpha PB$ in (9) cancel. Finally, $2\alpha QA$ in (9) cancels with the Q' term $-2\alpha QA$ in (19).

For the remaining Q' terms, first observe $\beta \sum_m m(m-1)B_m$ in (17) and $-\beta \sum_m m(m-1)B_m$ in (22) cancel, and $4\alpha QB$ in (15) and $-4\alpha QB$ in (20) cancel. Then last two terms for Q', βkR in (16) and $-\alpha Q$ in (21), cancel with (24) and (23) for R'. This completes the cancelations and shows $M' = 0$.   □

5   Simulation Results and Discussion

We use Mathematica to implement the dynamics of the example splicing system introduced in Sec. 3. Our code treats molecules up to 20 F-units long. The graphs given in this section show some of our simulation results. [1]

In Fig. 2, the left-hand graph shows the number of N^{10} molecules decreases shortly after the system is initialized. This is expected because initial molecules can be immediately cut to produce A^k-type molecules for $k < 10$. The right-hand graph shows an (expected) increase in the number A^i-type molecules for $2 \leq i \leq 9$. These molecules appear as the N^{10} molecules begin to be cut. It is interesting that all of these curves coincide, so only one curve appears. This is because there are exactly nine cutting sites in an N^{10} molecule, each is equally likely to be cut, and each cut produces one A^j and one A^k where $j + k = 10$. B-type molecules should appear once A molecules are present, because B's are generated when A molecules are cut. As expected, the left-hand graph in Fig. 3

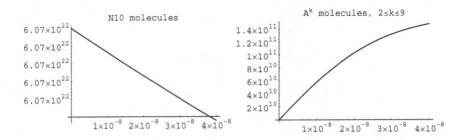

Fig. 2. The left-hand graph shows the initial decline in the number of N^{10} molecules (*black curve*) over time in seconds. The right-hand graph shows that the numbers of A^i molecules, $2 < i < 9$, initially increase at identical rates, thus their graphs coincide (*single curve*).

---

[1] In all graphs in this paper, the x-axis is time in seconds after the system is initialized, and the y-axis is the number of molecules present.

shows that B^k molecules with $k < 9$ appear just after A^i molecules for $1 \leq i \leq 9$ first appear in Fig. 2. Notice that the smaller the k, the more quickly the corresponding B^k curve rises. This makes sense because these B^k molecules are coming from cuts of the A^k molecules, $1 \leq k \leq 9$, which are present in equal numbers. There are, for example, more ways to generate a B^2 molecule than a B^8 molecule because any of the A^i molecules for $i > 2$ can be cut to create a B^2 molecule, whereas only an A^i for $i > 8$ can be cut to make a B^8 molecule. Similarly, N^k molecules for $k < 10$ appear after A molecules less than length 10 appear because an A^i can paste with an A^j for $i + j < 10$ to create such an N^k molecule. The right-hand graph in Fig. 3 shows this dynamic. Notice that the larger k is, the more quickly N_k increases. (Why?) Space constraints require us to omit a simulation of N^k molecules for $10 \leq k \leq 20$, but we can report that they also appear shortly after the A^i molecules for $1 \leq i \leq 9$ appear, in proportions that also make sense. Observe the very low numbers of B molecules in the left graph. Are those being generated and destroyed immediately thereafter? Or are most reactants that generate these B molecules involved in other reactions, leaving few to generate the B's? Deeper study is required.

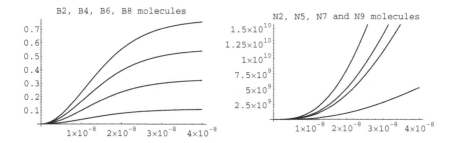

Fig. 3. The left-hand graph shows the increase in B^k molecules for $k = 2$ (*top curve*), $k = 4$ (*higher middle curve*), $k = 6$ (*lower middle curve*) and $k = 8$ (*lowest curve*). The right-hand graph shows the increase in N^j molecules for $j = 2$ (*lowest curve*), $j = 5$ (*lower middle curve*), $j = 7$ (*higher middle curve*), and $j = 9$ (*highest curve*).

The graphs in Fig. 4 show some longer molecules. The left graph shows the A^{11} molecules. An A^k molecule for $k \geq 10$ can arise from the paste of an A^i and B^j molecule if $i + j > 9$. From Figs. 2 and 3 we see such A and B molecules are present before the A^{11}'s appear. An N^j molecule with $j > 11$ can be cut to form an A^{11} and an A^{j-11} molecule. We reported above that N^j molecules for $10 \leq j \leq 18$ can (and do) arise once A^k molecules with $1 \leq k \leq 9$ are available. In Fig. 2 we see that such A molecules appear before the A^{11}'s appear here. The right-hand graph in Fig. 4 shows the N^{19} and N^{20} molecules. Either can result from the paste of two A-type molecules if A's longer than 9 are available. Because A^k molecules with $k > 9$ do not appear immediately, as seen in the Fig. 4, the time lag before the N^{19} and N^{20} molecules appear makes sense. The A^{11} molecules should (and do) appear before the N^k molecules for $k > 20$,

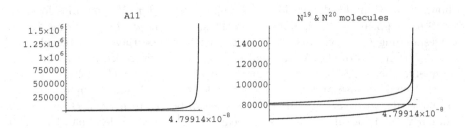

Fig. 4. The left-hand graph shows a rapid increase in the number of A^{11} molecules (*curve*) when they appear at time $t \approx 4.799 \times 10^{-8}$ seconds. The right graph shows that N^{19} (*top curve*) and N^{20} (*bottom curve*) molecules appear immediately after the A^{11} molecules appear.

because such an N molecule at this stage of the game requires the paste of two A molecules, with at least one longer than 11. In fact, no N^k molecules for $k > 20$ ever appear because our code only handles molecules up to length 20. Indeed, the simulation halts very shortly after the A^{11}'s appear. We interpret this to indicate that an N^k molecule for $k > 20$ was in fact generated.

6   Suggestions for Questions

While we think the splicing and limit languages given in Sect. 3 for Ex. 1 satisfy the definitions laid out in [3], it is still not clear how the relative numbers of the different molecules will evolve over an extended period of time. For example, while N^{10} molecules are destroyed when cut to form A molecules as long as A^9, some of these A molecules paste back together to recreate N^{10}. Others paste to form longer N molecules which again generate short A molecules that, in turn, replenish the N^{10}'s. Over the long run there are more ways to create an N^{10} than destroy one. We can easily count the finitely many ways to destroy one, whereas counting the number of ways to create one becomes ominous if there are A-type molecules of arbitrary length in the system. But the fact remains that any molecule can be destroyed at any time. The question is further complicated by the assumption that virtually infinitely many copies of each initial string is present at time $t = 0$. Would various upper bounds on initial quantities affect the long-range behavior of the system? Or would the limit languages have similar structures despite the amount of initial material? What is the best way to define and investigate this "structure?" A simultaneous graphical comparison of the quantities of certain molecular types and sizes at a fixed time t might be informative. Which molecules should we compare? Those of similar type, or perhaps those of similar size across all types? This problem begs for a more comprehensive simulation program that can run the system for many more time steps, and provide more graphical output. Finally, real experimental data from "the bench" would help us evaluate the viability of our model.

Acknowledgements. I extend my sincerest thanks to all the helpful and hard-working people who made DNA13 and this LNCS volume possible, especially Max Garzon.

References

1. Culik II, K., Harju, T.: Splicing semigroups of dominoes and DNA. Discrete Applied Mathematics 31, 261–277 (1991)
2. Freund, R., Csuhaj-Varjú, E., Wachtler, F.: Test tube systems with cutting/recombination operations. In [7]
3. Goode, E., Pixton, D.: Splicing to the limit. In: Jonoska, N., Păun, G., Rozenberg, G. (eds.) Aspects of Molecular Computing. LNCS, vol. 2950, pp. 189–201. Springer, Heidelberg (2003)
4. Head, T.: Formal language theory and DNA: An analysis of the generative capacity of specific recombinant behaviors. Bulletin of Mathematical Biology 49(6), 737–759 (1987)
5. Head, T., Păun, G., Pixton, D.: Generative mechanisms suggested by DNA recombination. In [14]
6. Mauri, G., Bonizzoni, P., DeFelice, C., Zizza, R.: DNA and circular splicing. In: Condon, A., Rozenberg, G. (eds.) DNA 2000. LNCS, vol. 2054, pp. 117–129. Springer, Heidelberg (2001)
7. Pacific Symposium on Biocomputing (1997), http://WWW-SMI.Stanford.EDU/people/altman/psb97/index.html
8. Păun, G.: Regular extended H systems are computationally universal. J. Autom. Lang. Comb. 1(1), 27–36 (1996)
9. Păun, G.: Controlled H systems and the Chomsky Hierarchy. Fundam. Inf. 30(1), 45–57 (1997)
10. Păun, G. (ed.): Computing with Bio-Molecules: Theory and Experiments. Springer, New York (1998)
11. Pixton, D.: Linear and circular splicing systems. In: INBS 1995. Proceedings of the First International Symposium on Intelligence in Neural and Biological Systems, p. 181. IEEE Computer Society, Washington (1995)
12. Pixton, D.: Regularity of splicing languages. Discrete Applied Mathematics 69(1-2), 101–124 (1996)
13. Pixton, D.: Splicing in abstract families of languages. Theoretical Computer Science 234, 135–166 (2000)
14. Rozenberg, G., Salomaa, A. (eds.): Handbook of Formal Languages. Springer, New York (1996)
15. Siromoney, R., Subramanian, K.G., Rajkumar Dare, V.: Circular DNA and splicing systems. In: ICPIA 1992. Proceedings of the Second International Conference on Parallel Image Analysis, pp. 260–273. Springer, London (1992)

Asynchronous Spiking Neural P Systems: Decidability and Undecidability

Matteo Cavaliere[1], Omer Egecioglu[2], Oscar H. Ibarra[2], Mihai Ionescu[3], Gheorghe Păun[4], and Sara Woodworth[2]

[1] Microsoft Research-University of Trento CoSBi, Italy
cavaliere@cosbi.eu
[2] Dept. of Computer Science, University of California, Santa Barbara, USA
{omer, ibarra, swood}@cs.ucsb.edu
[3] Research Group on Mathematical Linguistics, Universitat Rovira i Virgili, Tarragona, Spain
armandmihai.ionescu@urv.cat
[4] Institute of Mathematics of the Romanian Academy, Bucharest, Romania
george.paun@imar.ro, gpaun@us.es

Abstract. In search for "realistic" bio-inspired computing models, we consider asynchronous spiking neural P systems, in the hope to get a class of computing devices with decidable properties. However, although the non-synchronization is known in general to decrease the computing power, in the case of using extended rules (several spikes can be produced by a rule) we obtain again the equivalence with Turing machines (interpreted as generators of sets of vectors of numbers). The problem remains open for the case of restricted spiking neural P systems, whose rules can only produce one spike. On the other hand, we prove that asynchronous spiking neural P systems, with a specific way of halting, using extended rules and where each neuron is either bounded or unbounded, are equivalent to partially blind counter machines and, therefore, have many decidable properties.

1 Spiking Neural P Systems – An Informal Presentation

In the present paper we continue the investigation of spiking neural P systems (SN P systems, in short). A survey of results and the biological motivations for these systems can be found in [5] and [2]. In the meantime, two main research directions were particularly active in this area of membrane computing: looking for classes of systems with tractable (for instance, decidable) properties, and looking for the possibility of using SN P systems for efficiently solving computationally hard problems. Along the second research line are the investigations related to the possibility of simulating an SN P system by a Turing machine with a polynomial slowdown (preliminary results can be found in [3]) and those trying to improve the efficiency of SN P systems, e.g., by enhancing the parallelism of the system (see, for instance, [7]).

In this paper we report several recent results concerning the first topic mentioned above – specifically, removing the synchronization (common in many membrane computing models), calling them asynchronous SN P systems. These

M.H. Garzon and H. Yan (Eds.): DNA 13, LNCS 4848, pp. 246–255, 2008.

systems were introduced in [6] with the aim of incorporating, into membrane computing, specific ideas from spiking neurons, a field that is being heavily investigated in neural computing (see, e.g., [8]).

An SN P system consists of a set of *neurons* placed in the nodes of a directed graph and sending signals (*spikes*) along the arcs of the graph (they are called *synapses*). Thus, the architecture is that of a tissue-like P system, with only one kind of object present in the cells (the reader is referred to [10,11] for an introduction to membrane computing and to [12] for the up-to-date information about this research area). The objects evolve by means of *spiking rules* placed in the nodes and enabled when the (number of) spikes present in the nodes fulfill specified regular expressions. When a spiking rule is executed in a neuron, spikes are produced and sent to all neurons connected by an outgoing synapse from the neuron where the rule was applied.

Two main types of results were obtained for synchronous (i.e, with obligatory use of the rules) systems using standard rules (producing one spike): computational completeness ([6]) in the case when no bound was imposed on the number of spikes present in the system, and a characterization of semilinear sets of numbers in the case when a bound was imposed. Improvements in the form of the regular expressions and normal forms can be found in [4].

In the proofs of these results, the synchronization plays a crucial role, but both from a mathematical point of view and from a neuro-biological point of view it is rather natural to consider non-synchronized systems (even if a neuron has a rule enabled in a given time unit, this rule is not necessarily used).

The synchronization is in general a powerful feature, useful in controlling the work of a computing device. However, it turns out that the loss in power entailed by removing the synchronization is compensated in the case of SN P systems where extended rules (producing several spikes) are used: such systems are still equivalent with Turing machines.

On the other hand, we also show that a restriction which looks, at first sight, rather minor, has a crucial influence on the power of the systems and decreases their computing power: in particular, we identify a class of asynchronous SN P systems equivalent to partially blind counter machines (i.e., not computationally complete) and for which the configuration reachability, membership (in terms of generated vectors), emptiness, infiniteness, and disjointness problems can be decided.

2   SN P Systems – Formal Definitions

A *spiking neural P system* (in short, an SN P system), of degree $m \geq 1$, is a construct of the form $\Pi = (O, \sigma_1, \ldots, \sigma_m, syn, out)$, where:

1. $O = \{a\}$ is the singleton alphabet (a is called *spike*);
2. $\sigma_1, \ldots, \sigma_m$ are *neurons*, of the form $\sigma_i = (n_i, R_i), 1 \leq i \leq m$, where:
   a) $n_i \geq 0$ is the *initial number of spikes* contained by the neuron;
   b) R_i is a finite set of *extended rules* of the form $E/a^c \to a^p; d$, where E is a regular expression with a the only symbol used, $c \geq 1$, and $p, d \geq 0$, with $c \geq p$; if $p = 0$, then $d = 0$, too.

3. $syn \subseteq \{1, 2, \ldots, m\} \times \{1, 2, \ldots, m\}$ with $(i, i) \notin syn$ for $1 \leq i \leq m$ (*synapses*);
4. $out \in \{1, 2, \ldots, m\}$ indicates the *output neuron*.

A rule $E/a^c \rightarrow a^p; d$ with $p \geq 1$ is called *extended firing* (we also say *spiking*) *rule*; a rule $E/a^c \rightarrow a^p; d$ with $p = d = 0$ is written in the form $E/a^c \rightarrow \lambda$ and is called a *forgetting rule*. If $L(E) = \{a^c\}$, then the rules are written in the simplified form $a^c \rightarrow a^p; d$ and $a^c \rightarrow \lambda$. A rule of the type $E/a^c \rightarrow a; d$ and $a^c \rightarrow \lambda$ is said to be *restricted* (or *standard*).

A rule is *bounded* if it is of the form $a^i/a^c \rightarrow a^p; d$, where $1 \leq c \leq i, p \geq 0$, and $d \geq 0$. A neuron is *bounded* if it contains only bounded rules. A rule is called *unbounded* if is of the form $a^i(a^j)^*/a^c \rightarrow a^p; d$, where $i \geq 0, j \geq 1, c \geq 1, p \geq 0, d \geq 0$. (In all cases, we also assume $c \geq p$; this restriction rules out the possibility of "producing more than consuming", but it plays no role in arguments below and can be omitted.) A neuron is *unbounded* if it contains only unbounded rules. A neuron is *general* if it contains both bounded and unbounded rules. An SN P system is *bounded* if all the neurons in the system are bounded. It is *unbounded* if it has bounded *and* unbounded neurons. Finally, an SN P system is *general* if it has general neurons (i.e., it contains at least one neuron which has both bounded and unbounded rules).

If the neuron σ_i contains k spikes, $a^k \in L(E)$ and $k \geq c$, then the rule $E/a^c \rightarrow a^p; d \in R_i$ is enabled and it can be applied. This means that c spikes are consumed, $k - c$ spikes remain in the neuron, the neuron is fired, and it produces p spikes after d time units. If $d = 0$, then the spikes are emitted immediately, if $d = 1$, then the spikes are emitted in the next step, and so on. In the case $d \geq 1$, if the rule is used in step t, then in steps $t, t+1, t+2, \ldots, t+d-1$ the neuron is *closed*; this means that during these steps it uses no rule and it cannot receive new spikes (if a neuron has a synapse to a closed neuron and sends spikes along it, then the spikes are lost). In step $t + d$, the neuron spikes and becomes again open, hence can receive spikes (which can be used in step $t+d+1$). The p spikes emitted by a neuron σ_i are replicated and they go to all neurons σ_j such that $(i, j) \in syn$ (each σ_j receives p spikes). If the rule is a forgetting one of the form $E/a^c \rightarrow \lambda$ then, when it is applied, $c \geq 1$ spikes are removed.

In an asynchronous SN P system in each time unit *any neuron is free to use a rule or not* (a global clock, marking the time for all neurons, is considered). Hence, in each time unit, each neuron can use either zero or one rule. Even if enabled, a rule is not necessarily applied, the neuron can remain still not used in spite of the fact that it contains rules which are enabled by its contents. If the contents of the neuron is not changed, a rule which was enabled in a step t can fire later. If new spikes are received, then it is possible that other rules will be enabled – and applied or not.

It is important to point out that when a neuron spikes, its spikes immediately leave the neuron and reach the target neurons simultaneously (i.e., there is no time needed for passing along a synapse from one neuron to another neuron).

The *initial configuration* of the system is described by the numbers $n_1, \ldots, n_m$ representing the initial number of spikes present in each neuron. Using the rules as suggested above, we can define transitions among configurations. Any sequence of transitions starting in the initial configuration is called a *computation*. A computation is *successful* if it reaches a configuration where all bounded and unbounded neurons are open but none is fireable (i.e., the SN P system has halted). The *result* of a computation is defined here as the total number of spikes sent into the environment by the output neuron.

Successful computations which send no spike out can be considered as generating the number zero, but in what follows we adopt the convention to ignore number zero when comparing the computing power of two devices.

SN P systems can also be used for generating sets of vectors, by considering several output neurons, $\sigma_{i_1}, \ldots, \sigma_{i_k}$. In this case, the system is called a *k-output SN P system*. Here a vector of numbers, $(n_1, \ldots, n_k)$, is generated by counting the number of spikes sent out by neurons $\sigma_{i_1}, \ldots, \sigma_{i_k}$ respectively during a successful computation. We denote by $N_{gen}^{nsyn}(\Pi)$ $[Ps_{gen}^{nsyn}(\Pi)]$ the set [the set of vectors, resp.] of numbers generated in the non-synchronized way by a system Π, and by $NSpik_{tot}EP_m^{nsyn}(\alpha, del_d)$ $[PsSpik_{tot}EP_m^{nsyn}(\alpha, del_d)], \alpha \in \{gen, unb, boun\}, d \geq 0$, the family of such sets of numbers [sets of vectors of numbers, resp.] generated by systems of type α (*gen* stands for general, *unb* for unbounded, *boun* for bounded), with at most m neurons and rules having delay at most d. (The subscript *tot* reminds us of the fact that we count all spikes sent to the environment.)

A *0-delay SN P system* is one where the delay in all the rules of the neurons is zero. Because in this paper we always deal with 0-delay systems, the delay $(d = 0)$ is never specified in the rules. Because there is no confusion, in this paper, asynchronous SN P systems are often simply called SN P systems.

In the next section we present a module of the construction from the proof of the universality theorem, and this can illustrate and clarify the above definitions. On that occasion we also use the standard way to pictorially represent a configuration of an SN P system. Specifically, each neuron is represented by a "membrane", marked with a label and having inside both the current number of spikes (written explicitly, in the form a^n for n spikes present in a neuron) and the evolution rules. The synapses linking the neurons are represented by directed edges (arrows) between the membranes. The output neuron is identified by both its label, *out*, and pictorially by a short arrow exiting the membrane and pointing to the environment. Examples of SN P systems working in an asynchronous way can be found in the technical report [1].

3   Computational Completeness of General SN P Systems

We now show that the power of general neurons (with extended rules) can compensate the loss of power entailed by removing the synchronization.

Theorem 1. $NSpik_{tot}EP_*^{nsyn}(gen, del_0) = NRE.$

Proof. (sketch) We only prove that $NRE \subseteq Spik_{tot}EP_*^{nsyn}(gen, del_0)$ and to this aim, we use the characterization of NRE (i.e., the family of sets of numbers computed by Turing machines) by means of counter machines (abbreviated CM), [9]. Let $M = (m, H, l_0, l_h, I)$ be a counter machine with m counters, such that the result of a computation is the number stored in counter 1 and this counter is never decremented during the computation. We construct a spiking neural P system Π as follows.

For each counter r of M let t_r be the number of instructions of the form $l_i : (\text{SUB}(r), l_j, l_k)$, i.e., all SUB instructions acting on counter r (of course, if there is no such a SUB instruction, then $t_r = 0$, which is the case for $r = 1$). Denote by

$$T = 2 \cdot \max\{t_r \mid 1 \leq i \leq m\} + 1.$$

For each counter r of M we consider a neuron σ_r in Π whose contents correspond to the contents of the counter. Specifically, if the counter r holds the number $n \geq 0$, then the neuron σ_r will contain $3Tn$ spikes.

With each label l of an instruction in M we also associate a neuron σ_l. Initially, all these neurons are empty, with the exception of the neuron σ_{l_0} associated with the start label of M, which contains $3T$ spikes. This means that this neuron is "activated". During the computation, the neuron σ_l which receives $3T$ spikes will become active. Thus, simulating an instruction $l_i : (\text{OP}(r), l_j, l_k)$ of M means starting with neuron σ_{l_i} activated, operating the counter r as requested by OP, then introducing $3T$ spikes in one of the neurons $\sigma_{l_j}, \sigma_{l_k}$, which becomes in this way active. When activating the neuron σ_{l_h}, associated with the halting label of M, the computation taking place in the counter machine M is completely simulated in Π; we will then send to the environment a number of spikes equal to the number stored in the first counter of M. Neuron σ_1 is the output neuron of the system. Further neurons will be associated with the counters and the labels of M; all of them being initially empty.

The construction consists of modules simulating the ADD and SUB instructions of M, as well as a final module. We present here, in Figure 1, only the SUB module.

Let us start with $3T$ spikes in neuron σ_{l_i} and no spike in other neurons, except neurons associated with counters; assume that neuron σ_r holds a number of spikes of the form $3Tn$, $n \geq 0$. Assume also that this is the sth instruction of this type dealing with counter r, for $1 \leq s \leq t_r$, in a given enumeration of instructions (because l_i precisely identifies the instruction, it also identifies s).

Some time, neuron σ_{l_i} spikes and sends $3T - s$ spikes both to σ_r and to $\sigma_{i,0}$. These spikes can be forgotten in this latter neuron, because $2T < 3T - s < 4T$. At a certain time, also neuron σ_r will fire, and will send $2T + s$ or $3T + s$ spikes to neuron $\sigma_{i,0}$. If no spike is here, then no other action can be done, also these spikes will eventually be removed, and no continuation is possible (in particular, no spike is sent out of the system).

If neuron $\sigma_{i,0}$ does not forget the spikes received from σ_{l_i} (this is possible, because of the non-synchronized mode of using the rules), then eventually neuron σ_r will send here either $3T + s$ spikes – in the case where it contains more than $3T - s$ spikes (hence counter r is not empty), or $2T + s$ spikes – in the case where its only spikes are those received from σ_{l_i}. In either case, $\sigma_{i,0}$ accumulates more than $4T$ spikes, hence it cannot forget them.

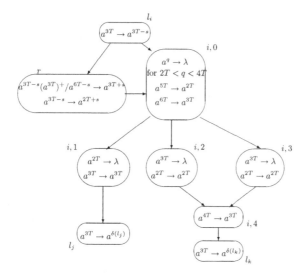

Fig. 1. Module SUB (simulating $l_i : (\text{SUB}(r), l_j, l_k)$)

Depending on the number of spikes accumulated, either $6T$ or $5T$, neuron $\sigma_{i,0}$ eventually spikes, sending $3T$ or $2T$ spikes, respectively, to $\sigma_{i,1}$, $\sigma_{i,2}$, and $\sigma_{i,3}$. The only possible continuation of neuron $\sigma_{i,1}$ is to activate neuron σ_{l_j} (precisely in the case where the first counter of M was not empty). Neurons $\sigma_{i,2}$ and $\sigma_{i,3}$ will eventually fire and either forget their spikes or send $4T$ spikes to neuron $\sigma_{i,4}$, which activates σ_{l_k} (in the case where the first counter of M was empty).

It is important to note that if any neuron $\sigma_{i,u}, u = 1, 2, 3$, skips using the rule which is enabled and receives further spikes, then no rule can be applied there anymore and the computation is blocked, without sending spikes out.

The simulation of the SUB instruction is correct in both cases, and no "wrong" computation is possible inside the module from Figure 1. What remains to examine is the possible interferences between modules, for instance, between neurons σ_r for which there are several SUB instructions, and this was the reason of considering the number T in writing the contents of neurons and the rules. Specifically, each σ_r for which there exist t_r SUB instructions can send spikes to all neurons $\sigma_{i,0}$ as in Figure 1. However, only one of these target neurons also receives spikes from a neuron σ_{l_i}, the one identifying the instruction which we want to simulate. By a careful analysis of the number of spikes a neuron can receive, the reader can check that the only computations in Π which can reach the neuron σ_{l_h} associated with the halting instruction of M are the computations which correctly simulate the instructions of M and correspond to halting computations in M.

In a similar way we construct the ADD and FIN modules of an SN P system Π (the reader can find the detailed construction in the technical report [1]). Hence $N_{gen}^{nsyn}(\Pi) = N(M)$.

Theorem 1 can be extended by allowing more output neurons and then simulating a k-output CM, producing in this way sets of vectors of natural numbers.

4   Characterization of Unbounded SN P Systems by Partially Blind Counter Machines

For the constructions in this section, we restrict the SN P systems syntactically to make checking a valid computation easier. Specifically, for an SN P system with unbounded neurons $\sigma_1, \ldots, \sigma_k$ (one of which is the output neuron) we assume as given non-negative integers $m_1, \ldots, m_k$, and for the rules in each σ_i we impose the following restriction: *If $m_i > 0$, then $a^{m_i} \notin L(E)$ for any regular expression E appearing in a rule of neuron σ_i.* This restriction guarantees that if neuron σ_i contains m_i spikes, then the neuron is not fireable. It follows that when the following conditions are met during a computation, the system has halted and the computation is valid: (1) All bounded neurons are open, but none is fireable, and (2) each σ_i contains *exactly* m_i spikes (hence none is fireable, too). This way of defining a successful computation, based on a vector $(m_1, \ldots, m_k)$, is called *μ-halting*. In the notation of the generated families we add the subscript μ to N or to Ps, in order to indicate the use of μ-halting.

A *partially blind k-output* CM (k-output PBCM) is a k-output CM, where the counters cannot be tested for zero. The counters can be incremented by 1 or decremented by 1, but if there is an attempt to decrement a zero counter, the computation aborts (i.e., the computation becomes invalid). Note that, as usual, the output counters are nondecreasing. Again, by definition, a successful generation of a k-tuple requires that the machine enters an accepting state with *all* non-output counters zero.

We denote by $NPBCM$ the set of numbers generated by PBCMs and by $PsPBCM$ the family of sets of vectors of numbers generated by using k-output PBCMs. It is known that k-output PBCMs can be simulated by Petri nets, and vice-versa. Hence, PBCMs are not universal.

Basic Construction: Let C be a partially blind counter. It is operated by a finite-state control. C can only store nonnegative integers. It can be incremented/decremented but when it is decremented and the resulting value become negative, the computation is aborted. Let i, j, d be given fixed nonnegative integers with $i \geq 0, j > 0, d > 0$. Initially, C is incremented (from zero) to some $m \geq 0$. Depending on the finite-state control (which is non-deterministic), one of the following operations is taken at each step: (1) C remains unchanged; (2) C is incremented by 1; (3) If the contents of C is of the form $i + kj$ (for some $k \geq 0$), then C is decremented by d. Note that in (3) we may not know whether $i + jk$ is greater than or equal to d, or what k is (the multiplicity of j), since we cannot test for zero. But if we know that C is of the form $i + jk$, when we subtract d from it and it becomes negative, it aborts and the computation is invalid, so we are safe. Note that if C contains $i + jk$ and is greater than or equal to d, then C will contain the correct value after the decrement of d. We can implement the computation using only C and the finite-state control as follows:

Case: $i < j$. Define a modulo-j counter to be a counter that can count from 0 to $j-1$. We can think of the modulo-j counter as an undirected circular graph with nodes $0, 1, \ldots, j-1$, where node s is connected to node $s+1$ for $0 \leq s \leq j-2$ and $j-1$ is connected to 0. Node s represents count s. We increment the modulo-j counter by going through the nodes in a "clockwise" direction. So, e.g., if the current node is s and we want to increment by 1, we go to $s+1$, provided $s \leq j-2$; if $s = j-1$, we go to node 0. Similarly, decrementing the modulo-j counter goes in the opposite direction, i.e., "counter-clockwise" – we go from s to $s-1$; if it is 0, we go to $s-1$.

The parameters of the machine are the triple (i, j, d) with $i \geq 0, j > 0, d > 0$. We associate with counter C a modulo-j counter, J, which is initially in node (count) 0. During the computation, we keep track of the current visited node of J. Whenever we increment/decrement C, we also increment/decrement J. Clearly, the requirement that the value of C has to be of the form $i + kj$ for some $k \geq 0$ in order to decrement by d translates to the J being in node i, which is easily checked.

Case: $i \geq j$. Suppose $i = r + sj$ where $s > 0$ and $0 \leq r < j$. There are two subcases: $d > i-j$ and $d \leq i-j$. We can show (we omit the "tricky" consruction here) that both subcases can also be implemented.

Using the above construction we get the following, rather surprising result.

Theorem 2. $N_\mu Spik_{tot} EP_*^{nsyn}(unb, del_0) = NPBCM$.

Proof. (sketch) We describe how a PBCM M simulates an unbounded 0-delay SN P system Π. Let B be the set of bounded neurons; assume that there are $g \geq 0$ such neurons. The bounded neurons can easily be simulated by M in its finite control. So we focus more on the simulation of the unbounded neurons. Let $\sigma_1, \ldots, \sigma_k$ be the unbounded neurons (one of which is the output neuron). M uses counters $C_1, \ldots, C_k$ to simulate the unbounded neurons. M also uses a nondecreasing counter C_0 to keep track of the spikes sent by the output neuron to the environment. Clearly, the operation of C_0 can easily be implemented by M. We introduce another counter, called ZERO (initially has value 0), whose purpose will become clear later.

Assume for the moment that each bounded neuron in B has only one rule, and each unbounded neuron σ_t ($1 \leq t \leq k$) has only one rule of the form $a^{i_t}(a^{j_t})^*/a^{d_t} \to a^{e_t}$. M incorporates in its finite control a modulo-j_t counter, J_t, associated with counter C_t, implemented by using the above basic construction. One step of Π is simulated in five steps by M as follows:

1. Non-deterministically choose a number $1 \leq p \leq g + k$.
2. Non-deterministically select a subset of size p of the neurons in $B \cup \{\sigma_1, \ldots, \sigma_k\}$.
3. Check if the chosen neurons are fireable. The neurons in B are easy to check, and the unbounded neurons can be checked using their associated J_t's (modulo-j_t counters). If at least one is not fireable, abort the computation by decrementing counter ZERO by 1.

4. Decrement the chosen unbounded counter by their d_t's and update their associated J_t's. The chosen bounded counters are also easily decremented by the amounts specified in their rules (in the finite control).

5. Increment the chosen bounded counters and unbounded counters by the total number of spikes sent to the corresponding neurons by their neighbors (again updating the associated J_t's of the chosen unbounded counters). Also, increment C_0 by the number of spikes the output neuron sends to the environment.

At some point, M non-deterministically guesses that Π has halted: It checks that all bounded neurons are open and none is fireable, and the unbounded neurons have their specified values of spikes. M can easily check the bounded neurons, since they are stored in the finite control. For the unbounded neurons, M decrements the corresponding counter by the specified number of spikes in that neuron. Clearly, $C_0 = x$ (for some number x) with all other counters zero if and only if the SN P system outputs x with all the neurons open and non-fireable (i.e., the system has halted) and the unbounded neurons containing their specified values.

It is straightforward to verify that the described construction generalizes to when the neurons have more than one rule. An unbounded neuron with m rules will have associated with it m modulo-j_{t_m} counters, one for each rule and during the computation, and these counters are operated in parallel to determine which rule can be fired. A bounded neuron with multiple rules is easily handled by the finite control. We then have to modify item 3 above to:

3'. Non-deterministically select a rule in each chosen neuron. Check if the chosen neurons with selected rules are fireable. The neurons in B are easy to check, and the unbounded neurons can be checked using the associated J_t's (modulo-j_t counters) for the chosen rules. If at least one is not fireable, abort the computation by decrementing counter ZERO by 1.

The proof of the converse, which we omit (for lack of space), is an intricate modification of the simulation in the proof of Theorem 1. Because each neuron can only have either bounded rules or unbounded rules (but *not* both), the simulation by PBCM is possible.

Theorem 2 can be generalized to the case with multiple outputs:

Theorem 3. $Ps_\mu Spik_{tot} EP_*^{nsyn}(unb, del_0) = PsPBCM.$

This is the best possible result we can obtain, since if we allow bounded rules and unbounded rules in the neurons, SN P systems become universal (Theorem 1).

It is known that PBCMs with only one output counter can only generate semilinear sets of numbers. Hence:

Corollary 1. *Unbounded 0-delay SN P systems with μ-halting can only generate semilinear sets of numbers.*

The results in the following corollary can be obtained using Theorem 3 and the fact that they hold for k-output PBCMs.

Corollary 2. *1. The sets of k-tuples generated by k-output unbounded 0-delay SN P systems with μ-halting is closed under union and intersection, but not under complementation.*

2. The membership, emptiness, infiniteness, disjointness, and reachability problems are decidable for k-output unbounded 0-delay SN P systems with μ-halting (for reachability, we do not need to define what is a halting configuration as we are not interested in tuples the system generates); but containment and equivalence are undecidable.

5   Final Remarks

Many issues remain to be investigated for asynchronous SN P systems. We only mention two of them: whether or not asynchronous SN P systems with standard rules (i.e., that can only produce one spike) are Turing complete and whether or not the decidability results proved in Section 4 can be proved by using the usual halting (i.e., by removing the μ-halting).

Acknowledgments. The work of O. Egecioglu, O.H. Ibarra, and S. Woodworth was supported in part by NSF Grants CCF-0430945 and CCF-0524136. The work of M. Ionescu was supported by "Formación de Profesorado Universitario" fellowship from MEC, Spain.

References

1. Cavaliere, M., Egecioglu, O., Ibarra, O.H., Woodworth, S., Ionescu, M., Păun, Gh.: Asynchronous spiking neural P systems. Tech. Report 9/2007 Microsoft Research - University of Trento, Centre for Computational and Systems Biology, http://www.cosbi.eu
2. Chen, H., Ionescu, M., Ishdorj, T.-O., Păun, A., Păun, Gh., Pérez-Jiménez, M.J.: Spiking neural P systems with extended rules: Universality and languages, Natural Computing (special issue devoted to DNA12 Conf.) (to appear)
3. Gutiérrez-Naranjo, M.A., et al.: Proceedings of Fifth Brainstorming Week on Membrane Computing, Fenix Editora, Sevilla (in press, 2007)
4. Ibarra, O.H., Păun, A., Păun, Gh., Rodríguez-Patón, A., Sosik, P., Woodworth, S.: Normal forms for spiking neural P systems. Theoretical Computer Sci. (to appear)
5. Ionescu, M., Păun, A., Păun, Gh., Pérez-Jiménez, M.J.: Computing with spiking neural P systems: Traces and small universal systems. In: Mao, C., Yokomori, T., Zhang, B.-T. (eds.) DNA Computing. LNCS, vol. 4287, pp. 32–42. Springer, Heidelberg (2006)
6. Ionescu, M., Păun, Gh., Yokomori, T.: Spiking neural P systems. Fundamenta Informaticae 71(2-3), 279–308 (2006)
7. Ionescu, M., Păun, Gh., Yokomori, T.: Spiking neural P systems with exhaustive use of rules. Intern. J. of Unconventional Computing (to appear)
8. Maass, W., Bishop, C. (eds.): Pulsed Neural Networks. MIT Press, Cambridge (1999)
9. Minsky, M.: Computation – Finite and Infinite Machines. Prentice-Hall, Englewood Cliffs (1967)
10. Păun, Gh.: Membrane Computing – An Introduction. Springer, Berlin (2002)
11. Păun, Gh.: Introduction to membrane computing. In: Ciobanu, G., Păun, Gh., Pérez-Jiménez, M.J. (eds.) Applications of Membrane Computing, Springer, Heidelberg (2006)
12. The P Systems Web Page, http://psystems.disco.unimib.it

On $5' \rightarrow 3'$ Sensing Watson-Crick Finite Automata

Benedek Nagy

Faculty of Informatics, University of Debrecen, Hungary
GRLMC, Rovira i Virgili University, Tarragona, Spain
`nbenedek@inf.unideb.hu`

Abstract. In this paper we introduce a variation of Watson-Crick automata in which both heads read the doubled DNA strand form 5' to 3'. The sensing version of these automata recognize exactly the linear context-free languages. The deterministic version is not so powerful, but all fixed-rated linear (for instance even-linear) languages can be accepted by them. Relation to other variations of Watson-Crick automata and pushdown automata are presented. The full-reading version of sensing $5' \rightarrow 3'$ automata recognizes non context-free languages as well.

1 Introduction

The theory of Watson-Crick (WK) automata is one of the main fields of DNA computing. Finite automata are well known and frequently used. They have a tape and, consequently, 1 input head reading the tape. The WK automata are such extension that work on WK tape, i.e. on double stranded tape; and it has an input head for each of the heads [3]. Several variations are presented in [7]. Their accepting power is intensively studied, several relations to formal language theory are shown. Formal languages are one of the bases of computer science. The class of linear languages is between the regular and context-free ones. A special subclass of linear grammars and languages, the even-linear one, – having rules in a symmetric shape – was investigated in [1]. In [2] the definition was extended to fix-rated linear languages. The terms k-linear grammars and k-regular languages were used. In the literature the terms k-linear grammars/languages are frequently used to refer metalinear grammars/languages [4], as they are extensions of the linear ones. Therefore we prefer the term fix-rated (k-rated) linear for those restricted linear grammars/languages that are used in [2]. The classes of k-rated linear languages are between the linear and regular ones.

In the nature the direction 5' to 3' is preferred, both DNA and RNA polymerase use this direction. The proteins get their meaning in this way and mRNA is read in this direction by ribosome. Therefore, it is a very natural idea to consider WK automata in which both of the heads are moving from the end 5' to 3', i.e. in the same direction concerning the structure of the DNA. In this paper we introduce and analyse these $5' \rightarrow 3'$ WK automata. The sensing version of these automata finishes the process of the input word when the heads meet. Several

M.H. Garzon and H. Yan (Eds.): DNA 13, LNCS 4848, pp. 256–262, 2008.

variations of these automata will be investigated. A class of $5' \to 3'$ WK automata recognizes exactly the linear languages. Another class recognizes exactly the even-linear languages. Moreover all k-rated linear languages can be accepted by deterministic $5' \to 3'$ WK automata. Their accepting powers comparing to the Chomsky hierarchy and other WK automata ([6,7]) are also presented.

2   Preliminaries

In this section we recall some well-known concepts of DNA computing and formal language theory ([7,4]). We also fix our notation.

Let V be an alphabet, i.e. a finite non-empty set of symbols (letters) and $\rho \subseteq V \times V$ be its complementary relation. For instance $V = \{A, C, G, T\}$ is usually used in DNA computing with the Watson-Crick complementary relation $\{(T, A), (A, T), (C, G), (G, C)\}$. The strings built up by complementer pairs of letters are double strands (of DNA). The sets of these strings are the languages.

A 5-tuple $A = (V, Q, s, F, \delta)$ is a finite state machine (finite automaton), with the (input) alphabet V, the finite (non-empty) set of states Q, the initial state $s \in Q$ and the set of final (accepting) states $F \subseteq Q$. The function δ is the transition function: $\delta : Q \times (V \cup \lambda) \to 2^Q$ / $\delta : Q \times V \to Q$ for non-deterministic/deterministic finite automaton (λ refers for the empty word). A word w is accepted by a finite automaton if there is a run starting with s, ending in a state in F and the symbols of the transitions of the path yield w.

A Watson-Crick finite automaton (WK automaton) is a finite automaton working on a Watson-Crick tape, that is a double stranded sequence (molecule) in which the lengths of the strands are equal and the elements of the strands are pairwise complements of each other: $\begin{bmatrix} a_1 \\ b_1 \end{bmatrix} \begin{bmatrix} a_2 \\ b_2 \end{bmatrix} \cdots \begin{bmatrix} a_n \\ b_n \end{bmatrix} = \begin{bmatrix} a_1\, a_2\, \dots\, a_n \\ b_1\, b_2\, \dots\, b_n \end{bmatrix}$ with $a_i, b_i \in V$ and $(a_i, b_i) \in \rho$ $(i = 1, ..., n)$. Formally a WK automaton is $M = (V, \rho, Q, s, F, \delta)$, where $\rho \subseteq V \times V$ is a symmetric relation, V, Q, s and F are the same as at finite automata, and the transition mapping $\delta : Q \times \binom{V \cup \{\lambda\}}{V \cup \{\lambda\}} \to 2^Q$.

The elementary difference between finite automata and WK automata besides the doubled tape is the number of heads. The WK automata scan separately each of the two strands, in a correlated manner.

Now we recall some language families related to the Chomsky hierarchy. A grammar is a construct $G = (N, V, S, H)$, where N, V are the non-terminal and terminal alphabets; $S \in N$ is the initial letter; H is a finite set of derivation rules. A rule is a pair written as $v \to w$ $(v \in (N \cup V)^* N(N \cup V)^*, w \in (N \cup V)^*)$. For the concepts of derivation $(\Rightarrow^*)$ we refer to [4]. The generated language $L = \{w | S \Rightarrow^* w \wedge w \in V^*\}$. Two grammars are equivalent if they generate the same language (mod λ). Due to the form of the rules we have the next classes

• context-sensitive (CS): $uAv \to uwv$ with $A \in N$ and $u, v, w \in (N \cup V)^*, w \neq \lambda$.
• context-free (CF): $A \to v$ with $A \in N$ and $v \in (N \cup V)^*$.
• linear (Lin): $A \to v$, $A \to vBw$; where $A, B \in N$ and $v, w \in V^*$.
• k-rated linear (k-Lin): linear and for each rule of the form $A \to vBw$: $k = \frac{|w|}{|v|}$ with a fixed rational number k ($|v|$ denotes the length of v).

- Specially with $k = 1$: even-linear (1-Lin) grammars.
- Specially with $k = 0$: regular (Reg) $A \to w$, $A \to wB$; with $A, B \in N$, $w \in V^*$.

The language family regular/linear etc. contains the languages that are generated by regular/linear etc. grammars. For various types of grammars various normal forms are introduced. Every linear grammar has an equivalent grammar with rules are in forms of $A \to aB, A \to Ba, A \to a$. Every even-linear language can be generated by rules of forms $A \to aBb, A \to a, A \to \lambda$. Every regular grammar has equivalent with rules of types $A \to aB, A \to a$ $(A, B \in N, a, b \in V)$.

The deterministic and non-deterministic finite automata recognize exactly the regular languages. For context-free and linear languages the pushdown and 1-turn pushdown automata fits. Their deterministic versions accept the deterministic context-free (dCF) and deterministic linear (dLin) languages.

3   The $5' \to 3'$ WK Automata

At the definition of WK automata we left open the interpretation of δ and the condition of acceptance of a word. Now we specialize them to get $5' \to 3'$ WK automata. In a $5' \to 3'$ WK automaton both heads start from the $5'$ end of the appropriate strand. Physically they read the double stranded sequence in opposite direction (Fig. 1 (a)). A $5' \to 3'$ WK automaton is sensing, if the heads sense that they are meeting. In sensing $5' \to 3'$ WK automata the process of the input string ends if for all pairs of the string one of the letters is read. In full reading version both heads read the whole strand from the end $5'$ to the end $3'$.

3.1   The Sensing $5' \to 3'$ WK Automata

The sensing $5' \to 3'$ automata ($5' \to 3'$ sWK) finish the reading of the input word, when the heads meet. In the last step both heads may step and then the automaton recognizes that all the input word is processed, since the heads meet. Otherwise we have the following agreement. In case only 1 pair of the input sequence is not being processed yet and so, both heads are at that position, they recognize each other and only one of the heads (the first one) can read the letter to finish processing the input word, the other head reads λ.

These automata can be represented graphically in a similar way as other finite automata. In transitions we put pair of symbols (a, b) meaning that the first head (upper strand) is reading symbol a, the second one (lower strand) is reading b and both are stepping (Fig. 1 (c)). Any (or both) of a and b can be λ.

Theorem 1. *The class of linear languages is exactly the one that is accepted by sensing WK finite automata.*

The proof is constructive in both directions. A WK automaton can be constructed for any linear grammar in a similar way as a finite automaton can be constructed from a regular grammar having rules in normal form. Starting from a linear grammar in normal form, the transitions are of the forms $B \in \delta(A, \binom{a}{\lambda})$ and $B \in \delta(A, \binom{\lambda}{a})$ $(A, B \in Q, a \in V)$. Similarly, one can construct a linear grammar based on the given automaton. As a special consequence of these constructions we define a 'normal form' for these automata. In simple, or shortly, $5' \rightarrow 3'$ sSWK automata at most 1 head moves at every transition.

Corollary 1. *For each $5' \rightarrow 3'$ sWK there is an equivalent $5' \rightarrow 3'$ sSWK.*

The transitions of a simple automata can be (a, λ) and (λ, a). Alternative notations $\rightarrow a$ and $\leftarrow a$ indicate the direction of the moving head (Fig. 1 (b)).

3.2   The Deterministic Sensing $5' \rightarrow 3'$ WK Automata

A WK automaton is deterministic if at each possible case of the triplets of actual state Q and letters under the heads $\binom{a}{b}$ there is at most 1 possible step (including the steps, where 1 of the heads does not read the letter).

Example 1. Let $L = \{w | w \in \{a, b\}^*, \ w = w^R\}$ be the language of palindromes. Fig. 1 (b) shows a deterministic simple $5' \rightarrow 3'$ sWK automaton accepting L.

The deterministic version of these automata is weaker than the non-deterministic one: $\{a^n b^n\} \cup \{a^{3n} b^n\}$ is accepted by a non-deterministic $5' \rightarrow 3'$ sWK trying both possibilities non-deterministically. With finite control it is impossible to accept this language in deterministic way. Now we consider the k-rated languages.

Theorem 2. *For any value of k, every k-rated linear language is accepted by deterministic $5' \rightarrow 3'$ sWK automata.*

Based on the normal form for the grammar the order of the steps of the heads can be fixed forehead. The proof that these languages can be accepted by deterministic automata is analogous with the set-construction (the proof of the equivalence of deterministic and non-deterministic finite automata). The class that is accepted by deterministic $5' \rightarrow 3'$ sWK automata strictly contains the k-rated linear languages, as the next example shows.

Example 2. The grammar $(\{S, A\}, \{a, b\}, S, \{S \rightarrow aAa, S \rightarrow bAbb, A \rightarrow aaaSb, A \rightarrow \lambda\})$ generates a linear language that is not k-rated linear (for any value of k), but it can be accepted by a deterministic $5' \rightarrow 3'$ sWK automaton.

3.3   The Both-Head Stepping $5' \rightarrow 3'$ sWK Automata

Let $5' \rightarrow 3'$ BWK denote those WK automata in which both heads must step at every transition step (but the case if only 1 letter is not processed yet in a sensing automaton). So, a $5' \rightarrow 3'$ sBWK automaton has transitions type $B \in \delta(A, \binom{a}{b})$ and possibly $C \in \delta(A, \binom{a}{\lambda})$, where C is a final state and there is not any transition from the final states which can be reached by such transitions. See Fig. 1 (c).

Theorem 3. *The even-linear languages are exactly those languages that*
i) are accepted by $5' \rightarrow 3'$ sBWK automata;
ii) are accepted by deterministic $5' \rightarrow 3'$ sBWK automata.

Case i) can be proven by the same constructions as in Theorem 1: the result is 1-linear. The automaton has transitions without finishing the input word only type $B \in \delta(A, \binom{a}{b})$. If the input has only 1 unread letter, then only the first head steps finishing the word and accepting it. The proof, that the deterministic version accepts all these languages goes in the same way as at Theorem 2.

3.4   The Full Reading $5' \rightarrow 3'$ sWK Automata

In the full reading variation of the $5' \rightarrow 3'$ sWK automata ($5' \rightarrow 3'$ fsWK) each head reads the whole word, but in different directions. It is easy to prove that

Lemma 1. *Every linear language is accepted by a $5' \rightarrow 3'$ fsWK automaton.* Furthermore, $5' \rightarrow 3'$ fsWK can accept more languages:

Example 3. The automaton shown in Fig. 1 (d) accepts the language $\{a^n b^n c^n\}$ if both heads read the whole input word. Moreover it works without sensing. This language is not even context-free; it cannot be accepted by $5' \rightarrow 3'$ sWK.

4   Comparison with Other Automata

Finite automaton (even with 2 heads) cannot recognize the language of correct bracket expressions. This language (DYCK) is a dCF language. The palindrome language is not a dCF language, but it is accepted by a deterministic $5' \rightarrow 3'$ WK finite automaton. Therefore, the classes of dCF languages and of languages accepted by deterministic $5' \rightarrow 3'$ WK automata are incomparable. Moreover the set of dLin languages is incomparable with the class of languages accepted by deterministic $5' \rightarrow 3'$ sWK automata: the language $\{a^n ba^{2n}\} \cup \{a^n ca^{3n}\}$ is only in the first class, while the language of palindromes only in the second one.

Usually in the literature the heads of the WK automata move in the same direction starting from the same end of the doubled sequence ([5,6,7]). The language $\{a^n b^n\}$ is accepted by the traditional and by the $5' \rightarrow 3'$ WK automata, moreover, the deterministic sensing version can be used. The $5' \rightarrow 3'$ automata accept the language $\{ww^R | w \in V^*\}$, that cannot be accepted by any 2-head finite automata with heads moving to the same direction. The 'copy'-language ($\{ww | w \in V^*\}$) can be recognized by a traditional non-deterministic WK automata, but it cannot be accepted by any $5' \rightarrow 3'$ WK finite machines. The languages $\{a^n b^n c^n\}$ and $\{a^n b^m c^n d^m\}$ can be accepted by traditional deterministic and by deterministic $5' \rightarrow 3'$ fsWK automata, as well.

5   Summary and Concluding Remarks

In this paper the sensing $5' \rightarrow 3'$ WK automata have been introduced. Full reading versions were also analysed. The simple version in which only 1 head can move in a step proved to be equivalently powerful to the non-restricted versions. Some subclasses such as deterministic and/or versions in which both heads step

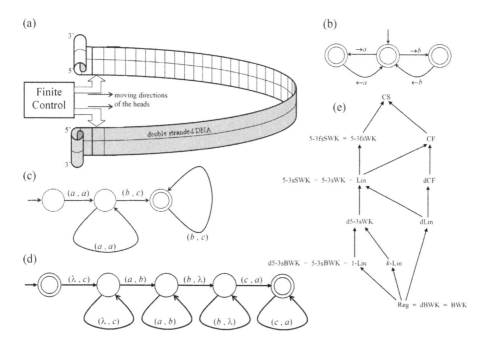

Fig. 1. Examples for $5' \to 3'$ WK automata and hierarchy of language classes

at the same time were presented. As a summary, Fig. 1 (e) shows the hierarchy of language families accepted by various $5' \to 3'$ sWK finite automata (d stands for deterministic, s for sensing, fs for full-reading sensing, B for both-head step at the same time, S for simple versions) and their relation to the classical classes. The arrows show strict inclusions, the nodes not having directed paths between them are representing incomparable classes. There is a future work to analyse the non-sensing and other (stateless, all final etc.) versions of $5' \to 3'$ WK.

Acknowledgements. The author thanks the comments of the reviewers and the support of DNA13, OTKA T049409 and of the [NKTH], Hungary.

References

1. Amar, V., Putzolu, G.R.: On a Family of Linear Grammars. Information and Control 7(3), 283–291 (1964)
2. Amar, V., Putzolu, G.R.: Generalizations of Regular Events. Information and Control 8(1), 56–63 (1965)
3. Freund, R., Paun, G., Rozenberg, G., Salomaa, A.: Watson-Crick finite automata. In: Third Annual DIMACS Symp. on DNA Based Computers, Philadelphia, pp. 305–317 (1997)
4. Hopcroft, J.E., Ullmann, J.D.: Introduction to Automata Theory, Languages, and Computation. Addison-Wesley, Reading (1979)

5. Hromkovic, J.: On one-way two-head deterministic finite state automata. Computers and Artificial Intelligence 4(6), 503–526 (1985)
6. Petre, E.: Watson-Crick-Automata. Journal of Automata, Languages and Combinatorics 8(1), 59–70 (2003)
7. Păun, G., Rozenberg, G., Salomaa, A.: DNA computing. In: New computing paradigms, Springer, Berlin (1998)

Equivalence in Template-Guided Recombination*

Michael Domaratzki

Department of Computer Science
University of Manitoba
Winnipeg, MB R3T 2N2 Canada
mdomarat@cs.umanitoba.ca

Abstract. We consider theoretical properties of the template-guided recombination operation. In particular, we consider the decidability of whether two sets of templates are equivalent, that is, whether their action is the same for all operands. We give a language-theoretic characterization of equivalence which leads to decidability results for common language classes. In particular, we show a positive answer for regular sets of templates. For context-free sets of templates, the answer is negative.

1 Introduction

The rearrangement of DNA in stichotrichous ciliates has received a significant amount of attention in the literature as a model of natural computing. Several potential formal models for the rearrangement have been proposed, including both intra-molecular and inter-molecular models. Ehrenfeucht *et al.* [7] give a detailed overview of ciliate DNA rearrangement and an investigation of one of the proposed models.

Template-guided recombination (TGR) is one of the formal models for recombination of DNA in stichotrichs. The model, proposed by Prescott *et al.* [9], has been the subject of much research in the literature [3,4,5,6,8]. Much of this work on TGR has focused on examining the closure properties of the operation. For example, McQuillan *et al.* [8] have recently shown that if a context-free language is iteratively operated upon with a regular set of templates (see Section 2 for definitions), then the resulting language is a context-free language which can be effectively constructed.

TGR specifies a set of templates which defines how the operation works: changes to the set of templates affect how the TGR operation functions on its operand, which represents the scrambled DNA in the ciliate. It is reasonable, therefore, to ask exactly what changes to the set of templates affect the operation of TGR. This is the question we address in this paper: given two sets of templates, do they define equivalent TGR operations? We give a natural condition on subwords of templates which exactly characterizes equivalence for template sets over an alphabet of at least three symbols.

From this characterization, we then establish decidability results: given two regular sets of templates, it is decidable whether they are equivalent. We also

---

* Research supported in part by a grant from NSERC.

M.H. Garzon and H. Yan (Eds.): DNA 13, LNCS 4848, pp. 263–272, 2008.

show an interesting universality result: determining whether a set of templates is equivalent to the universal set of all templates is not difficult, as it can be decided even for recursive sets. However, for alphabets of size at least three, there exists a fixed regular set of templates T_0 such that it is undecidable if a given context-free set of templates is equivalent to T_0.

As a proposed model for natural computing, understanding the equivalence of template sets is a critical prerequisite for understanding the potential for employing the natural computing power of ciliate DNA rearrangement. Under the hypothesis that a distinct set of DNA material (the templates) exactly guides rearrangement, a potential method for altering of the computational action of the rearrangement is a modification of the set of templates which are present during rearrangement.

With the results in this paper, we are able to determine exactly the situations in which modifying the set of templates modifies the computational process of rearrangement which occurs. For a recent survey of experimental results and hypotheses in identifying exogenic factors affecting ciliate DNA rearrangement, see Cavalcanti and Landweber [2]. Recent experimental results lend some support to the TGR model. Vijayan *et al.* [11] demonstrate that the addition of permuted RNA to the parental macronucleus does affect the rearrangement process during conjugation, and a modified micronucleus is produced.

Recently, Angeleska *et al.* [1] have reconsidered the TGR model by incorporating RNA templates (either single-stranded or double-stranded RNA). Their model does not incorporate any part of the RNA template into the rearranged DNA and reduces the number of required cuts to the DNA backbones. However, as noted by the authors, the new model does not have any impact when considered as an inter-molecular operation of formal languages as we do here.

2   Preliminary Definitions

We use the tools of formal language theory to study TGR. For additional background on formal languages, see Rozenberg and Salomaa [10]. Let Σ be a finite set of symbols, called *letters*; we call Σ an *alphabet*. Then Σ^* is the set of all finite sequences of letters from Σ, which are called *words*. The empty word ε is the empty sequence of letters. We denote by Σ^+ the set of non-empty words over Σ, i.e., $\Sigma^+ = \Sigma^* - \{\varepsilon\}$. The *length* of a word $w = w_1 w_2 \cdots w_n \in \Sigma^*$, where $w_i \in \Sigma$, is n, and is denoted by $|w|$.

A word $x \in \Sigma^*$ is a *prefix* of a word $y \in \Sigma^*$ if there exists $w \in \Sigma^*$ such that $y = xw$. Similarly, x is a *suffix* of y if there exists $u \in \Sigma^*$ such that $y = ux$. If $x \in \Sigma^*$, then pref(x) (resp., suff(x)) is the set of all prefixes (resp., suffixes) of x. We also use the notation first(x) and last(x) to denote the first and last letter of a non-empty word. That is, if $x \in \Sigma^+$ and $x = x_1 x_2$ where $x_1 \in \Sigma$ and $x_2 \in \Sigma^*$, then first(x) = x_1. Similarly, if $x = y_1 y_2$ where $y_1 \in \Sigma^*$ and $y_2 \in \Sigma$, then last(x) = y_2.

A *language* L is any subset of Σ^*. Given languages $L_1, L_2 \subseteq \Sigma^*$, their concatenation is defined by $L_1 L_2 = \{xy \; : \; x \in L_1, y \in L_2\}$. Given an alphabet Σ, we use the notation Σ^k to denote the set of all words in Σ^* of length k, while $\Sigma^{\geq k}$ (resp., $\Sigma^{\leq k}$) denotes the set of all words in Σ^* of length k or greater (resp., length k or less).

A *deterministic finite automaton* (DFA) is a five-tuple $M = (Q, \Sigma, \delta, q_0, F)$ where Q is the finite set of states, Σ is the alphabet, $\delta : Q \times \Sigma \to Q$ is the transition function, $q_0 \in Q$ is the start state, and $F \subseteq Q$ is the set of final states. We extend δ to $Q \times \Sigma^*$ in the usual way: $\delta(q, \varepsilon) = q$ for all $q \in Q$, while $\delta(q, wa) = \delta(\delta(q, w), a)$ for all $q \in Q$, $w \in \Sigma^*$ and $a \in \Sigma$. A word $w \in \Sigma^*$ is accepted by M if $\delta(q_0, w) \in F$. The *language accepted* by M, denoted $L(M)$, is the set of all words accepted by M, i.e., $L(M) = \{w \in \Sigma^* : \delta(q_0, w) \in F\}$. A language is called *regular* if it is accepted by some DFA. If L is a regular language, the *state complexity* of L, denoted $\mathrm{sc}(L)$, is the minimum number of states in any DFA which accepts L.

We assume the reader is familiar with the classes of context-free and recursive languages. A language is a *context-free* if it is generated by a context-free grammar. A language is *recursive* if it is accepted by a Turing machine which halts on all inputs. The classes of regular, context-free and recursive languages form a strict hierarchy of inclusions.

2.1 Template-Guided Recombination

We now give the formal definition of TGR, which was proposed by Prescott *et al.* [9] and first studied as a formal operation by Daley and McQuillan [4]. If $n_1, n_2 \geq 1$ and $x, y, z, t \in \Sigma^*$ are words, we denote by $(x, y) \vdash_{t, n_1, n_2} z$ the fact that we can write

$$x = u_1 \alpha \beta v_1 \tag{1}$$

$$y = v_2 \beta \gamma u_2 \tag{2}$$

$$z = u_1 \alpha \beta \gamma u_2 \tag{3}$$

$$t = \alpha \beta \gamma \tag{4}$$

with $\alpha, \beta, \gamma, u_1, u_2, v_1, v_2 \in \Sigma^*$, $|\alpha|, |\gamma| \geq n_1$ and $|\beta| = n_2$. If n_1, n_2 are understood, then we denote the relation $\vdash_{t, n_1, n_2}$ by $\vdash_t$. The word t is called the *template*.

Intuitively, x and y are the DNA strands which are to be recombined using the template t. The regions v_1 and v_2 represent the internal eliminated sequences (IESs) which do not form part of the final rearranged sequence, and β, which has a minimum length restriction, represents the pointer sequences in the ciliate DNA. Note that in the definition of $(x, y) \vdash_t z$, the words x and y are separate DNA sequences and so TGR is an inter-molecular model for ciliate DNA recombination. Recently, however, an intra-molecular TGR has been considered as well [3].

If $T, L \subseteq \Sigma^*$ are languages, then $\pitchfork_{T, n_1, n_2} (L)$ is defined by

$$\pitchfork_{T, n_1, n_2} (L) = \{z : \exists x, y \in L, t \in T \text{ such that } (x, y) \vdash_{t, n_1, n_2} z\}.$$

Again, we use the notation $\text{\reflectbox{m}}_T (L)$ if n_1, n_2 are understood or unimportant. The language T is the *set of templates*.

We require the following simple observation about TGR:

Observation 1. *If* $x, y, z, t \in \Sigma^*$ *such that* $(x, y) \vdash_{t,n_1,n_2} z$, *then* $|z| - (|x| + |y|) = -n_2 - (|v_1| + |v_2|)$, *where* v_1, v_2 *are as in (1)–(4).*

We now come to the definition of equivalence for sets of templates. Let $n_1, n_2 \geq 1$. For $T_1, T_2 \subseteq \Sigma^*$, we say that T_1 and T_2 are (n_1, n_2)-*equivalent*, denoted by $T_1 \equiv_{n_1,n_2} T_2$, if $\text{\reflectbox{m}}_{T_1} (L) = \text{\reflectbox{m}}_{T_2} (L)$ for all $L \subseteq \Sigma^*$. By $T_1 \sqsubseteq_{n_1,n_2} T_2$, we mean $\text{\reflectbox{m}}_{T_1} (L) \subseteq \text{\reflectbox{m}}_{T_2} (L)$ for all languages $L \subseteq \Sigma^*$. Note that $T_1 \equiv_{n_1,n_2} T_2$ if and only if $T_1 \sqsubseteq_{n_1,n_2} T_2$ and $T_2 \sqsubseteq_{n_1,n_2} T_1$ hold. We also note that $\equiv_{n_1,n_2}$ is an equivalence relation.

We consider the relationships between $\equiv_{n_1,n_2}$ and $\equiv_{n_1',n_2'}$ for different values of n_1, n_2, n_1', n_2'. We can show that these relations are incomparable.

Theorem 2. *Let* Σ *be an alphabet with size at least two and* $n_1, n_2 \geq 1$. *The relations* $\equiv_{n_1,n_2}$ *and* $\equiv_{n_1,n_2+1}$ *(resp.,* $\equiv_{n_1,n_2}$ *and* $\equiv_{n_1+1,n_2}$*) are incomparable.*

3   Language Theoretic Characterization

We can now give our main result, a language-theoretic characterization of equivalence of sets of templates. Let (C1) be the following condition:

$$\forall t, t_1, t_2 \in \Sigma^* \text{ with } |t| = 2n_1 + n_2, \tag{C1}$$
$$\text{if } t_1 t t_2 \in T_1 \text{ then } \exists t_1' \in \text{suff}(t_1), t_2' \in \text{pref}(t_2)(t_1' t t_2' \in T_2).$$

Condition (C1) is illustrated in Figure 1: for every subword t of length $2n_1 + n_2$ in a template in T_1, there must be an extension of t in T_2 which agrees with the template in T_1 on the subwords flanking t.

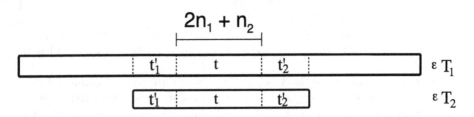

Fig. 1. Illustration of condition (C1)

Our main result uses condition (C1) to characterize equivalence of sets of templates:

Theorem 3. *Let* Σ *be an alphabet with* $|\Sigma| \geq 3$, $n_1, n_2 \geq 1$ *and* $T_1, T_2 \subseteq \Sigma^*$. *The condition (C1) holds if and only if* $T_1 \sqsubseteq_{n_1,n_2} T_2$.

Proof. ($\Rightarrow$): Suppose that (C1) holds. Let L be an arbitrary language and let $x \in \text{⋔}_{T_1}(L)$. Then there exist $y, z \in L, t \in T_1$ such that $(y, z) \vdash_t x$. Write x, y, z, t as

$$y = u_1 \alpha \beta v_1,$$
$$z = v_2 \beta \gamma u_2,$$
$$x = u_1 \alpha \beta \gamma u_2,$$
$$t = \alpha \beta \gamma,$$

where $|\alpha|, |\gamma| \geq n_1$, $|\beta| = n_2$ and $u_1, u_2, v_1, v_2 \in \Sigma^*$. Now write $\alpha = \alpha_1 \alpha_2$ and $\gamma = \gamma_1 \gamma_2$ where $|\alpha_2| = n_1$ and $|\gamma_1| = n_1$. Thus, $\alpha_2 \beta \gamma_1$ is a subword of t of length $2n_1 + n_2$. By (C1), let $\alpha_1' \in \text{suff}(\alpha_1)$ and $\gamma_2' \in \text{pref}(\gamma_2)$ be chosen so that $t' = \alpha_1' \alpha_2 \beta \gamma_1 \gamma_2' \in T_2$. Let $\alpha_1 = \alpha_1'' \alpha_1'$ and $\gamma_2 = \gamma_2' \gamma_2''$ for appropriate choices of α_1'', γ_2''. Note that

$$y = u_1 \alpha_1'' (\alpha_1' \alpha_2 \beta) v_1,$$
$$z = v_2 (\beta \gamma_1 \gamma_2') \gamma_2'' u_2,$$
$$x = u_1 \alpha_1'' \alpha_1' \alpha_2 \beta \gamma_1 \gamma_2' \gamma_2'' u_2.$$

Thus, $(y, z) \vdash_{t'} x$ and so $x \in \text{⋔}_{T_2}(L)$. We conclude that $T_1 \sqsubseteq_{n_1, n_2} T_2$.

($\Leftarrow$): Suppose for all $L \subseteq \Sigma^*$, we have $\text{⋔}_{T_1}(L) \subseteq \text{⋔}_{T_2}(L)$. Let $t, t_1, t_2 \in \Sigma^*$ with $|t| = 2n_1 + n_2$ and $t_1 t t_2 \in T_1$. Let $t_0 = t_1 t t_2$. Further, write $t = \alpha \beta \gamma$ where $|\alpha| = |\gamma| = n_1$ and $|\beta| = n_2$. Now, let $X_1, X_2 \in \Sigma$ be letters chosen so that they satisfy

$$X_1 \neq \text{first}(\gamma), \qquad X_2 \neq \text{last}(\alpha),$$
$$X_1 \neq \text{last}(\gamma t_2), \qquad X_2 \neq \text{first}(t_1 \alpha).$$

Note that this is possible since Σ has at least three letters.

Define the language $L \subseteq \Sigma^*$ as $L = \{t_1 \alpha \beta X_1, X_2 \beta \gamma t_2\}$. Note that $t_0 \in \text{⋔}_{T_1}(L) \subseteq \text{⋔}_{T_2}(L)$, as $(t_1 \alpha \beta X_1, X_2 \beta \gamma t_2) \vdash_{t_0} t_0$. Thus, there exist $x, y \in L$ and $t' \in T_2$ such that $(x, y) \vdash_{t'} t_0$. There are three cases, according to the choices for x, y.

(a) $x = t_1 \alpha \beta X_1, y = X_2 \beta \gamma t_2$. Thus, we must be able to write

$$x = t_1 \alpha \beta X_1 = u_1 \alpha' \beta' v_1,$$
$$y = X_2 \beta \gamma t_2 = v_2 \beta' \gamma' u_2,$$
$$t_0 = t_1 \alpha \beta \gamma t_2 = u_1 \alpha' \beta' \gamma' u_2,$$
$$t' = \alpha' \beta' \gamma',$$

where $|\alpha'|, |\gamma'| \geq n_1$, $|\beta'| = n_2$. Note that by Observation 1, $|t_1 \alpha \beta \gamma t_2| = |t_1 \alpha \beta X_1| + |X_2 \beta \gamma t_2| - n_2 - |v_1| - |v_2|$. Simplifying, we get that $|v_1| + |v_2| = 2$. We claim that $|v_1| = |v_2| = 1$. If not, then $|v_1| = 2$ and $|v_2| = 0$ or $|v_1| = 0$ and $|v_2| = 2$. We prove that the first case produces a contradiction; the second case is symmetrical.

If $|v_1| = 2$ and $|v_2| = 0$, then the equality $t_1\alpha\beta X_1 = u_1\alpha'\beta'v_1$ implies that $|u_1\alpha'\beta'| = |t_1\alpha\beta| - 1$ and (as $|\beta| = |\beta'| = n_2$) $|t_1\alpha| - 1 = |u_1\alpha'|$. Further $X_2\beta\gamma t_2 = \beta'\gamma'u_2$. Consider now that

$$t_1\alpha\beta\gamma t_2 = u_1\alpha'\beta'\gamma'u_2$$
$$= u_1\alpha'X_2\beta\gamma t_2$$

In this case, as $|t_1\alpha| - 1 = |u_1\alpha'|$, we have that $X_2 = \mathrm{last}(\alpha)$, a contradiction to our choice of X_2. (The case $|v_1| = 0$ and $|v_2| = 2$ produces the contradiction that $X_1 = \mathrm{first}(\gamma)$.)

Therefore, $|v_1| = |v_2| = 1$. Thus, $v_1 = X_1, v_2 = X_2$ and we get that $t_1\alpha\beta = u_1\alpha'\beta'$ and $\beta\gamma t_2 = \beta'\gamma'u_2$. We immediately conclude that $\beta = \beta'$ as both have length n_2. As $|\alpha'| \geq n_1 = |\alpha|$, the equality $t_1\alpha\beta = u_1\alpha'\beta'$ implies that there exists $t_1' \in \mathrm{suff}(t_1)$ such that $\alpha' = t_1'\alpha$. Similarly, $\gamma' = \gamma t_2'$ for some $t_2' \in \mathrm{pref}(t_2)$ by the equality $\beta\gamma t_2 = \beta'\gamma'u_2$. Finally, as $t' = \alpha'\beta'\gamma' = t_1'\alpha\beta\gamma t_2'$ and $t' \in T_2$, we note that condition (C1) holds, as required.

(b) $x = X_2\beta\gamma t_2$. Then regardless of the choice of $y \in L$, we have that

$$x = X_2\beta\gamma t_2 = u_1\alpha'\beta'v_1 \tag{5}$$
$$y = v_2\beta'\gamma'u_2 \tag{6}$$
$$t_0 = t_1\alpha\beta\gamma t_2 = u_1\alpha'\beta'\gamma'u_2 \tag{7}$$

Thus, equating (5) and (7), we get that $X_2 = \mathrm{first}(t_1\alpha)$, a contradiction.

(c) $y = t_1\alpha\beta X_1$. This is similar to case (b); we ultimately arrive at the contradiction $X_1 = \mathrm{last}(\gamma t_2)$.

We conclude that in all applicable cases, condition (C1) holds.   □

Example 1. Consider the following example with $n_1 = n_2 = 1$:

$$T_1 = \{baaab, caaac\},$$
$$T_2 = \{baaab, caa, aac\}.$$

Note that condition (C1) does *not* hold: for $t = aaa, t_1 = t_2 = c$, there is no prefix t_2' of t_2 and suffix t_1' of t_1 such that $t_1'aaat_2'$ is in T_2. Verifying Theorem 3, we note that $caaac \in \mathring{\mathrm{m}}_{T_1} (\{aac, caa\})$, but the same word is not in $\mathring{\mathrm{m}}_{T_2} (\{aac, caa\})$.

This example shows that condition (C1) cannot be replaced with the following more simple condition:

$$\forall t \in \mathrm{sub}_{2n_1+n_2}(T_1), \exists t_1', t_2' \in \Sigma^*(t_1'tt_2' \in T_2). \tag{8}$$

(here $\mathrm{sub}_m(L)$ is the set of all subwords of length m in L), since (8) *does* hold for the above sets T_1 and T_2. Intuitively, (8) is not an adequate formulation since it does not enforce that the chosen words t_1', t_2' agree with the regions surrounding the occurrence of t as a subword of length $2n_1 + n_2$ in a template in T_1.

We note that condition (C1) in Theorem 3 does not place any restrictions on templates in T_1 of length less than $2n_1+n_2$. Further, the extensions constructed (i.e., $t_1' t t_2'$ in (C1)) also have length at least $2n_1 + n_2$. Thus, there is no restriction on templates less than this critical length $2n_1 + n_2$. In other words, if $T_1 \equiv_{n_1, n_2} T_2$, then $T_1 \cap \Sigma^{\leq 2n_1 + n_2 - 1}$ and $T_2 \cap \Sigma^{\leq 2n_1 + n_2 - 1}$ can be modified completely arbitrarily and equivalence will still hold.

Finally, we do not know if the condition $|\Sigma| \geq 3$ in Theorem 3 can be improved to $|\Sigma| \geq 2$. However, the case of $|\Sigma| = 1$ is, as would be expected, trivial. In the case of a unary alphabet, we can replace condition (C1) by the following simpler condition:

$$\forall t \in T_1, |t| \geq 2n_1 + n_2, \exists t' \in T_2, 2n_1 + n_2 \leq |t'| \leq |t|.$$

We omit the proof.

4   Decidability Results

We now turn to employing Theorem 3 to demonstrate that we can determine algorithmically whether two sets of templates are equivalent. We first demonstrate that we can do so if the two sets of templates are regular. To establish this, we show that if T_1 and T_2 do not satisfy (C1), a bound on the length of a template in T_1 demonstrating this fact can be given:

Lemma 1. *Let* $T_1, T_2 \subseteq \Sigma^*$ *be regular sets of templates, with* $sc(T_i) = m_i$ *for* $i = 1, 2$. *If (C1) does not hold, then there exists* $t \in T_1$ *with* $|t| \leq m_1 2^{m_2} + 2n_1 + n_2$ *which witnesses this fact.*

Proof. Let $M_i = (Q_i, \Sigma, \delta_i, q_i, F_i)$ be DFAs with $|Q_i| = m_i$ and $L(M_i) = T_i$ for $i = 1, 2$.

The proof is by contradiction: Assume that (C1) does not hold. Let $t \in T_1$ be the shortest template that witnesses the fact that (C1) does not hold. Suppose that t has length strictly greater than $m_1 2^{m_2} + 2n_1 + n_2$. As (C1) does not hold, there exists a decomposition of t as $t = t_1 t' t_2$ such that $|t'| = 2n_1 + n_2$, and for all pairs (t_1', t_2') where $t_1' \in \text{suff}(t_1)$ and $t_2' \in \text{pref}(t_2)$, $t_1' t t_2' \notin T_2$.

By the length of t, we must have that either $|t_1| > m_1 2^{m_2 - 1}$ or $|t_2| > m_1 2^{m_2 - 1}$. Assume first that $|t_1| > m_1 2^{m_2 - 1}$. Let $k = |t_1|$ and $t_1 = \eta_1 \eta_2 \cdots \eta_k$ where $\eta_i \in \Sigma$ for all $1 \leq i \leq k$.

For all $1 \leq j \leq k$, let $\Pi_j \subseteq Q_2$ be the set of states

$$\Pi_j = \{\delta_2(q_2, s) \; : \; s \in \text{suff}(\eta_1 \cdots \eta_j)\}.$$

Note that

(a) $q_2 \in \Pi_j$ for all $1 \leq j \leq k$, since $\varepsilon \in \text{suff}(\eta_1 \cdots \eta_j)$.
(b) If $q \in \Pi_j$ and $t_2' \in \text{pref}(t_2)$, then $\delta_2(q, \eta_{j+1} \cdots \eta_k t' t_2') \notin F_2$; if this state were in F_2, then the subtemplate $\eta_i \cdots \eta_k t' t_2' \in T_2$ for some i with $1 \leq i \leq j+1$ (exactly the index i such that $\delta(q_2, \eta_i \cdots \eta_j) = q \in \Pi_j$).

By (a), there are at most $2^{m_2 - 1}$ possibilities for Π_j. Then considering all of the pairs $(\Pi_i, \delta_1(q_1, \eta_1 \cdots \eta_i))$ for all $1 \leq i \leq k$, as $k > m_1 2^{m_2 - 1}$, there must exist $1 \leq j < j' \leq k$ such that $(\Pi_j, \delta_1(q_1, \eta_1 \cdots \eta_j)) = (\Pi_{j'}, \delta_1(q_1, \eta_1 \cdots \eta_{j'}))$.

Claim. The template $t_0 = \eta_1\eta_2\cdots\eta_j\eta_{j'+1}\eta_{j'+2}\cdots\eta_k t't_2$ witnesses that (C1) does not hold.

Proof. First, $t_0 \in T_1$. To see this, note that $\delta_1(q_1, \eta_1\cdots\eta_j) = \delta_1(q_1, \eta_1\cdots\eta_{j'})$ by choice of j, j', and so substituting the prefix $\eta_1\cdots\eta_j$ for $\eta_1\cdots\eta_{j'}$ does not affect the finality of M_1 after reading the entire template, and t_0 is accepted by M_1.

Next, for each suffix t'' of $\eta_1\cdots\eta_j\eta_{j'+1}\cdots\eta_k$ and each prefix t_2' of t_2 we must have that $t''t't_2' \notin T_2$. For the suffixes of $\eta_{j'+1}\cdots\eta_k$ (and any prefix of t_2), this holds since they are also suffixes of t_1. Consider then a suffix of the form $\eta_i\cdots\eta_j\eta_{j'+1}\cdots\eta_k$ for some $1 \leq i \leq j$. Note that $\delta_2(q_2, \eta_i\cdots\eta_j) \in \Pi_j = \Pi_{j'}$. Thus, there exists a suffix $\eta_r\cdots\eta_{j'}$ of $\eta_1\cdots\eta_{j'}$ such that $\delta_2(q_2, \eta_i\cdots\eta_j) = \delta_2(q_2, \eta_r\cdots\eta_{j'})$. By (b) above, for all $t_2' \in \text{pref}(t_2)$, $\delta(q_2, \eta_i\cdots\eta_j\eta_{j'+1}\cdots\eta_k t't_2') = \delta(q_2, \eta_r\cdots\eta_{j'}\eta_{j'+1}\cdots\eta_k t't_2') \notin F_2$ and thus, $\eta_i\cdots\eta_j\eta_{j'+1}\cdots\eta_k t't_2' \notin T_2$ for any prefix t_2' of t_2, as required. $\square$

Now, as $j < j'$, we have that t_0 is shorter than t, contrary to our assumption that t was the shortest template in T_1 such that (C1) does not hold. The case where $|t_2| > m_1 2^{m_2-1}$ is similar. Thus, we must have that $|t| \leq m_1 2^{m_2} + 2n_1 + n_2$. $\square$

Corollary 1. *Let $n_1, n_2 \geq 1$ and $T_1, T_2 \subseteq \Sigma^*$ ($|\Sigma| \geq 3$) be effective regular sets of templates. Then it is decidable whether $T_1 \equiv_{n_1,n_2} T_2$.*

Proof. We can assume without loss of generality that $T_1, T_2 \subseteq \Sigma^{\geq 2n_1+n_2}$, as we have observed that templates of length less than this critical length do not affect equivalence.

By Theorem 3, $T_1 \equiv_{n_1,n_2} T_2$ if and only if (C1) holds twice, with T_1 and T_2 in both roles. To test (C1), it suffices to test all words up to the length given by Lemma 1. $\square$

Note that Corollary 1 is not an efficient algorithm: it requires checking an exponential number of templates up to a bound which is itself exponential in the size of the minimal DFA for T_1.

We note the following alternative proof for Corollary 1 which does not use Lemma 1, suggested to us by an anonymous referee. Let $t \in \Sigma^*$ and $T \subseteq \Sigma^*$ be arbitrary, and let $\# \notin \Sigma$. Define $T \ddagger t = \{t_1 \# t_2 : t_1 t t_2 \in T\}$. Note that if t is not a subword of t', then t' does not contribute anything to $T \ddagger t$. It is not difficult to demonstrate that $T \ddagger t$ is regular for all regular sets of templates T and all $t \in \Sigma^*$. We then note that

$$T_1 \sqsubseteq_{n_1,n_2} T_2 \iff \forall t \in \Sigma^{2n_1+n_2}, T_1 \ddagger t \subseteq \Sigma^*(T_2 \ddagger t)\Sigma^*.$$

That is, $T_1 \sqsubseteq_{n_1,n_2} T_2$ if and only if every word in $T_1 \ddagger t$ has a subword in $T_2 \ddagger t$. This subword must necessarily have an occurrence of $\#$, which has effectively replaced t, and so we capture (C1) exactly. Therefore, the process of testing the above condition for all words of length $2n_1 + n_2$ gives an alternate method of deciding whether $T_1 \sqsubseteq_{n_1,n_2} T_2$.

We can now give a somewhat surprising positive decidability result for recursive sets of templates. In particular, we can establish a universality equivalence result:

Theorem 4. *Let $n_1, n_2 \geq 1$ and Σ be an alphabet of size at least three. Given an effectively recursive set of templates $T \subseteq \Sigma^*$, we can determine whether $T \equiv_{n_1, n_2} \Sigma^*$.*

However, we also have the following result, which demonstrates that there is at least one regular set of templates such that determining equivalence for context-free sets of templates is undecidable:

Theorem 5. *Let Δ be an alphabet of size at least three and $n_1, n_2 \geq 1$. There exists a fixed regular set of templates $T_0 \subseteq \Delta^*$ such that the following problem is undecidable: Given a context-free set of templates $T \subseteq \Delta^*$, is $T \equiv_{n_1, n_2} T_0$?*

5   Conclusions

In this paper, we have considered equivalence of sets of templates. With a natural condition on extending subwords of the critical length $2n_1 + n_2$ in one set of templates to a template in the equivalent set, we have exactly characterized the equivalence of two sets of templates for alphabets of size three or more, which is sufficient for modelling biological processes. It is open whether the construction can be reduced to an alphabet of size two.

Using this characterization, we have shown that it is decidable whether two regular sets of templates are equivalent. This uses a result which establishes that if two regular sets of templates are not equivalent, a witness can be found within some finite bound. We have also established two other decidability results. First, deciding equivalence to the set of all possible templates is easier than might be expected: we can determine such an equivalence for recursive sets of templates. However, there exists a fixed regular set of templates T_0 such that it is undecidable whether a given context-free set of templates is equivalent to T_0.

We mention the problem of equivalence for iterated TGR, which has been defined as a formal operation by Daley and McQuillan [4]. Iterated TGR serves as a more realistic biological model of DNA rearrangement in ciliates. It is not difficult to show that if $T_1 \equiv_{n_1, n_2} T_2$, then the iterated TGR operations using T_1 and T_2 are also equivalent. Thus, equivalence of two sets of templates implies the equality of the corresponding iterated TGR operations using T_1 and T_2. However, the converse, i.e., whether equivalence of templates in iterated TGR implies equivalence for non-iterated TGR, is open and a topic for future research.

Acknowledgments

We thank the referees of DNA 13 for their helpful comments, and in particular, the suggested alternative proof of Corollary 1.

References

1. Angeleska, A., Jonoska, N., Saito, M., Landweber, L.: RNA-Guided DNA Assembly. Journal of Theoretical Biology 248(4), 706–720 (2007); Abstract appears in: Garzon, M., Yan, H. (eds.) DNA 13. LNCS, vol. 4848. Springer, Heidelberg (2007)
2. Cavalcanti, A., Landweber, L.: Insights into a biological computer: Detangling scrambled genes in ciliates. In: Chen, J., Jonoska, N., Rozenberg, G. (eds.) Nanotechnology: Science and Computation, pp. 349–360. Springer, Heidelberg (2006)
3. Daley, M., Domaratzki, M., Morris, A.: Intra-molecular template-guided recombination. International Journal of Foundations of Computer Science (to appear, 2007) Preliminary technical report,
   http://www.cs.acadiau.ca/research/technicalReports
4. Daley, M., McQuillan, I.: Template-guided DNA recombination. Theoretical Computer Science 330, 237–250 (2005)
5. Daley, M., McQuillan, I.: On computational properties of template-guided DNA recombination in ciliates. In: Carbone, A., Pierce, N.A. (eds.) DNA Computing. LNCS, vol. 3892, pp. 27–37. Springer, Heidelberg (2006)
6. Daley, M., McQuillan, I.: Useful templates and iterated template-guided DNA recombination in ciliates. Theory of Computing Systems 39, 619–633 (2006)
7. Ehrenfeucht, A., Harju, T., Petre, I., Prescott, D., Rozenberg, G.: Computation in Living Cells: Gene Assembly in Ciliates. Springer, Heidelberg (2004)
8. McQuillan, I., Salomaa, K., Daley, M.: Iterated TGR languages: Membership problem and effective closure properties. In: Chen, D.Z., Lee, D.T. (eds.) COCOON 2006. LNCS, vol. 4112, pp. 94–103. Springer, Heidelberg (2006)
9. Prescott, D., Ehrenfeucht, A., Rozenberg, G.: Template-guided recombination for IES elimination and unscrambling of genes in stichotrichous ciliates. Journal of Theoretical Biology 222, 323–330 (2003)
10. Rozenberg, G., Salomaa, A. (eds.): Handbook of Formal Languages. Springer, Heidelberg (1997)
11. Vijayan, V., Nowacki, M., Zhou, Y., Doak, T., Landweber, L.: Programming a Ciliate Computer: Template-Guided IN Vivo DNA Rearrangements in Oxytricha. In: Garzon, M., Yan, H. (eds.) DNA 13: Preliminary Proceedings, p. 172 (2007)

Watson-Crick Conjugate and Commutative Words

Lila Kari and Kalpana Mahalingam

University of Western Ontario,
Department of Computer Science,
London, ON, Canada N6A 5B7
{lila, kalpana}@csd.uwo.ca

Abstract. This paper is a theoretical study of notions in combinatorics of words motivated by information being encoded as DNA strands in DNA computing. We generalize the classical notions of conjugacy and commutativity of words to incorporate the notion of an involution function, a formalization of the Watson-Crick complementarity of DNA single-strands. We define and study properties of Watson-Crick conjugate and commutative words, as well as Watson-Crick palindromes. We obtain, for example, a complete characterization of the set of all words that are not Watson-Crick palindromes. Our results hold for more general functions, such as arbitrary morphic and antimorphic involutions. They generalize classical results in combinatorics of words, while formalizing concepts meaningful for DNA computing experiments.

1   Introduction

Theoretical DNA Computing is an area of biomolecular computing that has seen a surge of activity in recent years. It loosely encompasses contributions to fundamental research in computer science originated in or motivated by research in DNA computing. Examples are numerous and they include theoretical aspects of self-assembly [1], [20], DNA sequence design [11], [17], and mathematical properties of DNA-encoded information [10], [8].

This paper constitutes a contribution to the field of theoretical DNA computing by investigating a generalization of the classical notions of conjugacy and commutativity of words motivated by DNA-encoded information. The main idea is that information-encoding strings that are used in DNA computing experiments have an important property that differentiates them from their electronic computing counterparts. This property is the Watson-Crick complementarity between DNA single-strands that allows information-encoding strands to potentially interact. Mathematically, this translates into generalizing the identity function, which is the only one operating in the electronic realm, to an arbitrary involution function. An involution is a function θ such that θ^2 equals the identity. Given an alphabet Σ, an antimorphic involution, i.e., an involution θ with the additional property that $\theta(uv) = \theta(v)\theta(u)$ for all strings $u, v \in \Sigma^*$, is the mathematical notion that formalizes the Watson-Crick complementarity. Indeed,

M.H. Garzon and H. Yan (Eds.): DNA 13, LNCS 4848, pp. 273–283, 2008.
© Springer-Verlag Berlin Heidelberg 2008

an antimorphic involution captures the two main properties of the Watson-Crick complement of a DNA strand, namely its being the reverse (antimorphic property) complement (involution property) of the original strand. Replacing identity with involutions paves thus the way to concepts that are both meaningful formalizations of information-encoding DNA strands, and mathematically interesting generalizations of classical concepts in formal language theory, coding theory and combinatorics of words.

For example, using the concept of involutions one obtains generalizations of the classical notions of prefix codes, suffix codes and comma-free codes [12], [13]. In addition to being of theoretical interest, these notions prove to be meaningful in the context of DNA computing experiments. Indeed, if θ is the Watson-Crick involution, then a θ-sticky-free, or θ-overhang-free code is a set of words where no unwanted hybridizations of a certain type occur between DNA codewords. More recently, in [14] we extended the concept of bordered and unbordered words to involution-bordered and involution-unbordered words.

In this paper we extend the notions of conjugate and commutative words to Watson-Crick conjugate and Watson-Crick commutative words. Our results hold in a more general context where the function θ involved is an arbitrary morphic or antimorphic involution. To put these results in context, they augment studies of combinatorial properties of words which have meaningful applications in numerous other fields. For example, word properties such as periodicity and borderedness play a role in many areas including string searching algorithms [4,5,6], data compression [7,21] and in the study of coding properties of sets of words [2,19] as well as sequence assembly in computational biology [18]. Relevant to this paper, there are several classical results about conjugacy of words and words that commute [19]. In addition, in [3] the authors extend certain combinatorial properties of conjugacy of words to partial words with an arbitrary number of holes. An authoritative text on the study of combinatorial properties of strings would be [16].

The paper is organized as follows. We begin by reviewing basic concepts of combinatorics of words and the definition of θ-bordered and θ-unbordered words for an arbitrary morphic or antimorphic involution θ. In Section 2, we also define the concept of θ-conjugacy on words. If θ is the antimorphic Watson-Crick involution, this gives rise to the notion of Watson-Crick conjugate words. Figure 1 illustrates the interaction between two DNA strands u and v over the DNA alphabet $\Delta = \{A, C, G, T\}$ that are Watson-Crick conjugates to each other. We show that for a morphic involution θ, the θ-conjugacy on words is reflexive, symmetric and transitive. We also obtain several properties of θ-conjugate words including a general characterization of the words that are θ-conjugate in Proposition 1. These results generalize well-known properties of conjugate words [19].

In Section 3, we introduce the concept of θ-commutativity on words for an arbitrary morphic or antimorphic involution θ, and its particular case of Watson-Crick commutativity. Figure 3 illustrates the interaction between two DNA strands u and v that Watson-Crick commute. We obtain several properties of words that θ-commute, including their characterization (Proposition

3), and properties of the set $C_\theta(1)$ of words that cannot be written as a concatenation of two non-empty words x, y such that x θ-commutes with y. These properties generalize classical properties of words that commute, [19]. We define the notion of θ-palindrome that was obtained independently in [9]. Note that if θ is the Watson-Crick involution, then the notion of Watson-Crick palindromes coincides with the term "palindrome" as used by molecular biologists. We define a relation on words using the θ-commutativity and show that, for an antimorphic involution θ, the set of all θ-palindromes can be characterized using this relation.

2 Watson-Crick Conjugate Words

Before introducing the notion of Watson-Crick conjugate words, we review some basic concepts of combinatorics of words. An alphabet Σ is a finite non-empty set of symbols. A word u over Σ is a finite sequence of symbols in Σ. We denote by Σ^* the set of all words over Σ, including the empty word λ and, by Σ^+, the set of all non-empty words over Σ. We note that with the concatenation operation on words, Σ^* is the free monoid and Σ^+ is the free semigroup generated by Σ. For a word $w \in \Sigma^*$, the length of w is the number of symbols in w and is denoted by $|w|$. For a word w, the set of its prefixes/ suffixes are defined as follows: $Pref(w) = \{u \in \Sigma^+ | \exists v \in \Sigma^*, w = uv\}$ and $Suff(w) = \{u \in \Sigma^+ | \exists v \in \Sigma^*, w = vu\}$.

Bordered words were initially called "overlapping words" and unbordered words were called "non-overlapping words". For properties of bordered and unbordered words we refer the reader to [19,22]. In [14], we extended the concept of bordered words to involution bordered words. We now recall some notions defined and used in [22] and [14].

Definition 1. *Let θ be either a morphic or an antimorphic involution on Σ^*.*

1. *For $v, w \in \Sigma^*$, $w \leq_p v$ iff $v \in w\Sigma^*$.*
2. *For $v, w \in \Sigma^*$, $w \leq_s^\theta v$ iff $v \in \Sigma^*\theta(w)$.*
3. *$\leq_d^\theta = \leq_p \cap \leq_s^\theta$.*
4. *For $u \in \Sigma^*$, $v \in \Sigma^*$ is said to be a θ-border of u if $v \leq_d^\theta u$, i.e., $u = vx = y\theta(v)$.*
5. *For $w, v \in \Sigma^*$, $w <_p v$ iff $v \in w\Sigma^+$.*
6. *For $w, v \in \Sigma^*$, $w <_s^\theta v$ iff $v \in \Sigma^+\theta(w)$.*
7. *$<_d^\theta = <_p \cap <_s^\theta$.*
8. *For $u \in \Sigma^*$, $v \in \Sigma^*$ is said to be a proper θ-border of u if $v <_d^\theta u$.*
9. *For $u \in \Sigma^+$, define $L_d^\theta(u) = \{v | v \in \Sigma^*, v <_d^\theta u\}$.*
10. *$\nu_\theta(u) = |L_d^\theta(u)|$.*
11. *$D_\theta(i) = \{u | u \in \Sigma^+, \nu_\theta(u) = i\}$.*
12. *A word $u \in \Sigma^+$ is said to be θ-bordered if there exists $v \in \Sigma^+$ such that $v <_d^\theta u$, i.e., $u = vx = y\theta(v)$ for some $x, y \in \Sigma^+$.*
13. *A non-empty word which is not θ-bordered is called θ-unbordered.*

A word u in Σ^* is a conjugate of w in Σ^* if there exists $v \in \Sigma^*$ such that $uv = vw$. Note that conjugacy on words is an equivalence relation. In [3], the authors showed that conjugacy on partial words is reflexive and symmetric but not transitive. In this section we extend the concept of conjugacy of words to incorporate the notion of an involution function and show that θ-conjugacy on words is reflexive. We also show that θ-conjugacy on words is symmetric and transitive when θ is a morphic involution.

Definition 2. *Let θ be either a morphic or an antimorphic involution. A word u is a θ-conjugate of another word w if $uv = \theta(v)w$ for some $v \in \Sigma^*$.*

Example 1. Let $\Sigma = \{a, b\}$ and θ be an antimorphic involution which maps a to b and vice versa. Let $u = aba$ and $w = bab$. Then u is a θ-conjugate of w since $aba \cdot b = \theta(b) \cdot bab$.

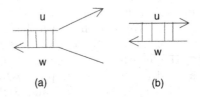

(a)   (b)

Fig. 1. If u is Watson-Crick conjugate of w, then u and the Watson-Crick complement of w overlap, resulting thus in one of the two intermolecular hybridizations shown above

For any DNA string u over the DNA alphabet $\Delta = \{A, G, C, T\}$, the Watson-Crick conjugates of u are defined as the DNA strings w such that $uv = \theta(v)w$ for some $v \in \Delta^*$. In this case, θ is the Watson-Crick involution which maps $A \mapsto T$, $C \mapsto G$ and viceversa such that θ is an antimorphic involution. In the following example we find all the Watson-Crick conjugates of a given DNA string.

Example 2. Let $\Delta = \{A, G, C, T\}$ be the DNA alphabet and let $u = ATAG$. Then the Watson-Crick Conjugates of u are given by $Conj_\theta(u) = \{ATAG, TAGT, AGAT, GTAT, CTAT\}$. For all $w \in Conj_\theta(u)$, there exists a $v \in \Sigma^*$ such that $uv = \theta(v)w$. These words v respectively are $T, AT, TAT, CTAT$.

The characterization of θ-conjugate words in Proposition 2 will show that if u and w are Watson-Crick conjugates, then u and the Watson-Crick complement of w overlap, hence forming the hybridization in Fig 1.

Note that for all $u \in \Sigma^*$, u is a θ-conjugate of u since $u\lambda = \theta(\lambda)u$. Also u is a θ-conjugate of $\theta(u)$ since $u\theta(u) = \theta(\theta(u))\theta(u)$ and hence for all $u, v \in \Sigma^*$, uv is a θ-conjugate to $v\theta(u)$ since $uv\theta(u) = \theta(\theta(u))v\theta(u)$. Even though we concentrate on Watson-Crick conjugates, we provide results that hold for any general morphic or an antimorphic involution. In the next lemma we show that the θ-conjugacy of words is transitive when θ is a morphic involution.

Lemma 1. *Let $u, v, w \in \Sigma^+$ such that u is a θ-conjugate of w and w is a θ-conjugate of v.*

1. *If θ is a morphic involution then u is a θ-conjugate of v.*
2. *If θ is an antimorphic involution then u is not necessarily a θ-conjugate of v.*

Proof. 1. Let θ be a morphic involution. Since u is a θ-conjugate of w and w is a θ-conjugate of v then there exists $r, s \in \Sigma^*$ such that $ur = \theta(r)w$ and $ws = \theta(s)v$ which implies that $urs = \theta(r)\theta(s)v$. Hence $urs = \theta(rs)v$ and u is a θ-conjugate of v.

2. Let θ be an antimorphic involution. Then u is not necessarily a θ-conjugate of v. For example let $\Sigma = \{a, b\}$ and $\theta(a) = b$ and let $u = aba$, $w = bab$ and $v = bba$. Note that aba is a θ-conjugate of bab since $aba \cdot b = \theta(b) \cdot bab$. Also bab is a θ-conjugate of bba since $bab \cdot ba = \theta(ba) \cdot bba$. Suppose there exist a $y \in \Sigma^*$ such that $aba \cdot y = \theta(y) \cdot bba$ then $\theta(y) = ax$ for some $x \in \Sigma^*$ which implies that $y = \theta(x)b$ which is not possible since y has to be of the form za. Hence the θ-conjugacy relation is not transitive for an antimorphic involution θ. □

Lemma 2. *Let $x, y \in \Sigma^*$ such that x is a θ-conjugate of y.*

1. *If θ is an antimorphic involution then for all $u \in \Sigma^*$ ux is a θ-conjugate of $y\theta(u)$.*
2. *If θ is a morphic involution then there exists a $u \in \Sigma^*$ such that ux is not a θ-conjugate of $y\theta(u)$.*

Proof. 1. Let θ be an antimorphic involution. Since x is a θ-conjugate of y there exists $v \in \Sigma^*$ such that $xv = \theta(v)y$ and hence $uxv\theta(u) = u\theta(v)y\theta(u)$. Take $r = v\theta(u)$, then $\theta(r) = \theta(v\theta(u)) = u\theta(v)$ which implies that $uxr = \theta(r)y\theta(u)$ hence ux is a θ-conjugate of $y\theta(u)$.

2. Let θ be a morphic involution and let $\Sigma = \{a, b\}$ such that $\theta(a) = b$. Note that for $x = abb$ and $y = bbb$, x is a θ-conjugate of y since $x \cdot b = \theta(b) \cdot bbb$. But for $u = ab$ ux is not a θ-conjugate of $y\theta(u)$. Also for $w = ux = ababb$, the set of all θ-conjugates is $C = \{babaa, bbaba, bbbab, abbba, babbb, ababb\}$ and clearly $y\theta(u) = bbbba \notin C$. □

Proposition 1. *Let u be a θ-conjugate of w such that $uv = \theta(v)w$ for some $v \in \Sigma^*$. Then for a morphic involution θ there exists $x, y \in \Sigma^*$ such that $u = xy$ and one of the following hold:*

1. $w = y\theta(x)$ *and* $v = (\theta(x)\theta(y)xy)^i\theta(x)$ *for some $i \geq 0$.*
2. $w = \theta(y)x$ *and* $v = (\theta(x)\theta(y)xy)^i\theta(x)\theta(y)x$ *for some $i \geq 0$.*

Proof. Let θ be a morphic involution. Given $uv = \theta(v)w$ for some $v \in \Sigma^*$. Then we either have $|u| < |v|$ or $|v| \leq |u|$. Suppose $|u| \geq |v|$ then $u = \theta(v)\alpha$ and $w = \alpha v$

for some $\alpha \in \Sigma^*$. Hence for $v = \theta(x)$, $u = xy$ and $w = y\theta(x)$. Assume that $|u| < |v|$. Then there exits $p_1 \in \Sigma^+$ such that $\theta(v) = up_1$ and $v = p_1w$. Hence $v = p_1w = \theta(u)\theta(p_1)$. Suppose $|u| < |p_1|$ then there exists $p_2 \in \Sigma^+$ such that $p_1 = \theta(u)p_2$ and $\theta(p_1) = p_2w$ and hence $u\theta(p_2) = p_2w$ and $v = \theta(p_2)\theta(w)w = \theta(u)u\theta(p_2)$. Continuing this way we can find a $p_n \in \Sigma^+$ such that $|u| > |p_k|$ and $v = a_j^n\theta(x_n)$ for $a_j = \theta(u)$ when j is odd and $a_j = u$ when j is even. When n is even, we have $n = 2k$ and $v = (\theta(u)u)^k\theta(x_{2k})$ with $u\theta(x_{2k}) = x_{2k}w$ which implies $u = x_{2k}r = xy$ and $w = r\theta(x_{2k}) = y\theta(x)$ and $v = (\theta(x)\theta(y)xy)^k\theta(x)$. When n is odd $n = 2k + 1$ for some k and $v = (\theta(u)u)^{2k-1}\theta(u)\theta(x_{2k+1})$ with $\theta(u)\theta(x_{2k+1}) = x_{2k+1}w$. Then we have $\theta(u) = x_{2k+1}r = \theta(x)\theta(y)$ and $w = r\theta(x_{2k+1}) = \theta(y)x$ and $v = (\theta(x)\theta(y)xy)^{2k-1}\theta(x)\theta(y)x$.   □

Corollary 1. *For a morphic involution θ on Σ^*, θ-conjugacy on words is a symmetric relation.*

Example 3. Let $\Sigma = \{a, b\}$ and let θ be a morphic involution which maps a to b and viceversa. From Proposition 1 for $w = ux = ababb$, the set of all θ-conjugates are $C = \{babaa, bbaba, bbbab, abbba, babbb, ababb\}$.

Proposition 2. *Let u be a θ-conjugate of w. Then for an antimorphic involution θ, there exists $x, y \in \Sigma^*$ such that either $u = xy$ and $w = y\theta(x)$ (Figure 1, (a)) or $w = \theta(u)$ (Figure 1, (b)).*

Corollary 2. *Let θ be either a morphic or an antimorphic involution and let u be a θ-conjugate of w for $u, w \in \Sigma^+$. Then either uw or wu is θ-bordered.*

Let u be a θ-conjugate of w. Then for an antimorphic involution θ, either uw or wu precisely form a hairpin-like structure. For example, choose a DNA string $u = ATAGCT$ and one of its Watson-Crick conjugates $w = GCTTAT$. Then $uw = ATAGCTGCTTAT = (ATA)GCTGCT\theta(ATA)$, as illustrated in Fig. 2.

Fig. 2. The DNA string $GCTTAT$ is a Watson-Crick conjugate of $ATAGCT$, and their catenation $ATAGCTGCTTAT$ forms a hairpin

3   Watson-Crick Commutative Words

Two words x and y are said to commute when $xy = yx$, [19]. In this section we define the concept of θ-commutative words and show that commutative words are a special case of θ-commutative words when θ is identity. We also introduce the θ-commutativity order and characterize the set of all θ-palindromes for an antimorphic involution θ.

Definition 3. *Let θ be either a morphic or an antimorphic involution.*

1. *For $x, y \in \Sigma^*$, x is said to θ-commute with y if $xy = \theta(y)x$.*
2. *We define the θ-commutativity order as $v \leq_c^\theta u$ iff $u = vx = \theta(x)v$ for some $x \in \Sigma^*$.*
3. *$L_c^\theta(u) = \{v | v \in \Sigma^*, v \leq_c^\theta u\}$.*
4. *$v_c^\theta(u) = |L_c^\theta(u)|$.*
5. *For $i \geq 1$, define $C_\theta(i) = \{u | u \in \Sigma^+, v_c^\theta(u) = i\}$.*
6. *A word $x \in \Sigma^*$ is called a θ-palindrome if $x = \theta(x)$.*

Suppose $uv = \theta(v)u$ holds. Then, if $v = \lambda$, then u is a θ-conjugate of u. (This also implies that u θ-commutes with λ.) Otherwise, it means that u θ-commutes with v. For any non-empty DNA strings u and v over the DNA alphabet $\Delta = \{A, G, C, T\}$, we say that u Watson-Crick commutes with v if $uv = \theta(v)u$ where θ is the Watson-Crick involution. The word $u \in \Delta^*$ is called a Watson-Crick palindrome if $u = \theta(u)$ where θ is the Watson-Crick involution. In what follows, we will show that for the Watson-Crick involution θ if u θ-commutes with v, then u is a Watson-Crick palindrome and either u is a prefix of $\theta(v)$ or $\theta(v)$ is a prefix of u.

Example 4. Consider a string $u = AGCT$ over the DNA alphabet Δ. Let θ be the Watson-Crick involution and $v = CTAGAGCT$. Then u θ-commutes with v since $uv = AGCT \cdot CTAGAGCT = \theta(CTAGAGCT) \cdot AGCT = \theta(v)u$.

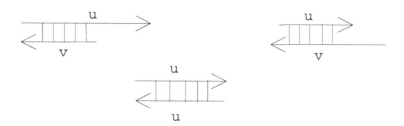

Fig. 3. If the DNA string u Watson-Crick commutes with v, then one of the intermolecular hybridizations (a) or (b) occurs and, in addition, u is a Watson-Crick palindrome (c).

If the word u Watson-Crick commutes with the word v, the characterization in Proposition 3 will show that u and v will form one of the hybridizations in Figure 3.

Observation 1 *Let θ be either a morphic or an antimorphic involution on Σ^*.*

1. *For all $u \in \Sigma^+$, $u \in L_c^\theta(u)$, i.e., $u \leq_c^\theta u$.*
2. *$C_\theta(1) = \{u \in \Sigma^+ | v \leq_c^\theta u \Leftrightarrow v = u\}$.*
3. *For all $u \in \Sigma^+$ such that u is a θ-palindrome we have $\lambda \in L_c^\theta(u)$.*
4. *For all $a \in \Sigma$ such that $a \neq \theta(a)$, $a^+ \subseteq C_\theta(1)$.*

Note that $C_\theta(1)$ is the set of all words that cannot be written as a catenation of two non-empty words x and y such that x θ-commutes with y. $C_\theta(1)$ is the set of all words u that have only one element in the set $L_c^\theta(u)$ namely u. In particular, θ-palindromes are not in $C_\theta(1)$. In the next lemma we show that for an antimorphic involution θ, the set $L_c^\theta(u)$ is a totally ordered set with respect to $\leq_c^\theta$.

Lemma 3. For an antimorphic involution θ and $u \in \Sigma^+$, $L_c^\theta(u)$ is a totally ordered set with $\leq_c^\theta$.

The proof technique of the following proposition is similar to that of Proposition 1 and hence we omit the proof.

Proposition 3. Let $u, v \in \Sigma^+$ such that u θ-commutes with v, i.e., $uv = \theta(v)u$.

1. If θ is an antimorphic involution then $u = x(yx)^i$, $v = yx$ where $i \geq 0$ (Figure 3 (a), or (b)) and u (Figure 3, (c)) as well as x, y are θ-palindromes, where $x \in \Sigma^+$, $y \in \Sigma^*$.
2. If θ is a morphic involution then $u = x(yx)^i$ and $v = yx$ where $yx = \theta(x)\theta(y)$ and $i \geq 0$ with $y \in \Sigma^*$ and $x \in \Sigma^+$.

It was shown in [15] that when $u = xy$ such that $x, y \in P_\theta$ and for an antimorphic involution θ, u can be written as $(\alpha\beta)^n$ with $x = (\alpha\beta)^i\alpha$ and $y = \beta(\alpha\beta)^{n-i-1}$. The authors also proved that for $u = xy = \theta(y)\theta(x)$ for a morphic involution θ, either $u = \alpha^m$ for $\alpha \in P_\theta$ or $u = [\alpha\theta(\alpha)]^n$ for some $\alpha \in \Sigma^+$. We use these results and Proposition 3 to deduce the following corollary.

Corollary 3. Let $u, v \in \Sigma^+$ such that u θ-commutes with v.

1. If θ is a morphic involution then one of the following hold:
   (a) $u = \alpha^m$ and $v = \alpha^n$ for some $m, n \geq 1$, $\alpha \in P_\theta$.
   (b) $u = \theta(\alpha)[\alpha\theta(\alpha)]^m$ and $v = [\alpha\theta(\alpha)]^n$ for some $n \geq 1$ and $k \geq 0$ with $\alpha \in \Sigma^+$.
2. If θ is an antimorphic involution, $u = \beta(\alpha\beta)^n$ and $v = (\alpha\beta)^m$ for some $\alpha, \beta \in P_\theta$ with $m \geq 1$ and $n \geq 0$.

Based on the definitions and the previous two results we have the following observation.

Lemma 4. Let $w \in \Sigma^+$ and θ be an antimorphic involution. Then w is a θ-palindrome iff there exists $v \in \Sigma^*$ such that $v \neq w$ and $v \leq_c^\theta w$.

Note that Lemma 4 states that, for an antimorphic involution θ, a word $w \in C_\theta(1)$ iff w is not a θ-palindrome, i.e., the set $L = \Sigma^* \setminus C_\theta(1)$ is the set of all θ-palindromes.

Note that for a word w which is not a θ-palindrome for an antimorphic involution θ, $L_c^\theta(w)$ may be an emptyset. For example, let $\Sigma = \{a, b\}$, and θ be an antimorphic involution that maps a to b and vice versa. Let $w = ababa$, then $\theta(w) = babab$. Clearly $w \neq \theta(w)$. Note that

$\quad - w = abab \cdot a \neq \theta(a) \cdot abab = babab$.
$\quad - w = aba \cdot ba \neq \theta(ba) \cdot aba = baaba$.

- $w = ab \cdot aba \neq \theta(aba) \cdot ab = babab$.
- $w = a \cdot baba \neq \theta(baba) \cdot a = babaa$.

Thus it is clear that for $w = ababa$ there does not exist a $v \in \Sigma^*$ such that $w = vx = \theta(x)v$ and thus $L_c^\theta(w) = \emptyset$.

Lemma 5. *Let θ be either a morphic or an antimorphic involution. For all $u \in \Sigma^+$, $\theta(L_c^\theta(u)) = L_c^\theta(\theta(u))$.*

Lemma 6. *Let θ be either a morphic or an antimorphic involution. Then for all $u \in C_\theta(1)$ we have $u^+ \subseteq C_\theta(1)$.*

It is shown in [14], that if u and v are θ-unbordered for an antimorphic involution θ, then for $u = u_1u_2$ such that $u_1, u_2 \in \Sigma^+$, u_1vu_2 is also θ-unbordered. But it is not true for words in $C_\theta(1)$. We illustrate it in the following example.

Example 5. Let $\Sigma = \{a, b\}$ and let θ be an antimorphic involution that maps a to b and vice versa. Note that $u = abb \in C_\theta(1)$ since $u \neq \theta(u)$ and $ab \cdot b \neq a \cdot ab$ and $a \cdot bb \neq aa \cdot a$. Let $v = a$ and $v \in C_\theta(1)$ since $v \neq \theta(v)$. But u_1vu_2 with $u_1, u_2 \in \Sigma^+$ is either $a \cdot a \cdot bb$ or $ab \cdot a \cdot b$. Note that both $aabb, abab \notin C_\theta(1)$ since $aabb = \theta(aabb)$ and $abab = \theta(abab)$.

Proposition 4. *Let $u, v \in C_\theta(1)$ and $\theta(Pref(u)) \cap Suff(v) = \emptyset$.*

1. *If θ is an antimorphic involution then $uv \in C_\theta(1)$.*
2. *If θ is a morphic involution then uv is not necessarily in $C_\theta(1)$.*

Proof. 1. Let θ be an antimorphic involution. Suppose for some $u, v \in C_\theta(1)$, $uv \notin C_\theta(1)$ then there exists $\alpha \in \Sigma^+$ such that $uv = p\alpha = \theta(\alpha)p$. Then we have the following cases. If $|\alpha| \leq |v|$ and $|\theta(\alpha)| \leq |u|$, we have $v = r\alpha$ and $u = \theta(\alpha)s$ then $\alpha \in \theta(Pref(u)) \cap Suff(v)$ a contradiction. If $|\alpha| \leq |v|$ and $|\theta(\alpha)| \leq |uv|$ we have $v = r\alpha$ and $\theta(\alpha) = us$ then $\alpha = \theta(s)\theta(u)$ and $v = r\theta(s)\theta(u)$ for some $r, s \in \Sigma^*$ which implies $\theta(u) \in \theta(Pref(u)) \cap Suff(v)$ which is a contradiction. If $|\alpha| \leq |uv|$ and $|\theta(\alpha)| \leq |u|$ we have $\alpha = rv$ and $u = \theta(\alpha)s$ then $u = \theta(v)\theta(r)s$ for some $r, s \in \Sigma^*$ which implies $\theta(v) \in Pref(u)$ and hence $v \in \theta(Pref(u)) \cap Suff(v)$ which is a contradiction. If $|\alpha| \leq |uv|$ and $|\theta(\alpha)| \leq |uv|$ we have $\alpha = rv$ and $\theta(\alpha) = us$ then $\alpha = rv = \theta(s)\theta(u)$. Then we have the following subcases:
   - If $|v| = |u|$ then $\theta(u) = v$.
   - If $|u| < |v|$ then $v = \beta\theta(u)$ for some $\beta \in \Sigma^+$ and $\theta(u) \in \theta(Pref(u)) \cap Suff(v)$.
   - If $|v| < |u|$ then $\theta(u) = \beta v$ for some $c \in \Sigma^+$ and $\theta(u_2)\theta(u_1) = \beta v$ with $u = u_1u_2$ and $|u_1| = |v|$ which implies $\theta(u_1) \in \theta(Pref(u)) \cap Suff(v)$.
   All the above cases arrive at a contradiction. Hence $uv \in C_\theta(1)$.
2. Let $u = ab$ and $v = a$ over the alphabet set $\Sigma = \{a, b\}$ and let θ be a morphic involution that maps a to b and vice versa. Then $\theta(u) = ba$ and $\theta(v) = b$. Note that $u, v \in C_\theta(1)$. Also $Pref(u) = \{a, ab\}$, $\theta(Pref(u)) = \{b, ba\}$ and $Suff(v) = \{a\}$. Note that $\theta(Pref(u)) \cap Suff(v) = \emptyset$. But $uv = aba \notin C_\theta(1)$ since $a \cdot ba = \theta(ba) \cdot a = aba$ which implies that $a \in L_c^\theta(uv)$.

□

Note that the converse of the statement 1 in Proposition 4 does not hold in general. Let $\Sigma = \{a, b\}$ and θ be an antimorphic involution that maps a to b and vice versa. Let $u = aba$ and $v = abb$. Then $\theta(u) = bab$ and $\theta(v) = aab$. Note that $u \neq \theta(u)$ and $v \neq \theta(v)$ and $u, v \in C_\theta(1)$. For $uv = abaabb$, $\theta(uv) = aabbab$ and

- $abaab \cdot b \neq \theta(b) \cdot abaab = aabaab.$
- $abaa \cdot bb \neq \theta(bb) \cdot abaa = aaabaa.$
- $aba \cdot abb \neq \theta(abb) \cdot aba = aababa.$
- $ab \cdot aabb \neq \theta(aabb) \cdot ab = aabbab.$
- $a \cdot baabb \neq \theta(baabb) \cdot a = aabbaa.$

Hence $uv \in C_\theta(1)$. But $\text{Pref}(u) = \{a, ab, aba\}$, $\text{Suff}(v) = \{b, bb, abb\}$ and $\theta(\text{Pref}(u)) = \{b, ab, bab\}$. Thus $b \in \theta(\text{Pref}(u)) \cap \text{Suff}(v) \neq \emptyset$.

Lemma 7. *Let θ be either a morphic involution or an antimorphic involution and let Σ be such that for all $a \in \Sigma$, $\theta(a) \neq a$. Then $D_\theta(1) \subseteq C_\theta(1)$.*

In [14], it was shown that for an antimorphic involution θ, the set of all θ-bordered words is regular. Note that from Lemma 4, $C_\theta(1)$ is the set of all non θ-palindromes for an antimorphic involution θ. We show using pumping lemma for regular languages that $\Sigma^* \setminus C_\theta(1)$ is not regular and hence $C_\theta(1)$ is not regular for an antimorphic involution θ.

Lemma 8. *When θ is an antimorphic involution, the set of all θ-palindrome words is not regular.*

Proof. Let $\Sigma = \{a, b\}$ and let θ be an antimorphic involution that maps $a \mapsto b$ and viceversa. Assume that the language L of all θ-palindromes is regular and let n be the constant given by the pumping lemma. Chose $w = a^n b^n$ and note that $w = \theta(w)$ and hence w is a θ-palindrome. Let $w = a^n b^n = xvy$ such that $|xv| \leq n$ and $|v| > 0$. Then $z = xv^i y$ contains more a's than b's for all i and hence z is not a θ-palindrome. Thus $L = \Sigma^* \setminus C_\theta(1)$ is not regular.   □

In our last proposition we construct a context-free grammar that generates the set of all θ-palindromes over a finite alphabet set for an antimorphic involution θ.

Proposition 5. *For an antimorphic involution θ, the set $L = \Sigma^* \setminus C_\theta(1)$ is context-free.*

Proof. Let Σ be a finite alphabet set and let $G = (\{X, Y\}, \Sigma, X, \mathcal{R})$ where $\mathcal{R} = \{X \to \lambda, Y \to \lambda, X \to a_i X \theta(a_i)$ for all $a_i \in \Sigma$ and $X \to b_i Y, Y \to b_i Y$ for all $b_i \in \Sigma$ such that $b_i = \theta(b_i)$ $\}$. It is easy to check that G generates the set of all θ-palindromes over Σ and G is context-free. Hence $L(G) = \Sigma^* \setminus C_\theta(1)$.   □

It is shown in Proposition 5.4 in [14] that for a morphic involution θ, the set of all θ-bordered words is not context-free. It is also clear from Proposition 5.4 in [14] that $L = \Sigma^* \setminus C_\theta(1)$ is not context-free when θ is a morphic involution.

Acknowledgment. Research supported by NSERC and Canada Research Chair grants for Lila Kari.

References

1. Adleman, L.: Towards a mathematical theory of self-assembly. Technical Report 00-722, Department of Computer Science, University of Southern California (2000)
2. Berstel, J., Perrin, D.: Theory of Codes. Academic Press, Orlando Florida (1985)
3. Blanchet-Sadri, F., Luhman, D.: Conjugacy on partial words. Theoretical Computer Science 289, 297–312 (2002)
4. Boyer, R., Moore, J.: A fast string searching algorithm. Communication of the ACM 20, 762–772 (1977)
5. Crochemore, M., Perrin, D.: Two-way string matching. Journal of Association of Computing Machinery 38, 651–675 (1991)
6. Crochmore, M., Rytter, W.: Jewels of Stringology. World Scientific (2003)
7. Crochmore, M., Mignosi, F., Restivo, A., Salemi, S.: Text compression using antidictionaries. In: Wiedermann, J., van Emde Boas, P., Nielsen, M. (eds.) ICALP 1999. LNCS, vol. 1644, pp. 261–270. Springer, Heidelberg (1999)
8. Daley, M., McQuillan, I.: On computational properties of template-guided DNA recombination. In: Carbone, A., Pierce, N.A. (eds.) DNA Computing. LNCS, vol. 3892, pp. 27–37. Springer, Heidelberg (2006)
9. de Luca, A., de Luca, A.: Pseudopalindrome closure operators in free monoids. Theoretical Computer Science 362, 282–300 (2006)
10. Domaratzki, M.: Hairpin structures defined by DNA trajectories. In: Mao, C., Yokomori, T. (eds.) DNA Computing. LNCS, vol. 4287, pp. 182–194. Springer, Heidelberg (2006)
11. Garzon, M., Phan, V., Roy, S., Neel, A.: In search of optimal codes for DNA computing. In: Mao, C., Yokomori, T. (eds.) DNA Computing. LNCS, vol. 4287, pp. 143–156. Springer, Heidelberg (2006)
12. Kari, L., Konstantinidis, S., Losseva, E., Wozniak, G.: Sticky-free and overhang-free DNA languages. Acta Informatica 40, 119–157 (2003)
13. Kari, L., Konstantinidis, S., Losseva, E., Sosik, P., Thierrin, G.: Hairpin structures in DNA words. In: Carbone, A., Pierce, N.A. (eds.) DNA Computing. LNCS, vol. 3892, pp. 158–170. Springer, Heidelberg (2006)
14. Kari, L., Mahalingam, K.: Involution bordered words. International Journal of Foundations of Computer Science (accepted, 2007)
   http://www.csd.uwo.ca/~lila/invbor.pdf
15. Kari, L., Mahalingam, K., Seki, S.: Language equations on Watson-Crick words, manuscript
16. Lothaire, M.: Combinatorics of Words. Cambridge University Press, Cambridge (1997)
17. Marathe, A., Condon, A., Corn, R.: On combinatorial DNA word design. In: Winfree, E., Gifford, D. (eds.) Proc. of DNA Based Computers 5, DIMACS Series in Discrete Math. and Theoretical Comp. Sci. pp. 75–89 (1999)
18. Margaritis, D., Skiena, S.: Reconstructing strings from substrings in rounds. In: Proceedings of the 36th Annual Symposium on Foundations of Computer Science, pp. 613–620 (1995)
19. Shyr, H.J.: Free Monoids and Languages. Hon Min Book Company (2001)
20. Soloveichik, D., Winfree, E.: Complexity of compact proofreading for self-assembled patterns. In: Carbone, A., Pierce, N.A. (eds.) DNA Computing. LNCS, vol. 3892, pp. 305–324. Springer, Heidelberg (2006)
21. Storer, J.A.: Data Compression: Methods and Theory. Computer Science Press, Rockville (1998)
22. Yu, S.S.: d-minimal languages. Discrete Applied Mathematics 89, 243–262 (1998)

DNA Coding Using the Subword Closure Operation*

Bo Cui and Stavros Konstantinidis

Department of Mathematics and Computing Science, Saint Mary's University,
Halifax, Nova Scotia, B3H 3C3 Canada
bo.cui@smu.ca, s.konstantinidis@smu.ca

Abstract. We investigate the problem of encoding arbitrary data into the words of a DNA language that is defined via the subword closure operation, which appears to be useful in various situations related to data encodings. We present a few theoretical results on the subword closure operation as well as an initial implementation of a web-system that computes DNA languages according to the user's parameters, and performs encodings of data into these languages.

1 Introduction

A language L is any set of words over some alphabet Σ. The prime example of alphabet in this paper is the DNA alphabet $\{a, c, g, t\}$. A *subword* of L is any word that occurs in some word of L; that is, u is a subword of L if there is a word of the form xuy in L. The expression $\mathrm{Sub}_k(L)$ denotes the set of all subwords of L of length k. We are interested in languages L whose subwords of length k, for some fixed parameter k, satisfy some desirable constraint. This topic is motivated by various questions related to DNA codes and combinatorial channels. Here we present a few theoretical results on the subword closure operation as well as an initial implementation of a web-system that computes DNA languages according to the user's parameters, and performs encodings of data into these languages.

Definition 1. *A subword constraint is any nonempty set S of words of length k, for some fixed $k \geq 1$. We say that a language L satisfies the subword constraint S if every subword of L of length k belongs to S; that is, $\mathrm{Sub}_k(L) \subseteq S$.*

The symbols Σ^*, Σ^m, $\Sigma^{\geq m}$ denote, respectively, the sets of all words; all words of length m; and all words of length at least m. In [10], a particular type of subword constraint was used to model bond-free DNA languages (see further below in this section), and the study of this constraint led to the technical concept of subword closure operation $\otimes$: $S^{\otimes} = \{w \in \Sigma^* : \mathrm{Sub}_k(w) \subseteq S\}$ Thus, $S^{\otimes}$ is the language of all words w such that every subword of w of length k belongs to S.

We list now a few examples to demonstrate that the concepts of subword constraint and subword closure operation are interesting objects of study.

---

* Research supported by NSERC.

M.H. Garzon and H. Yan (Eds.): DNA 13, LNCS 4848, pp. 284–289, 2008.

In [6], a language L is called a θ-k-code, where θ is an involution, if $\theta(x) \neq y$ for any subwords x, y in $\mathrm{Sub}_k(L)$. The relationship $\theta(x) = y$ indicates that the molecules corresponding to x and y can form chemical bonds between them. The θ-k-code property is meant to ensure that DNA strands cannot form unwanted hybridizations during DNA computations, and has been tested successfully in practice [12]. In [10], the concept of θ-k-code is extended to the Hamming bond-free property: $H(\theta(x), y) > d$ for any subwords x, y in $\mathrm{Sub}_k(L)$, where H is the Hamming distance function between words. When $d = 0$ the Hamming bond-free property coincides with the θ-k-code property. Obviously, if a language L is Hamming bond-free then $L \subseteq S^\otimes$, for some subword constraint S with $H(\theta(u), v) > d$ for all u, v in S.

The gc-ratio of words that represent DNA molecules is an important parameter related to the melting temperature of these molecules [11]. Let S be the set of all DNA words of length k such that the ratio of g and c's over k is about 50%. Then a language L has a gc-ratio of about 50% for subwords of length k, if it satisfies the subword constraint S.

Another example where the subword constraint and operation are relevant is when describing a combinatorial channel [9] via a set of edit strings. There are probably other situations where the subword constraint and closure concepts might be useful.

We note that there are other methods and systems available addressing problems related to DNA coding – see for instance [14,11,13]. However, these approaches are not comparable to ours as they address different DNA properties (other than the bond-free property discussed here), or they do not address the general encoding problem for arbitrary subword constraints.

The paper is organized as follows. The next section presents some theoretical results on the subword closure $S^\otimes$. In Section 3, we propose a method of encoding arbitrary data words into the words of such a language L. The method is designed for any S independently of the application. Section 4 contains simple algorithms and heuristics for the method of Section 3. Finally, Section 5 gives some details of a web-system that implements certain aspects of the method of Section 3.

2   Some Results on the Subword Closure Operation

The density of a language L is the function that returns, for each nonnegative integer n, the number of words in L of length n [15]. We use the expression $|L(n)|$ for the density value on n. This quantity is related to the efficiency (in terms of information capacity) of the language when it is used for encoding data. The next result provides an exact formula for the density of the language $S^\otimes$ for any subword constraint S.

Theorem 1. *For any subword constraint S of some length k, the density function $|S^\otimes(l)|$, for $l > k$, is given by the following recursive formula*

$$|S^\otimes(l)| = \sum_{a \in \Sigma, v_1 a \in \mathrm{Suf}_{k-1}(S)} \sum_{b \in \Sigma_{v_1 a}} |S^\otimes_{bv_1}(l-1)|,$$

where $\Sigma_v = \{b \in \Sigma : bv \in S\}$ and $S_v^\otimes(l)$ is the set of all words in $S^\otimes$ of length l that end with v, for any word v of length $k - 1$.

In many cases, a language is required to satisfy a certain type of subword constraint as opposed to a particular constraint S. For example, if a language L should be Hamming bond-free for some parameters d, k, there are many different choices for the subword constraint S such that $L \subseteq S^\otimes$. The next result (Theorem 2) characterizes languages that are maximal with respect to a set of constraints: Let k be a positive integer. A *k-subword property* $\mathcal{P}_k$ is a set of subword constraints of length k such that, if $S \in \mathcal{P}_k$, then $S' \in \mathcal{P}_k$ for any subset S' of S. A language L is *maximal* with respect to $\mathcal{P}_k$ if L satisfies a subword constraint in $\mathcal{P}_k$ and there is no language L' that satisfies a subword constraint in $\mathcal{P}_k$ and properly contains L.

The next result uses the ordinary concept of maximality: a maximal set of $\mathcal{P}_k$ is any set S in $\mathcal{P}_k$ which is not a proper subset of another set in $\mathcal{P}_k$.

Theorem 2. *Let $\mathcal{P}_k$ be a k-subword property. The set of all languages that are maximal with respect to $\mathcal{P}_k$ is equal to $\{S^\otimes : S \text{ is a maximal set of } \mathcal{P}_k\}$.*

It is interesting to note that in [10] a specific instance of the above result was obtained for the case of Hamming bond-free languages using a longer proof.

3   Encoding Data into $S^\otimes$

In this section we propose a method of encoding words of Σ^* into the language $S^\otimes$, where S is a subword constraint of some length k. In [10] it is shown that, given a trie T accepting the set S, we can construct a deterministic finite automaton $\mathrm{trie}(S)^\otimes$ accepting $S^\otimes$ whose number of states is equal to the number of states in T. However, it is not clear how one, in general, can encode arbitrary words into $S^\otimes$ using the automaton $\mathrm{trie}(S)^\otimes$. The approach we take here is as follows. We define a subset $B(l)$ of $S^\otimes$ whose words are of some fixed length $l \geq k$ such that $B(l)^* \subseteq S^\otimes$, that is, if we concatenate zero or more words of $B(l)$, the resulting word is in $S^\otimes$ – hence, it satisfies the subword constraint S. In practice, the cardinality $|B(l)|$ of $B(l)$ should be sufficiently small so that $B(l)$ can fit into a look-up memory table. Let m be the smallest positive integer such that $|\Sigma|^m \leq |B(l)|$, that is, $m = \lfloor \log_{|\Sigma|} |B(l)| \rfloor$. Then we can define any injective mapping enc : $\Sigma^m \to B(l)$, and we can encode any arbitrary word of the form $w_1 \cdots w_n$, with each $w_i \in \Sigma^m$, as $\mathrm{enc}(w_1) \cdots \mathrm{enc}(w_n)$ in $S^\otimes$. Conversely, the encoded word can be decoded uniquely as $w_1 \cdots w_n$. The encoding and decoding processes can be done via the look-up table that implements the mapping enc.

Theorem 3. *Let S be a subword constraint of some length k. Consider the following non-deterministic method: (1) Pick a nonempty subset S_e of S. (2) Let $S_b = \cap_{u \in S_e} S_b(u)$, where $S_b(u) = \{z \in S : uz \in S^\otimes\}$, for any word u. (3) Let $B = S^\otimes \cap S_b \Sigma^* \cap \Sigma^* S_e$; that is, B consists of all words in $S^\otimes$ that begin with a word in S_b and end with a word in S_e. (4) Define, for any $l \geq k$, the set $B(l) = B \cap \Sigma^l$. The following statements hold true.*

1. *The catenation closure of B is a subset of $S^{\otimes}$; that is, $B^* \subseteq S^{\otimes}$.*
2. *There is $l \geq k$ such that $B(l)$ is nonempty if and only if $S^{\otimes}$ is infinite.*
3. *If the automaton $\mathrm{trie}(S)^{\otimes}$ has a pair of 2-way communicating cycles (see the Appendix) then the set $B(l)$ can be chosen to be arbitrarily large.*

The above method would work in practice if the set $B(l)$ contains at least two elements – in this case it is not difficult to see that, for any n, there is an l_n such that $B(l_n)$ contains at least n elements. As stated above, a sufficient condition is that $\mathrm{trie}(S)^{\otimes}$ has a pair of 2-way communicating cycles. We conjecture that no such set $B(l)$ can be found when there is no pair of 2-way communicating cycles.

4   The Method of Theorem 3

The method of Theorem 3 is nondeterministic and depends on the choice of the initial subset S_e of S, which must be nonempty. Clearly there are $2^{|S|} - 1$ choices. A good choice would be one that produces a large set $B(l)$. Intuitively, when S_e is chosen to be large then S_b will be small and vice versa. However, it is not obvious how one can address mathematically this tradeoff. Our experimentation suggests that the following two guidelines are usually helpful: (i) If we pick a word u to be in S_e then we should also pick from S words whose suffix of length $k - 1$ is equal to the suffix of u of length $k - 1$. (ii) List all the sets $S_b(u)$ in decreasing order of cardinality, say $S_b(u_1), \ldots, S_b(u_n)$. Pick words in S_e such that, for these words, the corresponding sets are listed continuously, i.e., as in $S_b(u_i), \ldots, S_b(u_{i+t})$.

With these guidelines, we have a few choices for S_e. For each choice, we can calculate the size of $B(l)$ and pick the choice with the largest size of $B(l)$. Table 1 shows an example using $S = \{aaa, caa, gaa, aca, cca, gca, aga, cga, gga, aac, cac, aag\}$. Note that Choice 6 does not follow the second guideline, so the size of $B(10)$ for this choice is much smaller than the other choices.

5   Implemented System

In this section we describe our progress to the design and implementation of a web system for generating languages satisfying arbitrary subword constraints according to the method of Theorem 3. The system also allows one to enter an arbitrary data word over the DNA alphabet or the keyboard alphabet, and returns the encoded DNA word that satisfies the given subword constraints. Although the system is intended to work for arbitrary subword constraints, the examples in this paper are related to DNA-related constraints. The web interface can be found at http://cs.smu.ca/~b_cui/DNA13/ Currently the system is designed for fast response time and, therefore, it accepts only small values of k – the length of the subword constraint. The implementation language is C++ including libraries of the Grail project for finite state machines [4]. The flow of control of the system is given in Figure 1. A user can enter the subword constraint S via a text file, or via a special interface for the case of DNA constraints. The

Table 1. Comparison of the sizes of a few $B(10)$'s arising from different choices of S_e's

Choices	S_e	S_b	$B(10)$
Choice 1	$\{aaa, caa, gaa, aca, cca, gca,$ $aga, cga, gga, aac, cac, aag\}$	$\{aaa, aac, aag\}$	150
Choice 2	$\{aaa, caa, gaa, aca, cca, gca,$ $aga, cga, gga, aac, cac\}$	$\{aaa, aca, aac, aag\}$	180
Choice 3	$\{aaa, caa, gaa, aca, cca,$ $gca, aga, cga, gga\}$	$\{aaa, aca, aga, aac, aag\}$	150
Choice 4	$\{aaa, caa, gaa, aca, cca, gca\}$	$\{aaa, caa, aca, aga,$ $aac, aag, cac\}$	180
Choice 5	$\{aaa, caa, gaa\}$	$\{aaa, caa, gaa, aca,$ $aga, aac, aag, cac\}$	150
Choice 6*	$\{aaa, caa, gaa, aga, cga, gga\}$	$\{aaa, aca, aga, aac, aag\}$	111

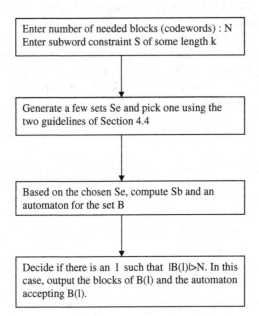

Fig. 1. The flow chart of the system

choice of l is based on Theorem 1. If the system decides that no set $B(l)$ exists for the given parameters, this simply means that the heuristic guidelines of Section 5 did not work. We refer the reader to [2] for further details on the system.

Appendix

For a deterministic automaton A, $[p_1, v_1, \ldots, p_n, v_n, p_{n+1}]$ denotes the path that starts at state p_1, forms the word v_1 between the states p_1 and p_2, then the word v_2 between the states p_2 and p_3, etc. When $p_{n+1} = p_1$ then the path is called a *cycle*. Two cycles $[p_1, v_1, p_1]$ and $[p_2, v_2, p_2]$ are called *equivalent* if they can be written in the form $[p_1, x_1, p_2, x_2, p_1]$ and $[p_2, x_2, p_1, x_1, p_2]$, respectively. The automaton A is said to have a pair of communicating cycles, if there are two non-equivalent cycles $[p_1, v_1, p_1]$ and $[p_2, v_2, p_2]$ in A such that there is a path of the form $[p_1, u_1, p_2]$. The pair of communicating cycles is called 2-way communicating if, in addition, there is a path in A of the form $[p_2, u_2, p_1]$.

References

1. Chen, J., Reif, J.H. (eds.): DNA Computing. LNCS, vol. 2943. Springer, Heidelberg (2004)
2. Cui, B.: Encoding methods for DNA languages. MSc Thesis, Dept. Math. and Computing Science, Saint Mary's University, Halifax, Canada (2007)
3. Ferretti, C., Mauri, G., Zandron, C. (eds.): DNA Computing. LNCS, vol. 3384, pp. 7–10. Springer, Heidelberg (2005)
4. Grail+: Department of Computer Science. University of Western Ontario, London, Canada. http://www.csd.uwo.ca/Research/grail/
5. Hagiya, M., Ohuchi, A. (eds.): DNA Computing. LNCS, vol. 2568. Springer, Heidelberg (2003)
6. Jonoska, N., Mahalingam, K.: Languages of DNA based code words. In: [1], 58–68.
7. Jonoska, N., Seeman, N.C. (eds.): DNA Computing. LNCS, vol. 2340. Springer, Heidelberg (2002)
8. Kari, L., Kitto, R., Thierrin, G.: Codes, involutions and DNA encoding. In: Brauer, W., Ehrig, H., Karhumäki, J., Salomaa, A. (eds.) Formal and Natural Computing. LNCS, vol. 2300, pp. 376–393. Springer, Heidelberg (2002)
9. Kari, L., Konstantinidis, S.: Descriptional complexity of error/edit systems. Journal of Automata, Languages and Combinatorics 9(2/3), 293–309 (2004)
10. Kari, L., Konstantinidis, S., Sosí, k.P.: Bond-free languages: formalizations, maximality and construction methods. Intern. Journal of Foundations of Computer Science 16(5), 1039–1070 (2005) (Conference version in [3], 169–181)
11. Kobayashi, S., Kondo, T., Arita, M.: On template method for DNA sequence design. In: [5], pp. 205–214
12. Mahalingam, K.: Involution codes: With application to DNA strand design. PhD Thesis, Dept. Mathematics, University of South Florida, Florida, USA (2004)
13. Mauri, G., Ferretti, C.: Word Design for Molecular Computing: A Survey. In: [1], pp. 37–46
14. Tanaka, F., Nakatsugawa, M., Yamamoto, M., Shiba, T., Ohuchi, A.: Developing support system for sequence design in DNA computing. In: [7], pp. 129–137
15. Yu, S.: Regular Languages. In: Rozenberg, G., Salomaa, A. (eds.) Handbook of Formal Languages, vol. 1, pp. 41–110. Springer, Berlin (1997)

Author Index

Lecture Notes in Computer Science

Sublibrary 1: Theoretical Computer Science and General Issues

For information about Vols. 1– 4588
please contact your bookseller or Springer

Vol. 4707: O. Gervasi, M.L. Gavrilova (Eds.), Computational Science and Its Applications – ICCSA 2007, Part III. XXIV, 1205 pages. 2007.

Vol. 4706: O. Gervasi, M.L. Gavrilova (Eds.), Computational Science and Its Applications – ICCSA 2007, Part II. XXIII, 1129 pages. 2007.

Vol. 4705: O. Gervasi, M.L. Gavrilova (Eds.), Computational Science and Its Applications – ICCSA 2007, Part I. XLIV, 1169 pages. 2007.

Vol. 4703: L. Caires, V.T. Vasconcelos (Eds.), CONCUR 2007 – Concurrency Theory. XIII, 507 pages. 2007.

Vol. 4700: C.B. Jones, Z. Liu, J. Woodcock (Eds.), Formal Methods and Hybrid Real-Time Systems. XVI, 539 pages. 2007.

Vol. 4699: B. Kågström, E. Elmroth, J. Dongarra, J. Waśniewski (Eds.), Applied Parallel Computing. XXIX, 1192 pages. 2007.

Vol. 4698: L. Arge, M. Hoffmann, E. Welzl (Eds.), Algorithms – ESA 2007. XV, 769 pages. 2007.

Vol. 4697: L. Choi, Y. Paek, S. Cho (Eds.), Advances in Computer Systems Architecture. XIII, 400 pages. 2007.

Vol. 4688: K. Li, M. Fei, G.W. Irwin, S. Ma (Eds.), Bio-Inspired Computational Intelligence and Applications. XIX, 805 pages. 2007.

Vol. 4684: L. Kang, Y. Liu, S. Zeng (Eds.), Evolvable Systems: From Biology to Hardware. XIV, 446 pages. 2007.

Vol. 4683: L. Kang, Y. Liu, S. Zeng (Eds.), Advances in Computation and Intelligence. XVII, 663 pages. 2007.

Vol. 4681: D.-S. Huang, L. Heutte, M. Loog (Eds.), Advanced Intelligent Computing Theories and Applications. XXVI, 1379 pages. 2007.

Vol. 4672: K. Li, C. Jesshope, H. Jin, J.-L. Gaudiot (Eds.), Network and Parallel Computing. XVIII, 558 pages. 2007.

Vol. 4671: V.E. Malyshkin (Ed.), Parallel Computing Technologies. XIV, 635 pages. 2007.

Vol. 4669: J.M. de Sá, L.A. Alexandre, W. Duch, D. Mandic (Eds.), Artificial Neural Networks – ICANN 2007, Part II. XXXI, 990 pages. 2007.

Vol. 4668: J.M. de Sá, L.A. Alexandre, W. Duch, D. Mandic (Eds.), Artificial Neural Networks – ICANN 2007, Part I. XXXI, 978 pages. 2007.

Vol. 4666: M.E. Davies, C.J. James, S.A. Abdallah, M.D. Plumbley (Eds.), Independent Component Analysis and Blind Signal Separation. XIX, 847 pages. 2007.

Vol. 4665: J. Hromkovič, R. Královič, M. Nunkesser, P. Widmayer (Eds.), Stochastic Algorithms: Foundations and Applications. X, 167 pages. 2007.

Vol. 4664: J. Durand-Lose, M. Margenstern (Eds.), Machines, Computations, and Universality. X, 325 pages. 2007.

Vol. 4661: U. Montanari, D. Sannella, R. Bruni (Eds.), Trustworthy Global Computing. X, 339 pages. 2007.

Vol. 4649: V. Diekert, M.V. Volkov, A. Voronkov (Eds.), Computer Science – Theory and Applications. XIII, 420 pages. 2007.

Vol. 4647: R. Martin, M.A. Sabin, J.R. Winkler (Eds.), Mathematics of Surfaces XII. IX, 509 pages. 2007.

Vol. 4646: J. Duparc, T.A. Henzinger (Eds.), Computer Science Logic. XIV, 600 pages. 2007.

Vol. 4644: N. Azémard, L. Svensson (Eds.), Integrated Circuit and System Design. XIV, 583 pages. 2007.

Vol. 4641: A.-M. Kermarrec, L. Bougé, T. Priol (Eds.), Euro-Par 2007 Parallel Processing. XXVII, 974 pages. 2007.

Vol. 4639: E. Csuhaj-Varjú, Z. Ésik (Eds.), Fundamentals of Computation Theory. XIV, 508 pages. 2007.

Vol. 4638: T. Stützle, M. Birattari, H. H. Hoos (Eds.), Engineering Stochastic Local Search Algorithms. X, 223 pages. 2007.

Vol. 4630: H.J. van den Herik, P. Ciancarini, H.H.L.M.(J.) Donkers (Eds.), Computers and Games. XII, 283 pages. 2007.

Vol. 4628: L.N. de Castro, F.J. Von Zuben, H. Knidel (Eds.), Artificial Immune Systems. XII, 438 pages. 2007.

Vol. 4627: M. Charikar, K. Jansen, O. Reingold, J.D.P. Rolim (Eds.), Approximation, Randomization, and Combinatorial Optimization. XII, 626 pages. 2007.

Vol. 4624: T. Mossakowski, U. Montanari, M. Haveraaen (Eds.), Algebra and Coalgebra in Computer Science. XI, 463 pages. 2007.

Vol. 4623: M. Collard (Ed.), Ontologies-Based Databases and Information Systems. X, 153 pages. 2007.

Vol. 4621: D. Wagner, R. Wattenhofer (Eds.), Algorithms for Sensor and Ad Hoc Networks. XIII, 415 pages. 2007.

Vol. 4619: F. Dehne, J.-R. Sack, N. Zeh (Eds.), Algorithms and Data Structures. XVI, 662 pages. 2007.

Vol. 4618: S.G. Akl, C.S. Calude, M.J. Dinneen, G. Rozenberg, H.T. Wareham (Eds.), Unconventional Computation. X, 243 pages. 2007.

Vol. 4616: A.W.M. Dress, Y. Xu, B. Zhu (Eds.), Combinatorial Optimization and Applications. XI, 390 pages. 2007.

Vol. 4614: B. Chen, M. Paterson, G. Zhang (Eds.), Combinatorics, Algorithms, Probabilistic and Experimental Methodologies. XII, 530 pages. 2007.

Vol. 4613: F.P. Preparata, Q. Fang (Eds.), Frontiers in Algorithmics. XI, 348 pages. 2007.

Vol. 4600: H. Comon-Lundh, C. Kirchner, H. Kirchner (Eds.), Rewriting, Computation and Proof. XVI, 273 pages. 2007.

Vol. 4599: S. Vassiliadis, M. Bereković, T.D. Hämäläinen (Eds.), Embedded Computer Systems: Architectures, Modeling, and Simulation. XVIII, 466 pages. 2007.

Vol. 4598: G. Lin (Ed.), Computing and Combinatorics. XII, 570 pages. 2007.

Vol. 4596: L. Arge, C. Cachin, T. Jurdziński, A. Tarlecki (Eds.), Automata, Languages and Programming. XVII, 953 pages. 2007.

Vol. 4595: D. Bošnački, S. Edelkamp (Eds.), Model Checking Software. X, 285 pages. 2007.

Vol. 4590: W. Damm, H. Hermanns (Eds.), Computer Aided Verification. XV, 562 pages. 2007.